Product Design for Engineers

Devdas Shetty
University of the District of Columbia
Washington, D.C.

CENGAGE
Learning™

Australia • Brazil • Japan • Korea • Mexico • Singapore • Spain • United Kingdom • United States

Product Design For Engineers, First Edition
Devdas Shetty

Product Director, Global Engineering:
   Timothy L. Anderson
Senior Content Developer: Mona Zeftel
Media Assistant: Ashley Kaupert
Product Assistants: Alexander Sham,
   Teresa Versaggi
Marketing Manager: Kristin Stine
Director, Content and Media Production:
   Sharon L. Smith
Content Project Manager: D. Jean Buttrom
Production Service: SPi Global
Compositor: SPi Global
Senior Art Director: Michelle Kunkler
Cover and Internal Design:
   Red Hangar Design, LLC
Cover Images:
   Blueprint background–
   © ildogesto/ShutterStock.com
   Jet Engine–
   © X-RAY pictures/ShutterStock.com
   Auto Assembly–
   © iurii/ShutterStock.com
   Vases CAD–
   © RATOCA/ShutterStock.com
   Robot Welder–
   © iStockphoto.com/Fertnig
   Auto 3D Model–
   © iStockphoto.com/pagadesign
   Vacuum–
   © iStockphoto.com/amriphoto
Intellectual Property
   Analyst: Christine Myaskovsky
   Project Manager: Sarah Shainwald
Text and Image Permissions Researcher:
   Kristiina Paul
Senior Manufacturing Planner: Doug Wilke

For product information and technology assistance, contact us at **Cengage Learning Customer & Sales Support, 1-800-354-9706**.

For permission to use material from this text or product, submit all requests online at **www.cengage.com/permissions**.

Further permissions questions can be emailed to **permissionrequest@cengage.com**.

Library of Congress Control Number: 2015931883
ISBN: 978-1-133-96204-5

Cengage Learning
20 Channel Center Street
Boston, MA 02210
USA

Cengage Learning is a leading provider of customized learning solutions with employees residing in nearly 40 different countries and sales in more than 125 countries around the world. Find your local representative at **www.cengage.com**.

Cengage Learning products are represented in Canada by Nelson Education Ltd.

To learn more about Cengage Learning Solutions, visit **www.cengage.com/engineering**.

Purchase any of our products at your local college store or at our preferred online store **www.cengagebrain.com**.

Design element: © Shutterstock/Ildogesto

Unless otherwise noted, all items © Cengage Learning

Printed in the United States of America
Print Number: 01   Print Year: 2015

To my wife, Sandya, and sons, Jagat and Nandan, for their love and support.

ABOUT THE AUTHOR

Dr. Devdas Shetty serves as Dean of the School of Engineering and Applied Sciences at the University of the District of Columbia, where he is also Professor of Mechanical Engineering. He previously served as Dean of Engineering at Lawrence Technological Institute, Southfield, Michigan, and Dean of Research at the University of Hartford, Connecticut. At the University of Hartford, he was the original holder of the distinguished Vernon D. Roosa Endowed Professorship. As Director of Hartford's Engineering Applications Center, he set up partnerships with more than 50 industries. He also held positions at the Albert Nerkin School of Engineering at the Cooper Union for the Advancement of Science and Art in New York City.

The author of three books and more than 225 scientific articles and papers, Dr. Shetty's textbooks on mechatronics and product design are widely used around the world. His work has been cited for contributions to the understanding of surface measurement, intellectual achievements in mechatronics, and contributions to product design. He has five patents for inventions that involve interdisciplinary areas of mechanical engineering, design, and computer science.

Dr. Shetty has led several successful multi institutional engineering projects. In partnership with Albert Einstein College, he invented the mechatronics process for supporting patients with ambulatory systems for rehabilitation. In partnership with Armament Research, Development and Engineering Center (ARDEC), he led a multi-university industry team for the successful design and testing of a hybrid projectile. He established research and academic programs in Laser Manufacturing in collaboration with Connecticut Center for Advance Technology (CCAT) under National Aerospace Leadership Initiative (NALI). Major honors include the James Frances Bent Award for Creativity, the Edward S. Roth National Award for Manufacturing from the Society of Manufacturing Engineers, the American Society of Mechanical Engineer Faculty Award, and the Society of Manufacturing Engineers Honor Award. He is an elected member of the Connecticut Academy of Science and Engineering (CASE). He also is the author of *Mechatronics System Design*, published by Cengage Learning, now in its second edition.

BRIEF CONTENTS

CONTENTS

PREFACE

The objective in writing *Product Design for Engineers* is to familiarize students with concepts, techniques, and tools that encourage creativity and innovation in product design. Its purpose is also to address the lack of engineering-oriented product design books that can familiarize both product designers and students with the concepts, techniques, and tools of new product design and development. This book provides a comprehensive coverage of many areas so that students can understand the range of engineering disciplines in improving product creation. Students are introduced to the creative problem-solving method for product redesign techniques and success through case studies that explore issues of promising products, design for function, design for assembly, disassembly, quality, maintainability, and sustainability. The book's interdisciplinary approach, step-by-step coverage, and helpful illustrations provide mechanical, industrial, aerospace, manufacturing, and automobile engineering students with everything they need to design cost-effective, innovative products that meet customer needs.

As competition grows from world-class manufacturers, engineering designers must critically examine their product design strategies. Competing in a global market requires the adaptation of modern technology to yield flexible, multifunctional products that are better, cheaper, and more intelligent than others. New product development procedures today use increasingly sophisticated solutions to streamline and speed product development as well as improve overall product quality.

Many companies expect engineers to be competent with the procedures involved in product creation. Engineering careers now involve designing, analyzing, prototyping, and testing a wide variety of products irrespective of the field. There is a need to educate students not only to produce design concepts but also to take a global view of emerging trends in digital manufacturing, virtual prototyping, failure mode analysis, life cycle evaluation, managing innovation, entrepreneurship, and design for supply chain. Students must be exposed to issues of design for assembly, disassembly, reliability, and maintainability.

Several engineering programs in the United States are re-examining the design content in their curricula to incorporate vertical integration of design through large projects between the first year and the final year of their programs. More universities are introducing interdisciplinary courses taught across disciplines. This book will enable students to become familiar with the steps in creative product design starting with concepts through production and marketing. Students are also introduced to different product redesign techniques using case studies.

This book is intended for use by upper-level undergraduate and graduate students in mechanical, industrial, manufacturing, aerospace, and biomedical engineering and in engineering management. The book is also useful in engineering technology and business management–related programs. It is designed to serve as a text for stand-alone product design courses, design for manufacturing courses, and interdisciplinary courses dealing with modeling and product prototyping. It can also be used as the main reference material in professional courses and continuous improvement programs conducted in industries.

Product design is the methodology of creating a new product. A very broad concept, it is essentially the efficient and effective generation and development of ideas. This book provides systematic approaches to enable students to learn how to create smarter design.

The real challenge in writing this book was to connect complex and seemingly independent topics. Its major organizational feature is the use of case studies in each chapter to discuss various tools and successful project implementation. End-of-chapter exercises can be useful for instructors.

Chapter Content

The chapters are organized to take students through the various steps of creative product design starting with problem definition and following through customer needs, function diagram, house of quality, design for assembly, disassembly, optimization, and production layout with value stream mapping.

Chapter 1 covers the need for better design methodology. It considers the product development process and discusses reasons why some good products have failed. The chapter examines the systematic product creation process and the contribution of digital tools in creating competitive products.

Chapter 2 is devoted to conceptual phases of the product design process. Various techniques are discussed to show how a design team uses customer needs to formulate product specifications and generate concepts for detailed design. The main benefit of an organized process is to ensure that the product is focused on customer needs without ignoring any primary requirements.

Chapter 3 identifies the way the designer evaluates concepts at the generation stage with respect to customer needs and identifies the relative strengths and weaknesses of proposed concepts as well as the best candidate for further examination. The chapter also examines how the designer defines the real problem, decomposes the problem into subproblems, gathers information, and examines patents and published literature at the concept generation stage.

Chapter 4 addresses the issues of product function. A product's functional design, which addresses upstream phases, involves the generation and synthesis of ideas and performance. It consists of defining a set of subfunctions and their interrelationships to determine a general product's architecture. Product architecture evolves during concept development, resulting in the arrangement of physical elements into building blocks.

Chapter 5 examines the methods used to evaluate the product design processes by highlighting their weaknesses and strengths. The main focus is on design for assembly. Several methods are discussed with examples of the calculation of product efficiency: analyzing a design, investigating how to process and assembling the design components, reduce cost, and increase product quality. The procedures outlined here evaluate an initial design's efficiency and examine its validity by using comparative tables.

Chapter 6 discusses factors concerning a product's disassembly and maintenance. The principles of design for disassembly identify the ease with which a product can be assembled, disassembled, maintained, serviced, and recycled. This chapter considers the disassembly of the individual components of a large product and identifies how this methodology can lead to a more efficient product. Because the creation of product architecture is a dominant activity, the team responsible for it must have input from the design, manufacturing, marketing, and supplier areas.

Chapter 7 examines areas that critically impact cost-effective product development both for new and in-process products. Designers should be aware of the manufacturing consequences of their decisions because minor design strategies during the early stages often prevent major problems later. Value stream mapping shows a systematic process of the flow of material and information and can be introduced both in existing production systems and while creating a new product. This chapter provides a clear explanation of the philosophies of the Toyota production system, including the pull system, kanban, just in time, and standard work. The chapter's case study examines the implementation of lean philosophies to develop continuous improvement and reduce non-value-adding processes.

Chapter 8 discusses the basic elements that greatly affect the outcome of the development process. Sustainability in product design embraces a broad view of product development by considering a product's full life cycle and the impact of its design, manufacture, use, and retirement. Key elements of robust design, optimization, and failure mode evaluation are presented.

Chapter 9 discusses how the emerging emphasis on product reliability and the desire to reduce product development time have focused on the use of software tools for design and production. Interactive modeling is crucial to the design process, and it can occur in an environment that combines real and virtual objects. The chapter also examines virtual prototyping tools that help companies take new products to market as well as the role of additive manufacturing that emphasizes quickly creating output in the form of a prototype. The chapter also discusses the emerging trend of additive manufacturing and its application in the creation of new products in different industries.

Chapter 10 includes two additional case studies that address design methodologies outlined in the previous nine chapters. The first case study describes the design approach for developing a surface inspection instrument for the aerospace industry. The second one is on unmanned aerial systems. This case study investigates the design, simulation, and prototyping of a transformable hybrid projectile used for defense.

In the past 25 years, enormous changes have occurred in the practice of engineering enterprises in the United States and worldwide. Paramount among these has been the evolution of the Internet, which has enabled the creation of designs that were previously not feasible. The result of the concurrent introduction of the principles of lean thinking, supply chain management, and virtual reality has been the explosion of new products and the ability to create an infinite variety of variations of existing products.

The instructor's website for this book also includes **Lecture Note PowerPoints** and an **Instructor's Solution Manual**.

MindTap Online Course and Reader

This textbook is also available online through Cengage Learning's **MindTap**, a personalized learning program that can be purchased separately or as an addition to the textbook. Students who purchase the MindTap have access to the book's MindTap Reader and are able to complete homework and assessment material online or on desktops, laptops, and iPads. Instructors who use a Learning Management System (such as Blackboard or Moodle) for tracking course content, assignments, and grading can seamlessly access the MindTap suite of content and assessments for this course.

With MindTap, instructors can:

- Personalize the Learning Path to match the course syllabus by rearranging content or appending original material to the online content
- Connect a Learning Management System portal to the online course and Reader
- Customize online assessments and assignments
- Track student progress and comprehension
- Promote student engagement through interactivity and exercises

Additionally, students can listen to the text through ReadSpeaker, take notes, create their own flashcards, highlight content for easy reference, and check their understanding of the material through practice quizzes and gradable homework.

Acknowledgments

I acknowledge the helpful suggestions made by reviewers during the course of this project with gratitude. They include Professors John L. Evans from Auburn University, Jin S. Kang from the U.S. Naval Academy, Katie Grantham from the Missouri University of Science and Technology, Christopher Kitts from Santa Clara University, Yuyi Lin from the University of Missouri, Ramin Sedaghati from Concordia University, and Mustapha S. Fofana from Worcester Polytechnic Institute.

Others who have contributed by reviewing this manuscript include Professor Claudio Campana of the University of Hartford and Dr. Jiajun Xu of the University of the District of Columbia. Dr. Neelesh Kumar of Central Scientific Instruments Organization and Dr. Arun Giriyapur of BVB College of Engineering and Technology in India offered valuable suggestions throughout this project.

It gives me great pleasure to acknowledge the support and help of Naresh Poudel, who reviewed the manuscript and assisted in finalizing it.

The frequent encouragement and support by Provost Rachel Petty and Chairs Dr. Samuel Lakeou, Dr. Pradeep Behera, Dr. Segun Adebayo, and Dr. Lily Liang of the University of District of Columbia is greatly appreciated.

I acknowledge the contributions from Matt Morrison (OTIS elevators) and from Neil MacDonald, Daniel Ly, Silvia Pellegrini, Luke Ionno, Tim Maver, Chris Bepko, and Vishweshwaran Coimbatore, Swetha Seethamraju, Ali Garrash, and Aaron Beal through their innovative theses and projects. I am indebted to many of my past graduate students who are now successful professionals and professors. They include Ken Rawolle, Kran Kolluri, Nilesh Dave, Troy Chicone, Rob Choquette of United Technologies Corporation, and Dr. Elizabeth Cudney. I also wish to acknowledge the several hundred students from the classes in which I have tested the teaching material.

Special thanks go to Dean Lou Manzione, Dr. Ivana Milanovic, Dr. Abby Ilumoka, Dr. Tom Eppes, and Dr. Saeid Moslehpour of the University of Hartford for their support and contributions over time and to Leon Manole from ARDEC for his sponsorship and encouragement. I gratefully acknowledge the valuable support given by Dr. Avital Fast, Professor Emeritus, Albert Einstein College of Medicine, New York City, during our collaboration.

Most of the material presented in this book is a collection of many years of research and teaching at the Cooper Union, New York City; University of Hartford, Connecticut; Lawrence Technological University, Michigan; and the University of the District of Columbia in Washington, D.C., as well as the insight gained by working closely with industries such as United Technologies Corporation and government agencies such as the National Science Foundation and the Armament Research, Development, and Engineering Center (ARDEC) at Picatinny Arsenal, New Jersey.

The staff at Cengage Learning was outstanding as usual. I would like to recognize senior content developer Mona Zeftel, content project manager Jean Buttrom, and Timothy Anderson for their excellent work.

CHAPTER 1
Product Design Process

OBJECTIVES

The process of creating a product is influenced by the process of designing, identifying the quality needs, and determining the overall cost. An organization has greatest control over a product at the early stages of its creation when the organization determines the market, factory and operational costs, and life cycle. The goal is to develop a reliable product that is best for performing its function, manufacturing, assembling, and servicing. The first chapter considers the product development process and discusses reasons some good products have failed. This chapter examines the systematic product creation process, the simulation and modeling, and the contribution of digital tools in creating competitive products. This chapter also discusses product design in a global environment and concludes with case studies.

1.1 Introduction

Competing in a global market requires the adaptation of modern technology to yield flexible, multifunctional products that are better, less expensive, and more intelligent. The challenge facing us today is how to improve the product development systems.

As we experience more and more competitive pressure from world-class manufacturers, it becomes important that we critically examine product design strategies.

Successful companies such as Microsoft, 3M, Apple, and Toyota continuously engage in product innovation because each company has its own culture of design approach and production strategy. Done properly, these processes affect every aspect of product development, launch, and service. By applying a design-oriented approach, these companies have developed innovative solutions, and their senior management is aligned with the design team to avoid pitfalls. These companies are able to create materials that align technical and nontechnical participants and understand how people interact with technology, how to manage innovation, and how to evaluate product effectiveness.

When using a toaster, microwave, or any other product, consumers are not aware of the complexities that went into its creation. Designing a product involves a constant decision-making process, which includes problem solving in a sequential fashion and analyzing the constraints at each step. Human beings are special types of designers, and our design philosophy influences our own lives and environment. In general, design represents an answer to a problem—one that has visible form, shape, and function.

Design Intent

Various professions define *design* differently; business professionals, physicians, architects, and engineers have their own unique views on the nature of design and their own personal experience with its application. However, we limit our discussion to the professional design as practiced by engineers. Design provides a set of rules for reorganizing the elements of creation toward some greater purpose, known as *design intent*, which has ethical and moral dimensions. In the larger context, design is, in fact, any purposeful, "thought-out" activity. It is a way of doing things characterized by decision making. A typical product may be made up of many technical and nontechnical components that factor into its design. The product may not succeed, however, if it lacks balance between design and other important factors such as ergonomics, aesthetics, and safety.

1.1.1 **Product Realization Process**

The ideal goal of design is to develop a reliable product that is best for functioning, manufacturing, assembling, and servicing—which is difficult to achieve. Thus, the design process must be managed in steps that ultimately must be integrated successfully.

Refer to **Figure 1.1** for the structure of a design process that starts with a need statement and proceeds through need establishment, concept development, detailed design, testing, and production. The figure shows different stages of a typical product's life cycle, including

- use identification,
- technical research,
- design,
- development,
- market testing,
- quality assurance, and
- servicing.

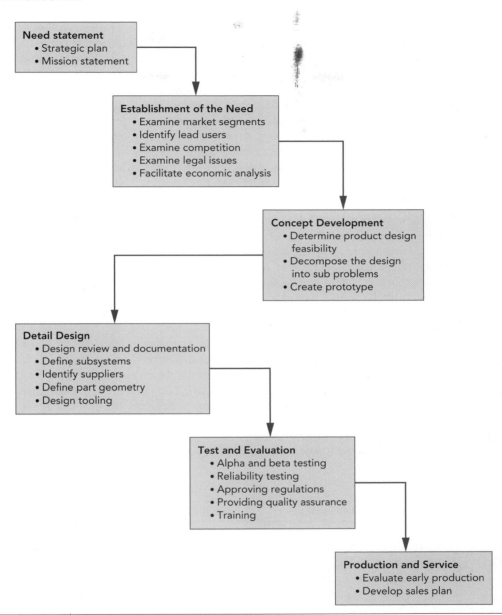

FIGURE 1.1 General Structure of the Design Process

The speed of a product's development process, market testing, and production are important factors in its life cycle. In the first stage, the organization's recognition of market need for the product initiates the process. Next is the task of identifying product use; this step involves planning the item based on the organization's strategic goals, which include completing feasibility study; performing research and development, prototyping;

testing the market; developing manufacturing and marketing processes. Products that become well defined in the early design stages have higher chances of success than those that lack such planning. Many studies have shown that conducting early design work, preparing sketches and prototypes, performing simulation studies, estimating costs, and talking to potential customers reduce the uncertainties in a new product launch.

The process of creating a product and its overall cost are influenced greatly by the design process. An organization has greatest control over a product in the early stages of its creation when market interest, factory cost, operational cost, and life cycle are determined. At this stage, a product's status can be unstable while the organization tries to optimize the product's distinctiveness to ensure high market acceptance.

1.1.2 **Characteristics of a New Product Launch**

The design and development of a product invariably involves considerable investment of time, effort, and money. It is essential that a new product be thoroughly examined and reviewed before it is presented to the public. A company's reputation and finance rest on the launching of each of its products. The fundamental questions a product designer must consider relate to the why, what, how, who, and when.

See **Table 1.1** for five key characteristics of a new product launch.

Why?

Why involves the company's business-approach strategies and its evaluation of a potential product's success and failure. Success does not depend on the design, technology, and marketing alone but also on realistic planning. Among several reasons for launching a product, profit is usually the main factor. The company's financial situation, including its cash inflow and outflow and net payback time for investments, is one of the issues that influences the decision. However, a successful new product could translate into enhancing a company's prestige or even maintaining its survival. Also important is the creation of a comprehensive survey translating customers' requirements,

	Strategic Objective	Financial Consideration	Market Consideration
Why	Survival of the company New business opportunity	Financial reward	Synergistic impact
What	Product definition Adoption or modification of existing design	Joint venture Collaboration Licensing	Projected product demand
How	Strengths and weaknesses Physical infrastructure	Financial benchmark	Market response
Who	Structural decision making Human consideration	Accountability and responsibility	Information security Information dissemination
When	Timing	Money and investment	Market window

TABLE 1.1 Five Key Characteristics of a New Product Launch

concept definitions, feasibility documentation, and prototype demonstration. Along with market research, these are the essential elements of technical and commercial implementation.

What?

What pertains to the product under consideration and its market. The market segment influences a product's function, while the technology segment determines the manner in which it can function. Market segments are dynamic and constantly evolve. They must be well specified before a product is planned and resources are allocated. At this stage, designers should consider adaption or modification of an existing product.

Scientific study of marketing research can provide quantitative data analyzing past demand and making predictions about future demand. This research can provide information on the trends of technology, the influence of the governmental and local restrictions, and the impact of industrial restructuring in a particular field. A major factor at this early stage is the importance of choosing among various product ideas; the ideas must be compared and ranked based on business history, technological infrastructure, and marketing thrust.

How?

How relates to aspects of using the existing scientific and technical capabilities and providing the mechanism for implementation. Product implementation is, to a large extent, based on a company's strengths, weaknesses, and basic capabilities, particularly human and financial resources and could be related to its physical infrastructure. Basic exploratory activity is very important before moving forward with technical developments. Commercial development is the heart of the entire process; it involves establishing a list of strategies, analyzing market competition, benchmarking with respect to other companies, pricing, and market distribution. Obtaining and enforcing intellectual property rights and trademarks must be systematically planned and conducted in addition to the product's technical and business-related developments. It is also possible to create a product through collaboration, licensing, or joint venture from external agencies.

Who?

Who defines the accountability and responsibility for the new product development. No matter what the organizational structure, the people involved in product development must understand the institutional setup and layout. Many product failures have been attributed to breakdowns in communication and inadequate follow-up at critical junctures. Effective communication and information dissemination of business objectives, goals, work functions of various participants, and financial and legal issues throughout the company are critical. Responsible information retrieval and delivery are crucial in new product development.

When?

When requires the product developer to be sensitive to the timing of a product launch. Time is critical because it is closely related to money, investment, and window of market opportunity. A product that enters the market before or after its time will certainly not meet the company's profit goal.

1.1.3 Reasons That Promising Products Fail

Why should products with high expectations fail despite clear signals of market need? One reason may be that any innovative product, especially if it has advanced technological content, meets customer resistance and sells slowly until consumers perceive it as safe. Neil Rackham (1998) suggests that the main problem is the way in which highly innovative products are launched to the sales force, which, in turn, influences the manner in which the product is sold.

Steps for introducing a new product to the sales force are given in **Figure 1.2**. The sales force learns product-centered information about product capabilities and communicates this information to customers in the same way it was communicated to them. Focusing on all of the new features that make a product highly innovative draws the attention of the sales force from the most important issue in the process: customer needs. Launching becomes product centered instead of being customer centered, thereby decreasing customer interest. A more customer-oriented approach that focuses on a product in terms of problems it solves for the customer is more successful.

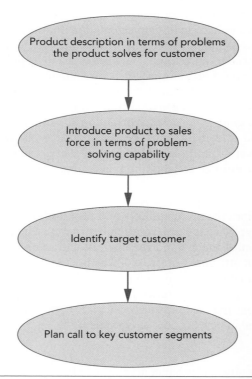

FIGURE 1.2 Steps for Introducing a Product to the Sales Force

The following two examples show the impact caused by the customer viewpoint.

EXAMPLE 1.1

NEW COKE: CUSTOMER VIEW

In 1985, the Coca-Cola Company replaced its original formula and brand Coke with a sweeter New Coke. It failed because Coca-Cola did not adequately research the taste preferences and emotional attachment of its most loyal customers to the original Coke formula. The product launch team did not consider the fact that brand loyalty was actually to the *brand*. The U.S. public's reaction to the New Coke was negative, and it was a major marketing failure. New Coke was unsuccessful not because it tasted any better or worse than original Coke but because the millions of loyal customers believed that Coke had completely replaced their old favorite with something else. For customers, it did not matter at all whether New Coke was better or worse; it was simply different and not wanted regardless of what marketing research had recommended. Eventually, after the restoration of Coke (with the name change to Coke Classic), the company resumed its normal lead as the dominant soda company in the United States.

EXAMPLE 1.2

BETAMAX VIDEO

In 1975, Japan's Sony Company introduced the first reasonably priced home videotape system. Utilizing the Betamax format for the tape cassettes, the company pioneered a special tape recording system. The original tape cassettes had a maximum recording time of one hour. In 1976, JVC Company introduced a similar product utilizing the VHS format. These tapes were slightly larger, and the screen resolution was improved, but the tape drive mechanism was noisier. The big difference was that JVC offered a two-hour recording time. What consumers quickly realized was that the VHS tapes could record feature films but the Betamax could not. Although Sony defended the Betamax format as technically superior and countered with a two-hour capacity with slightly degraded resolution, matching that of VHS, the new Betamax had no market penetration.

Sony's delay in addressing the two-hour taping seems to be what doomed the product in the U.S. and European markets, and by extension, globally. By 1980, Sony's initial market of 100% had dropped to less than 30% and by 1986 was down to around 7%. Sony licensed the format to allow other companies to produce Betamax machines, but by 1984, 40 companies were producing VHS machines compared to only 12 producing Betamax. The end came in 2002 when Sony produced the last Betamax machine en.wikipedia.org/wiki/Betamax. The Betamax format failed because Sony initially did not recognize the importance of recording features for consumers.

1.2 **Building Blocks of New Product Design**

A company should have a systematic design process that it communicates to the design team. The actual process by which product designers implement their tasks and responsibilities is typically determined by the individuals involved. The abilities, approaches, habits, and degree of documentation of these employees can be unique and randomly acquired, thus causing difficulties for one person to follow up after another's work.

Strategic Plan

The strategic planning process is an organization's major vehicle for policy deployment and its execution. It consists of creating a mission statement, performing situation analysis, identifying objectives and strategies, preparing action plans, and implementing policy issues. A strategic plan:

1. Identifies the company's mission
2. Identifies its strengths and weaknesses and establishes long-range business objectives
3. Selects the market segments to pursue
4. Formalizes the process for selecting products for development
5. Selects the products and identifies the strategic and tactical issues that the company must resolve to facilitate success
6. Projects the financial returns expected from the selected market and products

A company's mission statement should provide a clear product and market focus and identify the sources of opportunities. The strategic plan also should include a situation analysis that addresses changes in the company, changes in the market place and competition, the company's technical readiness, and pertinent standards and regulations. Whereas market analysis focuses on market share, competition, and customer perception, a company's internal analysis considers the strengths, weaknesses, threats, and growth opportunities for different products. Based on these considerations, management outlines the product objectives.

Development Process

A well-planned product development process enables companies to select, propose, design, develop, and market new products effectively. A systematic product design and development process are key elements of an organization's infrastructure. Such a process can provide a stable structure for strategic planning, decision making, operation, effective communication, implementation, and control. See **Figure 1.3** for the components of product development.

Companies around the world have improved production process structure by studying the manufacturing processes and fluctuations in workloads and identifying techniques to reduce variation and eliminate bottlenecks. Paul Adler has studied a dozen companies that have started applying the process management techniques to product development, including Raychem, Motorola, Harley-Davidson, Hewlett-Packard,

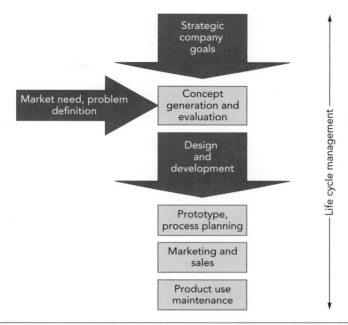

FIGURE 1.3 Components of Product Development

General Electric, AT&T, Ford, General Motors, and NEC. The results of this study have shown three significant observations:

1. Product development projects are accomplished quickly if an organization takes on fewer projects at a time.
2. Investment made by a company to relieve project bottlenecks can lead to early market launch.
3. Eliminating unnecessary variation in workloads and work processes removes distractions and delays, thereby freeing the organization to focus on key areas of development.

Product Definition

Customer satisfaction is essential in a product's success and must be considered in product planning. Converting customer inputs into specific product functions, features, and specifications is necessary. The "voice of the customer" must be translated into product requirements that meet their needs and expectations. These expectations can be summarized as high quality and reliability, low maintenance, and low price.

The product development process uses customer expectations as input and concurrent engineering as a design approach. The voice of customers regarding their needs create increased pressure to offer products of ever higher quality to the market in ever shorter times. The company's customers and shareholders scrutinize product price/performance ratios carefully. The traditional approach to product design and development reduces a company's ability to compete effectively on the global market. This approach has the following weaknesses:

1. Insufficient product definition
2. Inadequate studies of the influence of design on manufacturing and assembly
3. Lack of clear guidelines on how to develop the product in detail before production
4. Inadequate cost analysis
5. Changes occurring during the design process

The appropriate response to these weaknesses is addressed by using concurrent approach in the early stage of product development.

The response requires a new approach and change in a company's culture that initiates product design and associate processes, which is not the case in the traditional serial approach. Studies show that more time spent early in the design process is more than compensated by the time saved when prototyping occurs. The concurrent approach not only increases product quality and customer satisfaction but also reduces the time to launch the product.

Key concepts of new product development include customer orientation, major decision making, early concurrent development of product design and production processes, and use of cross functional teams and efficient design and manufacturing techniques.

1.2.1 Concurrent Engineering Approach to Product

Concurrent engineering is a design approach that merges product design and manufacturing. It is based on the idea that people can do a better job if they cooperate to achieve a common goal. As the term *concurrent engineering* means, product design and process planning take place at the same time. Concurrent engineering requires design teams to create a product that addresses:

- Robustness
- Design for manufacture, maintenance, and disposal
- Reliability and sustainability
- Environmental concerns

The characteristics of concurrent engineering are:

- More precise definition of the product without late changes
- Manufacture and assembly design undertaken in the early design stage
- Well-defined product development
- More accurate cost estimates

A team must develop sound insight into the nature of the activities required to make a product. Decision making is delegated to people doing the work commensurate with their responsibilities while managers function as facilitators. Therefore, most of the decisions relating to design are made early in the process by a team that consists of experts from the different stages of a product's life, including the marketing, maintenance, and service stages. While individually performing their respective functional design responsibilities, cross functional team members work very closely with other team members. For this reason, good communication is essential to make a company's operations effective.

Using concurrent engineering principles as a guide, the designed product is likely to have four basic characteristics:

- High quality
- Low cost
- Reduced time to market
- Higher customer satisfaction

See **Figure 1.4** for the basic concurrent engineering model. Concurrent design improves the quality of early design-related decisions and has a significant impact on a product's life cycle. It is well suited for team-oriented project management that emphasizes collective decision making. Successful implementation of concurrent engineering is possible by coordinating optimal exchange of information and addressing organizational barriers.

Concurrent engineering has been influenced partly by the recognition that decisions affecting many of the high costs in manufacturing are decided at the product design stage itself. The cost of design component is generally less than 5% of the total budget of a project, but its influence on total costs is significant (see **Table 1.2** and **Figure 1.4**).

The decisions made in the design stage can result in profound effects on product cost (see **Figure 1.5**). Thus, concurrent engineering considers customer need, life cycle management, quality, and reliability.

Design for Manufacturing Intent

A major element in the product design schedule is the development of a design for manufacturing (DFM) intent. DFM is a technique for developing a product that meets the desired performance specifications while optimizing the design through the production system. DFM uses the concurrent engineering approach that is based on a fundamentally different way of considering how the products are conceived, produced, and supported. It follows a procedure that is intended to help designers consider all

	Initial Project Cost (percentage)	Influence on Overall Cost (percentage)
Design	5	70
Materials	50	20
Labor	15	5
Overhead	30	5

TABLE 1.2 Design Versus Other Product Costs

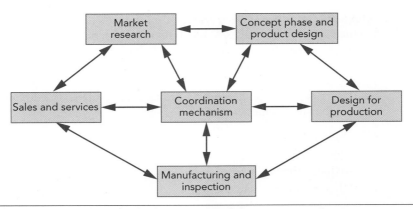

FIGURE 1.4 Basic Concurrent Engineering Model

elements of the product system life cycle from conception through disposal, including quality, cost, and user requirements. Design for Manufacturability (DFM) methods solve problems of improper product definition and address the issues of building standard and reliable products for industry.

Additional Considerations

General considerations in product design include cost, function, market design, factory elements, and supply chain. Cost is an obvious consideration as seen in **Figure 1.5**. Market design considerations involve items such as customers' needs, the breadth of the product line, product customization, upgrading, time to market, and the need for future designs. Factory considerations that must be planned are delivery, quality and reliability, ease of assembly, ability to test, ease of service, shipping, and repair. Supply chain considerations (discussed later) include optimization in product manufacturing, cost delivery, service, and safety.

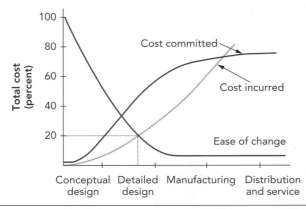

FIGURE 1.5 Percentage of Costs of a Product's Life Cycle Committed during the Design Stage

1.3 **Digital Tools for Product Design**

Product innovation is the primary instrument that companies as well as nations use to drive international competition. Sustained success depends upon continuous improvement in products and processes. Since the customers are aware of their needs, the current trend is toward mass-customized and less service-intensive products. The emphasis on reliability and the need to reduce product development time have increased the use of technology and software tools in design and manufacturing. Global economic pressures led manufacturers to slash budgets and cut costs wherever possible. To compress new product development time, most major automotive, aerospace, and other manufacturers deploy increasingly sophisticated product life cycle management (PLM) solutions to streamline and speed product development. Digital manufacturing tools help manufacturers, particularly those in increasingly price-conscious industries, deal with the pressures of getting new products to market fast and at low prices.

With new digital tools and improved three-dimensional (3-D) visualizations, manufacturers can digitally design and validate full factories in a fraction of the time previously required. The impact of the digital age in product design and manufacturing has driven the creation of electronically stored information to never before seen levels. A few areas seeing a major increase in storage requirements include:

- Complex design files
- Simulation data for design optimization
- Design and estimation information shared between different departments
- Parts and assembly files
- Data created on the plant floor

In addition, many firms are encouraging employees to collaborate on projects from different locations and on multiple devices. As a result, many organizations are using data management tools to manage their digital information and enable collaboration efforts. The companies have to deal with big data that involves transaction processing logs, application data, database backup files, and an increasing variety of structured and unstructured data. As software companies take their solutions into the cloud, the complexity of operations dealing with big (and secure) data from large customer bases interacting online is significant. All of the critical data being created and stored needs to be protected so that it can be restored in the event of an unforeseen disaster.

For example, in the concept phase, a 3-D computer-aided design (CAD) can be used to generate data for 3-D rapid prototyping technology, allowing designers to examine a product's shape and parameters such as stress, vibration, heat, and flow in the design analysis phase. Changes can be integrated by automatically converting data from 3-D computer-aided design (3-D CAD) systems.

1.3.1 **Simulation and Modeling Tools**

Engineers need effective software tools to comprehensively design, model, synthesize, and analyze a design before a prototype is fabricated; experimental evaluation of torque and stress for a mechanical product cannot be performed until a working prototype has been completed. Thus, engineers need a rapid way to fabricate a working prototype. Simulating complex systems allows designers to develop a system involving virtual product design procedures without finalizing the hardware.

The simulation procedure can be a "what-if" scenario when the hardware needed to manufacture a product does not exist. Two critical issues to consider are speed and accuracy. The trade-off between these two is necessary.

Interactive modeling is crucial to the design process, and it can occur in a mixed environment where real and virtual objects are combined. The key aspect of virtual environments is that the visual representation of system partitioning and interaction lends itself to electromechanical products. Virtual simulations enable everyone to work on development before the first prototype is completed. Engineers can validate the entire operating cycle for a machine by driving the simulation with control system logic and timing. With industries leading all-digital design, validation, and commissioning of factory automation devices, virtual simulation of factory floor layout has become important.

Computer-aided manufacturing (CAM) systems can generate manufacturing process data directly from solid models without human intervention. Modern materials-based finite element modeling (FEM) and manufacturing simulation software provide solutions to optimize manufacturing processes. Simulation software tools enable detailed analyses of manufacturing processes such as milling, turning, and grinding. Doing so allows designers to improve production rates, part quality, and tool performance while decreasing the need for physical testing.

Digital tools are also available to predict force and temperature; they have the ability to predict the production tool wear and analyze the 3-D residual stresses exerted on a product. DFM methodologies generate feedback on the consequences of design decisions in areas such as product assembly/disassembly and maintenance. In the rapid prototyping phase, modern software tools can create precision 3-D prototyped parts directly from a 3-D CAD system. Different software tools for use from conception to prototype are available (**Figure 1.6**). Integration of these software tools can reduce development cost, shorten development times, and improve reliability.

An example of a virtual product model that is available to production workers on the manufacturing side and engineering design is provided in **Figure 1.7**. Increased interaction with the 3-D models enables more users, whether initial designers or shop floor engineers, to access the model. Interactive modeling technologies allow people to work with the geometry in much more free-form manner.

Figure 1.8 is an example of a platform used in many industrial applications. In this case, there are effectively two models: the simulated physical model and the application model. The physics-based model accounts for a simulated product. The application model interacts with this environment to simulate the real-world application. Simulink and MATLAB are model-based development tools that assist in the creation of an application model. The basic design represented in the physical world by CAD tools (such as CATIA, Creo, and SolidWorks) has advanced simulation tools, although their orientation is toward physical construction rather than process control integration.

The simulation platform can examine stress under dynamic loading conditions and can address nonlinear analysis such as deflection and impact with flexible materials

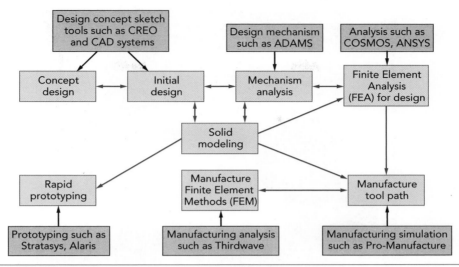

FIGURE 1.6 Software Tools for Each Phase of the Design Spectrum

such as foam, rubber, and plastic. In many cases, simulation and analysis of physical entities are useful in a design that does not include a computer-based controller.

1.3.2 **Right the First Time by Using Virtual Simulation**

Interactive modeling is crucial to the design process, and it can occur in a mixed environment where real and virtual objects are combined. A real robotic arm may be coupled with a virtual assembly line involved in laser welding end plates to determine

Chuck Rausin/Shutterstock.com

FIGURE 1.7 Prototyping a Mechanical Interface Using a 3-D CAD System

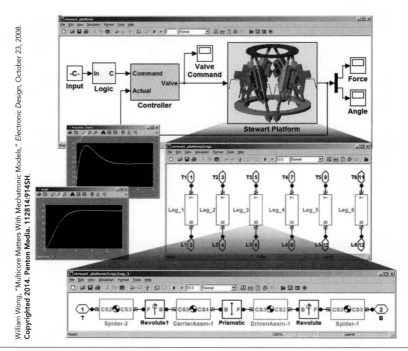

FIGURE 1.8 Simulink Model of a Platform

whether the hand on the robotic arm can reorient an object. The key is getting the virtual objects and their control counterparts to interact with real objects by using computer code running on remote devices. In an integrated product creation process, a small mechanical change may increase the mass of the part; it also may affect how fast the control system ramps up motor speed and how long the part is held in place before returning. Several product creation systems also use software that routes, tracks, and shares work. Most common are workflow tools that automatically route work packages, warn about deadlines, and notify the right people of changes. Many companies use product data management tools to manage multidisciplinary bills of materials.

Multi-domain tools such as Simscape tie together the electronics, mechanical drive elements, and mechanical and hydraulic tools. By reducing the amount of expertise required for developing multidisciplinary products, developers can spend time and effort on other areas in which they are skilled. Likewise, having a model environment permits a better interaction of ideas between design teams.

As application models become more complex, they offer the availability of design verifier and assertion blocks, providing the advantage of being able to look within a model to determine whether a specific object is used (**Figure 1.9**).

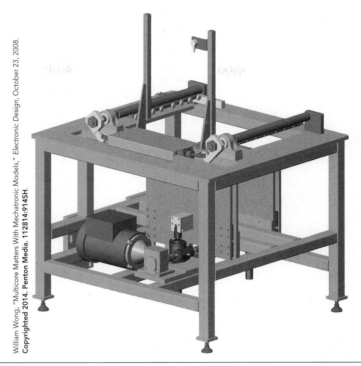

William Wong, "Multicore Matters With Mechatronic Models," *Electronic Design*, October 23, 2008. Copyrighted 2014. **Penton Media. 112814:914SH.**

FIGURE 1.9 Assembly Line Design Using CAD Models

1.3.3 **Evaluating Trade-Offs by Simulation and Modeling**

Virtual prototyping helps designers reduce risk by identifying system-level problems, finding interdependencies, and evaluating performance trade-offs. Simulations enable all functional areas to work on development before the first prototype is completed. Engineers can use force and torque data from simulations for stress and strain analysis to validate whether mechanical components are strong enough to handle the load during operation (**Figure 1.10**). They can validate a machine's entire operating cycle by simulating control system logic. They can calculate a realistic estimate for cycle time performance, which is typically the top performance indicator for a mechanical design, and compare force and torque data with the realistic limitations of transmission components and motors. This information can help identify flaws and drive design alterations from within the CAD environment. Simulations also simplify evaluating engineering trade-offs between different conceptual designs.

Trade-Off Opportunities in Design

Consider the analysis of the torque load for the bottom lead screw actuator of a machine tool design. If a person violates the limits specified by the lead screw supplier, the mechanical transmission parts may not last for their rated life cycles. The use of

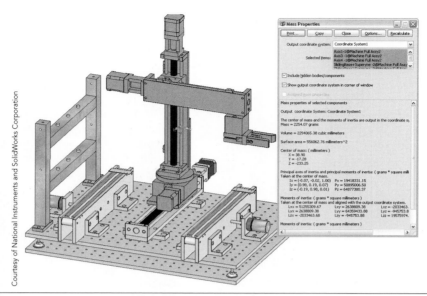

Courtesy of National Instruments and SolidWorks Corporation

FIGURE 1.10 Simulation Verification of Torque Load

simulation software can help find the mass of all the components mounted on the lead screw and determine the resulting center of mass. It can do this by creating a reference coordinate system located at the center of the lead screw table and calculating the mass properties with respect to that coordinate system. With this information, a user can calculate the static torque on the lead screw because of the gravity caused by the overhanging load. Evaluating the dynamic torque induced by the motion is important because it tends to be much higher than the static torque load. Realistic motion profiles help to simulate inverse vehicle dynamics, which can provide more accurate torque and velocity requirements based on the transmission's motion profiles and the mass, friction, and gear ratio properties.

At times, a designer may consider compliance issues when designing assemblies, but incorrect assumptions about operational forces and torques can lead to errors. The flexibility of mechanical transmission components, such as the connecting rods and couplings, causes rotational compliance or linear compliances. Another phenomenon is backlash caused by the clearance between mating components such as gear teeth during change of direction. Compliance and backlash issues can make the proportional-integral-derivative control system difficult to tune. By incorporating onboard prognostic and diagnostic capabilities into product designs, manufacturers are able to make a repair in other locations and substantially reduce maintenance costs.

1.3.4 Hardware-in-the-Loop (HIL) Simulation

Hardware-in-the-loop (HIL) simulation has become an integral part of the current virtual product development process. This testing reassures designers that any assumptions made about a model are correct. PC-based integration of systems benefits from various software packages that use graphical programming to create virtual instrumentation. HIL simulation is also a cost-effective method to perform system tests in a virtual environment. It demonstrates a level of interaction with the modeling of a system that is not possible when a code is directly ported to the final target platform.

HIL facilitates the replacement of conventional mechanical motion control devices with digital devices. Mechanical systems are increasingly being controlled by sophisticated electric motor drives that have digital intelligence from software that runs on an embedded processor. Making correct electromechanical designs requires multidisciplinary teamwork and superb communication among team members. A decision such as choosing the characteristics of a lead screw actuator has a cascading effect throughout the design and can influence the performance of other systems. Integrating motion simulation with CAD simplifies design because the simulation uses information, such as assembly mates, couplings, and material mass properties, that already exists in the CAD model. Adding a high-level functional block language for programming the motion profiles provides easy access to control those assemblies.

In HIL simulation testing, mathematical models replace most of the components of the system environment, and the components to be tested are inserted into the closed loop. If any assumptions are incorrect, the designer has the opportunity to continue optimizing the design before committing to the final hardware platform. Two methods are currently used to accomplish HIL simulation testing. One method utilizes the virtual instrumentation-based user interface coupled with standard data acquisition and control interface. The actual plant environment is used instead of the plant simulation model so that actual sensors and actuators between the plant and the interface are connected. In the prototyping step, many of the model's noncomputer subsystems are replaced with actual hardware. Sensors and actuators provide the interface signals necessary to connect the hardware subsystems back to the model. The resulting block diagram representation of both hardware and software is shown in **Figure 1.11 (b)**.

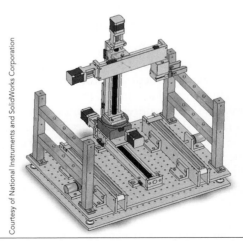

Courtesy of National Instruments and SolidWorks Corporation

FIGURE 1.11 (a) Evaluating Trade-Offs in a CAD Environment

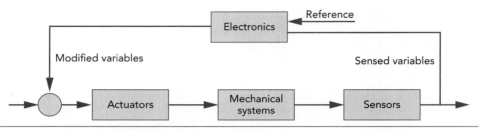

FIGURE 1.11 (b) Hardware in the Loop (HIL)

Because the real part of the model inherently evolves in real time while the mathematical part evolves in simulated time, the two parts must be synchronized. This process of fusing and synchronizing model, sensor, and actuator information is called *real-time interfacing* or *HIL simulation* and is essential in the modeling and simulation environment.

Another method for accomplishing HIL testing involves cross-compiling the control algorithm to target an embedded real-time processor platform. The embedded processor platform is a digital signal processor with input/output (I/O) that is customized for embedded system products. The cross-compiled code is then downloaded onto the embedded processor, sensors are connected to the inputs of the embedded processor board, and actuators are connected to the outputs of the embedded processor board. In other words, this concept called *virtual machine prototyping* brings together motion control software and simulation tools to create a virtual model of an electromechanical product that is in use.

1.3.5 **Simulated Models from the Virtual World to the Real World**

Virtual prototyping (VP) is rapidly gaining importance as the engineering practice of choice to shorten the design cycle and aid rapid product development. The enabling trends for the adoption and rapid proliferation of the VP methodology include:

- Availability of low-cost PC-based parametric simulation/analysis tools
- Integration of multiphysical simulations into a unified environment

Virtual prototype development allows the use of commercially off-the-shelf components that reduce the need for custom-built structures. The specific description and dimensions allow construction of the needed component that is then inserted into the model. This "virtual assembly" process points out any areas where off-the-shelf components do meet requirements and custom fabrication is needed. The model then supports the development of specifications and drawings for custom components. These drawings, specifications, and instructions can be sent to the fabricator accompanied by the 3-D drawings to ensure that the custom or modified items are fabricated correctly.

See **Figure 1.12 (a)** for an example of a conceptual design of a hydraulic valve used in the control industry, **Figure 1.12 (b)** for the design of a wheelchair with an embedded safety mechanism to prevent tilt during uneven motion, and **Figure 1.12 (c)** for an example of the design of a dispensing mechanism for adhesive/glue.

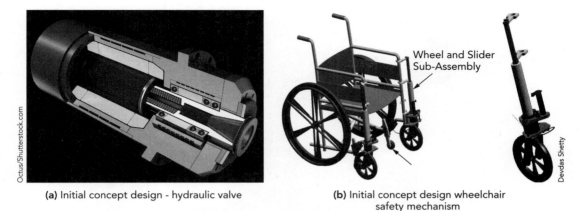

(a) Initial concept design - hydraulic valve

Octus/Shutterstock.com

(b) Initial concept design wheelchair safety mechanism

Wheel and Slider Sub-Assembly

Devdas Shetty

(c) Adhesive dispensing device

Devdas Shetty

FIGURE 1.12 Virtual Design Models

A "walk-through" of a virtual facility via a computer monitor helps developers to identify a model's existing problems and to determine any potential failure areas and more efficient equipment placement. Subsystems can be relocated and moved around in the computer for test fitting and impact analysis. By studying and manipulating the model, developers can experiment with static equipment placement, such as support structures, walkways, plumbing, and electrical distribution systems, to determine optimum results. Computer simulation can currently be used to calculate a system's geometric, kinematic, and dynamic responses and to visualize the results in a 3-D interactive graphical virtual environment. Furthermore, VP can run a wide variety of test suites on computer models without running the risks of overtesting (and possibly wearing out components), thus aiding in the redesign process to achieve the desired performance (**Figure 1.13**). Virtual prototypes can additionally facilitate the involvement of management, sales personnel, and consumers early in the design. Thus, VP has gained acceptance as the method of choice for designing mechanical system products in several leading manufacturing industry sectors.

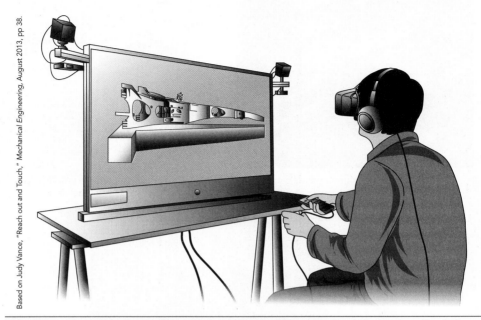

Based on Judy Vance, "Reach out and Touch," *Mechanical Engineering*, August 2013, pp 38.

FIGURE 1.13 Comparison of Simulated Model of Machine Tool Frame in a Real-Time Environment

1.3.6 **Virtual Machine Prototyping**

Designing around the full product life cycle encourages engineers to make prudent decisions about component design and materials—choices that can offer payback when it is time to recycle or pursue "aftermarkets." If in-service data are collected from the product itself or field service organizations, designers can obtain a realistic perspective on product performance by comparing the actual data to original expectations. Because component development is no longer imbedded within a particular product's development program, engineers are free to choose between externally and internally developed components.

Product development programs in effect become product integration programs. By integrating pre-developed components, manufacturers can respond quickly to market opportunities. Increasing the percentage of designs for a carryover component that can be reused in another product not only reduces cycle time but also lowers risk. Some manufacturers have employed this practice in a full-scale program-development platform that manages each product as a generation within a product family. Skilled integrators also incorporate customer feedback, not just as a one-time exercise but also as an ongoing, parallel activity.

1.3.7 **Virtual Prototyping Parallels Production**

Virtual prototyping is one of the most convenient methods of communication between a designer and customers. The ability to turn 360 degrees and flip end over end makes it possible for designers to see a product from every possible angle. For example, during the development of a vehicle, automotive industries can use collaborative platforms to

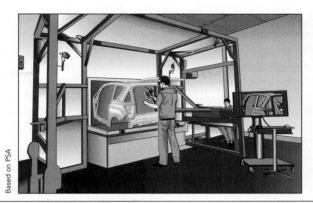

Based on PSA

FIGURE 1.14 INCA 6D, Haptic Device

perform tests, especially to validate the design of interiors (**Figure 1.14**). In this example, a product data model is employed to build a computational prototype on which operations and analysis can be performed as a real-world physical representation. Product realization activities are first performed with respect to the virtual world where all necessary product data and manufacturing processes are modeled. The interaction between the design world and the virtual world comprises a virtual design environment that integrates product definition, design engineering, and manufacturing.

1.3.8 Design for Supply Chain Efficiency

In the past, producers worked relentlessly to improve manufacturing efficiencies because nearly three-quarters of the cost was incurred during the production process. Companies have realized that by designing and testing the entire production process digitally and the supply chain simultaneously, they can drastically reduce production time. The use of supply chain effectively in the virtual prototyping stage contributes to the better understanding of the product. Digital processes can also increase the likelihood of being able to manufacture a product as designed. The product design freeze can, in effect, become the manufacturing design freeze. As an example, many complex products, such as telecommunications systems, radio frequency-based systems (e.g., radars and wireless base stations), and advanced consumer electronics, include sophisticated test sets to ensure quality manufacturing and performance.

1.4 Critical Factors That Influence Supply Chains in a Global Environment

The growth in globalization, and the additional management challenges it brings, has motivated both practitioner and academic interest in global supply chain management. Supply chain management chains transcend national boundaries, imposing the challenges of globalization on designers who design supply chains for existing and new products. A supply chain design problem comprises the decisions regarding the number and location of production facilities, the amount of capacity at each facility, the assignment of each market region to one or more locations, and supplier selection for sub-assemblies, components and materials. Critical virtual reality (VR) factors influence the product supply chain (**Table 1.3**). Globalization has enabled many supply organizations to offer products worldwide, and VR can present their products in a 3-D environment. The supply chain must be flexible enough to accommodate transient conditions such as new contracts, currency fluctuations, natural disasters, and capacity

Critical Factors	Influence on Product Supply Chain	Virtual Reality Result
Product release	Using the supply chain effectively enables companies to remain flexible by varying product releases. In attempting to shrink the overall development cycle, manufacturers have learned the importance of using supply chains and decide when and how they can "freeze" a product's design.	Collaboration from a single virtual product definition can enable geographically dispersed teams to be informed and to coordinate on design. A designer can work faster without waiting for complete information, and one in a different location working on a related part can suggest a design. The designer can insert a virtual placeholder for an incomplete design. Companies can incorporate supplier product-related insight to optimize their designs, increasing their product appeal to the market.
Critical path management	Products and processes are less likely to change during their design and can be interchangeable with one another.	The aerospace industry uses materials that take substantial time in creating complex shapes. By using digital prototypes in design, manufacturers have the confidence to postpone creation of physical prototypes until tools are in place. By using qualified suppliers and postponing the creation of tooling to a later stage in the cycle, the automotive industry has demonstrated savings.
Suppliers as partners	Because they are pre-identified, suppliers can provide systems at a lower price as the result of spreading research development costs across multiple product generations.	Experts from supplier organizations are brought in to participate in real-time exchange of expertise. The product that reaches the marketplace is the culmination of the impact of all tools, processes, and collaborators. Partners can discuss product design regardless of their location. Automotive industries obtain up to 80% of their products from subsystem suppliers. To achieve such gains, collaboration with supplier is essential.

(continued)

Critical Factors	Influence on Product Supply Chain	Virtual Reality Result
Consumer feedback	Industries not only gather consumer requirements early but also develop systematic ways to incorporate feedback throughout the product development cycle. Monitoring a company's market pulse prepares for market acceptance at product launch.	Through virtual technologies, companies are finding new ways to confirm that consumer requirements are being met. Automotive manufacturers invite customers to "virtual car clinics" where future buyers browse through electronic showrooms to view virtual vehicles and give feedback.
Product life cycle management	A supply chain's efficiency influences an organization's future maintenance cost. For example, products such as airplanes are designed to allow for continuous overhaul and repair.	With a virtual product, manufacturers can electronically simulate routine service activities. The cost of maintaining a jet engine can exceed its initial purchase price. Customers often weigh service costs as part of their purchase decision. Early detection of maintenance challenges allows product makers to adapt the product design while it is still fluid.
Security	Companies that work with multiple suppliers must establish environments that adequately protect data without hiding information from those who need it.	Network connectivity, especially over the Internet, introduces an element of risk.

TABLE 1.3 Critical Factors Influencing a Product Supply Chain and Virtual Reality

loadings at suppliers. In many cases, product design and, in particular, supplies for a product can be provided by different links to make the supply chain flexible. A subsystem that uses particularly expensive tests that are unlikely to be replicated at three of four production locations worldwide is not attractive to a product designer. Virtual prototyping and teleconferencing offer companies instant communication, allowing designers to evaluate subsystems thoroughly to decide which ones provide the greatest supply chain flexibility.

With the implementation of new or significantly improved products, operational processes, marketing methods in business practice, workplace organization has had a more serious impact on our lives and environment than ever before. It has also been vital to the development of new production processes that have lowered manufacturing costs, improved quality, and driven economies of scale. At the same time, these technological innovations have led to higher rates of productivity, and therefore, higher return rates on investment. This book mainly looks at the methodologies needed to adopt design-oriented approach in a company and to introduce processes and/or products which add value, lower costs, and raise productivity. See **Table 1.4** for a list of the tools and techniques discussed in the chapters of this book.

Product Development Phase	Tools and Methods	Chapter
Product Design Process		
Introduction	Building Blocks	1
Customer Focus	Market Studies Voice of the Customer Case Studies	2
Creative Concept Generation and Evaluation	Creative Concepts Axiomatic Design Structured Innovation TRIZ Case Studies	3
Product Configuration and Design for Function	Design for Function Developing Product Architecture Case Studies	4
Design Evaluation – Assessing Design Assembly	Design for Assembly Design for Assembly Examples	5
Product Evaluation – Assessing Design for Disassembly and Maintenance	Design for Disassembly (DFD) Assessing DFD Design for Maintenance Case Studies	6
Product Architecture – The Impact on Manufacturing	Systematic Process Selection Streamlining Production Design for Life Cycle Design for Supply Chain Value Stream Mapping Case Studies	7
Sustainable Product Design through Reliability	Sustainable Design Model Robust Design Tools for Optimum Design Evaluation of Failure Modes Root Cause Analysis Case Studies	8
Digital Manufacturing and Virtual Product Prototyping	Digital Manufacturing Virtual Model Creation Product Prototyping Additive Manufacturing Case Studies	9
Additional Case Studies	Putting It Together	10

TABLE 1.4 Tools and Techniques Discussed in Chapters

CASE STUDY 1.1
Why Promising Products Failed

Automotive Examples

The Ford Edsel

The Ford Motor Company introduced the Edsel (**Figure 1.15**) in 1958 as a cutting-edge and innovative product. The company's marketing of Edsel led customers to expect something extraordinary. The car was priced higher than other Ford models, although it looked similar to the existing ones. In spite of the heavy marketing, sales were poor. In addition, the market economy had been better when the company initiated the project. By the time the Edsel came to the market, the U.S. economy was in a recession. Ford considered the Edsel to be a luxury car, but consumers wanted an "economic" car, so the car failed. The president of Ford convinced the board to drop the Edsel.

This example shows that the customer needs and requirements must be a key component of product innovation.

The Tucker 48

In 1948, Preston Tucker introduced a new sedan to the market (**Figure 1.16**). It had a number of advanced features for safety and performance, many of which have become standard items on nearly every modern car. Using surplus helicopter engines and building in a surplus WWII factory, the company raised enough capital to build only 51 cars. All initial capital was used for research and development instead of production launch. Although market interest in the vehicle was high, the product failed because of financial problems and the news media's negative publicity. A significant factor in the company's demise in 1949 was the focus on too many innovative features in the new vehicle.

© Bettmann/CORBIS

FIGURE 1.15 The Ford Edsel

FIGURE 1.16 The Tucker 48

CASE STUDY 1.2
Examples of Ready-to-Use Products

Maxwell House Ready-to-Drink Coffee

Maxwell House Coffee introduced its ready-to-drink coffee in 1990. The idea behind this product was to provide consumers the ability to have their coffee ready to use whenever and wherever they wanted. The promotional materials promised the drink would be a "convenient new way to enjoy the rich taste of Maxwell House Coffee." The reasons for failure could have been consumers' inability to heat the coffee in its original container in a microwave, forcing users to pour coffee into a mug before heating the beverage. Consumers, it turned out, seemed happier pouring their coffee from a coffee pot than from a cold carton. Thus, the product failed to provide the expected functions of portability, quickness, and ease of use, and consumers decided to continue making their coffee in a pot.

McDonald's Arch Deluxe

In 1996, McDonald's launched a new burger, the Arch Deluxe, that was created to capture the market of adults with sophisticated taste. Because McDonald's is a fast-food company, customers were not as interested about sophistication as McDonald's marketing team had hoped. The main reason behind this failure was the lack of participation by the appropriate market segment. After one year, this new product was taken out of McDonald's menu.

Bengay Aspirin

At its introduction, Bengay aspirin was expected to be successful by having a well-known name behind it. Bengay is primarily known for its burning sensation to relieve pain upon contact with skin but was notorious for its strong unpleasant odor. In this case, having a well-known brand closely tied to a single product image and linked to the new product hindered the product's acceptance.

CASE STUDY 1.3
Creation of a Successful Product by Virtual Reality

Virtual Reality Creation of Ferrari

Ferrari, a widely known producer of sports and race cars, uses virtual design and prototyping in its design and testing activities that support the development of the vehicles' physical components (**Figure 1.17**). These areas include packaging design and assembly, which require a trade-off between weight and dimension, layout optimization, and component mounting. Developers can analyze numerous designs via a computer screen. Simulations of cockpit airflow using virtual smoke, tank refueling sequences, rollover testing, and roll bar activation enable the analysis of designs that is especially important for race cars.

The use of virtual reality for design analysis through simulations includes changing the lateral acceleration for better drivability and varying the car's steering and longitudinal acceleration. The company also uses virtual reality simulations for analyzing high-speed braking, studying the relationship between braking-force distribution between the front and rear along the track distance, and testing two- and four-wheel drives at different tracks with varying surface roughness. The simulation provides ample data on engine control system integration and four-wheel drive torque distribution.

Virtual prototyping technology is useful in evaluating hybrid technology and battery packs for emission reduction. Ferrari uses the virtual prototyping to design the control of each subsystem independently rather than a trial and error approach. The driving simulator can provide visualizations of most driver activities, including having the vehicle driver inside the loop and evaluating lap time. In addition, prototyping technology can allow designers to study the vehicle's setup and reaction times. Design management teams have realized that virtual prototyping provides human–machine

FIGURE 1.17 Ferrari Racing Car

interaction and ergonomics for superior mechatronics product creation. Management has realized that virtual prototyping can reduce the cycle time from initial concept development to design, prototyping, and testing by 30%, resulting in a 30% reduction in cost. According to CEO of Ferrari, Dr. Amedeo Felisa, the philosophy of virtual prototyping has been engrained into the fabric of the Ferrari company and can be viewed as an irreversible process.

Ferrari uses virtual prototyping in the following areas:

- Cash flow analysis
- Comprehensive packaging design and assembly requirement
- Race car layout optimization
- Vehicle fluid dynamics analysis
- Air flow simulation with smoke
- Fluid dynamics for system cooling and evaluation
- Simulation studies in vehicle tank refueling
- Cockpit airflow analysis
- Vehicle dynamics and rollover testing
- Evaluation of vehicle dynamics for hybrid emission reduction technology
- High-speed braking analysis
- Control system integration and simulation
- Dashboard rendering

In this example, the hands-on design creation, analysis of vehicle dynamics, engine testing, control system simulation, safety analysis, and examination of all aspects of virtual product development experiments were possible. With the virtual experiments, the designers working in a team were able to share results on their experiments. In addition, designers can ask each other questions about the results. Experiments show that the existing methodologies can be effectively used to validate the design and reduce the time from conception to market.

Conclusion

In today's highly customized and quality-focused product design and manufacturing, virtual prototyping has become an important and useful tool in science and engineering. VR applications cover a wide range of industrial areas from product design to analysis and product prototyping to manufacturing. The design and manufacturing of a product can be viewed, evaluated, and improved in a virtual environment before its prototype is made, which is an enormous cost saving.

With the virtual product development, the designers working in a team can share results on their experiments with each other and ask questions about the results. As an experiment progresses, periodic response profiles are constructed and posted on the server.

REFERENCES

1. Devdas Shetty and Ric Kolk, *Mechatronics System Design* (Toronto: Cengage Learning International, 2012).
2. Devdas Shetty, *Design for Product Success Society of Manufacturing Engineers* (Dearborn, MI: SME, 2003).
3. Neil Rackham, "From Experience: Why Bad Things Happen to Good New Products," *Journal of Product Innovation Management* 15, no. 3 (1998), 201–20.
4. Avi Mandelbaum, Vien Nguyen, and Elizabeth Schnerer. "Getting the Most of Your Product Development Process," *Harvard Business Review* (March-April 1996), 134–51.
5. Robert L. Nagel, Kenneth Perry, Robert B. Stone, and Daniel A. McAdams, "Function-Based Design Process for an Intelligent Ground Vehicle Vision System," *Journal of Electronic Imaging* 19, no. 4 (2010).
6. William Wong, "Simulink Model of a Platform," *Electronic Design* (October 2008).
7. Weiqun Cao et al., "Digital Product Development in a Distributed Virtual Environment," *Proceedings SPIE* 5444 (no. 322) (2004).
8. A. Ali, Z. Chen, and J. Lee, "Web-Enabled Platform for Distributed and Dynamic Decision Making Systems," *International Journal of Advanced Manufacturing Technology* 38, no. 11 (2007).
9. Brian MacCleery and Nipun Mathur, "Right the First Time," *Mechanical Engineering* (June 2008).
10. David Nelson and Elaine Cohen, "Optimization-Based Virtual Surface Contact Manipulation at Force Control Rates," Proc. IEEE Virtual Reality (March 2000).
11. David Johnson and Elaine Cohen, "Specialized Normal Cone Hierarchies," *Proceedings of 2001 ACM Symposium on Interactive 3D Graphics, Research Triangle Park, NC, March 2001.*
12. G. Caruso, S. Polistina, and M. Bordegoni, "Collaborative Mixed-Reality Environment to Support the Industrial Product Development," *Proceedings of ASME World Conference on Innovative Virtual Reality, June 2011, Milan, Italy.*
13. Aberdeen Group, *System Design: New Product Development for Mechatronics* (Boston, MA: Author, 2008).
14. N. Shyamsunder and R. Gadh, "Collaborative Virtual Prototype of Product Assemblies over the Internet," *Computer-Aided Design* 34, no. 10 (2002), 755–68.
15. R. Bedini et al., "From Traditional to Virtual Design of Machine Tools, a Long Way to Go—Problem Identification and Validation," Paper Presented at the conference of the International Mechanical Engineers (IMECE) (November 2006).
16. Devdas Shetty, "Virtual Product Design Using Innovative Mechatronics Techniques for Global Supply Chain," ASME 2011 International Mechanical Engineering Congress & Exposition, IMECE 2011-64228, Denver, CO, November 2011.
17. T.V. Thompson II and E. Cohen, "Direct Haptic Rendering of Complex Trimmed NURBS Models," Proceedings of ASME Dynamic Systems and Control, Nashville, TN November 14–16, 1999.
18. Ian Gibson, David Rosen, and Brent Stucker, *Additive Manufacturing Technologies* (New York: Springer, 2010).
19. mathworks.com/products/simscape/demos.html mathworks.com/company/events/webinars/wbnr42107.html?id=42107&p1=688660407&p2=688660425.
20. Judy Vance, "The Hands on Control of Virtual Reality Gives Designers Increasing Power to Make Better Decisions," *Mechanical Engineering*, August 2013.
21. MSC software, www.mscsoftware.com/info mathworks.com/products/simscape/demos.html mathworks.com/company/events/webinars/wbnr42107.html?id=42107&p1=688660407&p2=688660425.
22. Brian MacCleery and Nipun Mathur, "Right the first time," *Mechanical Engineering*, June 2008.
23. Paul Adler, Avi Mandelbaum, Vien Nguyen and Elizabeth Schnerer, "Getting the Most of your Product Development Process", *Harvard Business Review*, March-April: pp 134–151.

EXERCISES

1.1. Suggest possible improvements you would like to see made to a common household item.

1.2. List several tools and techniques required for product development. How would you use them effectively?

1.3. List two examples of promising products that have failed and briefly describe the reasons for it.

1.4. Examine the 1998 case study of Breakfast Mates from Kellogg. Identify the reasons for the product to be withdrawn.

1.5. What role does research play in product design? Indicate four sources of research and the desired information that you can expect from each source.

1.6. With the help of sketches, show the basic concepts that can be used in the development of a product. What strategies would you use to have an effective design as speedily as possible?

1.7. Explain how you would select a team to undertake the product design for a multicomponent product.

1.8. Draw the functional diagram of an entire product. With examples, illustrate the difference between constraints and engineering specifications.

1.9. From the perspective of engineering design, what are the differences and similarities between a system and a product?

1.10. Using web-based research, identify five corporations and their product lines. In what way do these products support each corporation's strategy?

CHAPTER 2
Customer Focus

OBJECTIVES

The process of creating a product is also influenced by the proper determination of customer requirements. Global competitiveness has brought greater focus on customers' views. Gathering information from customers involves getting information from both internal customers and external customers. The process involves both collecting the data and then interpzs customer needs to formulate product specifications, generating concepts for detailed design. The main benefit of an organized process is to ensure that the product is focused on customer needs and that no primary requirements are ignored.

2.1 Introduction: Customer Focus

During the past two decades, the emergence of global economy along with strong global competitiveness and markets influenced by customer preferences about products in addition to technological change has caused major shift in quality. The quality management practices of Japanese and U.S. companies provided an opportunity to influence the cost and lead times of new products. Early quality initiatives focused on reducing the process variability in manufacturing; later efforts were focused on using concurrent engineering philosophy for product and process development. The advantages that come from cutting time to market and continuously developing quality products are so great that the balance in some sectors is shifting in favor of companies that adopt new strategies. Companies that introduce new products and react quickly to external changes are racing ahead of their competitors.

One essential element affecting the entire development process is the proper determination of customer needs. Worldwide competitiveness has brought increased focus on customers' needs and opinions. Customer needs-analysis projects customers' future needs, not merely current ones identified by the marketing department. Encouraging members of the product design team to participate in the customer needs-analysis can enhance their creative contributions. It enables them to see opportunities that they might not see by simply reading a market report. Internal customers are people connected with corporate management, manufacturing personnel, and sales and field service people. External customers are the end users.

The Quality Function Deployment (QFD) methodology is ideally suited for supporting a company's total quality initiative. It provides a framework for product or program design, which starts with the customers to find out what they want. This methodology identifies a mechanism by which an organization can respond to customer needs. QFD methods are employed by some of the larger U.S. companies as a way to obtain better quality products in a shorter cycle time.

2.1.1 Quality for Product Success

Companies can survive only by providing quality. If a company's objective is to bring new or improved products to market with low cost and high quality, the QFD procedure is a vehicle for achieving this. QFD is an organized procedure for early product design. Basically, this technique involves incorporating customer requirements and expectations into product design and then monitoring them through the stages of design, planning, and manufacturing. More precisely, QFD utilizes a conceptual map or group of matrices known as the *house of quality* to relate customer attributes to product specifications and design operations.

In QFD, the term *quality* takes on a much broader meaning than the more conventional definition in design and projection, which is limited to a product's adherence to its manufacturer's specification. The basic definition of QFD, which is translated from Japanese words "HIN SHITSU KI NU TEN KAI" is "a system for translating customer requirements at each stage from research, product development to engineering and manufacturing to marketing/sales and distribution." Ultimately, customers receive a product that meets their demands with the minimal amount of design changes.

What makes QFD a powerful tool in the manufacturing sector is the direct input from the customer and the interdepartmental communication between marketing, engineering, manufacturing, and management. In a typical industry, the interdisciplinary group usually referred to as the "quality team" is charged with the responsibility to integrate customer requirements into product design. This group usually meets on a regular basis and conducts brainstorming sessions. The end result is the development of a quality system that delivers a product of specified quality standards.

QFD is totally driven by the concept of quality and results in the best possible product to market. It requires a paradigm shift from traditional manufacturing quality control to product design quality control. The old paradigm is quality control by inspecting physical products through observation and measurements; this is *inspected-in quality*. But the new QFD paradigm is to design quality into the products and their manufacturing processes so that products are produced error free; this is referred to as *designed-in quality* (**Figure 2.1**).

The advantages of using QFD are

- Reduction of product design time
- Cost reduction
- Early exposure of design trade-offs

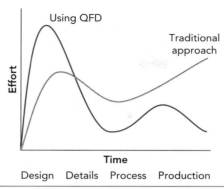

FIGURE 2.1 Relationship between Effort and Time

- Written documentation of design decisions
- Reduction in errors and corrections to design
- Clarity for decisions
- Incorporation of a collective experience of a multifunctional team capable of guiding sound decisions

QFD is very essential in current industrial climate and as customer demands vary with changing times. It needs to be incorporated in the design accordingly and accurately. It is the only comprehensive quality system aimed specifically at satisfying the customer. Furthermore, QFD allows customers to prioritize their requirements with a comparative analysis of current designs and identify those features that will lead to the greatest competitive advantage.

The four phases of QFD according to Akao (1997) and Sullivan (1986) are:

- Product planning phase
- Part deployment phase
- Process deployment phase
- Product deployment phase

In the *product planning phase*, customers' preferred attributes are identified based on surveys, interviews, observations, field contacts, focus groups, employee feedback, publications, and sales records and are then converted into product characteristics. A relationship matrix between the customer requirements and the product characteristics is drawn. This matrix of data from market evaluation includes customer-expressed importance ratings and data on competitors' products. The matrix has information on current product strengths and weaknesses, measurable targets to be achieved and selling points. In the *part deployment phase*, the product characteristics are translated into a component's characteristics. At this stage, the characteristics of the final product are converted into component part details. In the *process deployment phase*, the process plan for the manufacture of the component, subassembly, and assembly are identified as well as the quality parameters. In the *production deployment phase*, the output from the process

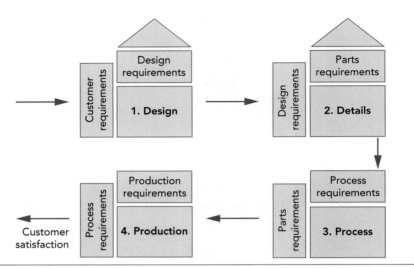

FIGURE 2.2 Four Phases of QFD

deployment charts provides a measure of critical product and process parameters. At this stage, production operations for all critical components are identified (**Figure 2.2**).

2.2 How to Build a House of Quality for a Product

The central body of the house of quality consists of *whats* (customer needs), *hows* (what manufacturer controls), and a matrix of relationship between the *whats* and *hows*. See **Figure 2.3** for the essential elements of a general house of quality. The customer requirements/needs/wants are listed horizontally and are known as *whats*. The counterpart technical characteristics are listed vertically and are known as *hows*. The interrelationship matrix (*whats* against the *hows*) is shown where the horizontal and vertical axes meet. The roof of the matrix shows the correlation between the *hows* and *hows*. The bottom of the house of quality gives an indication of the technical characteristics against the technical bench marking ("*hows*" against the "*how muchs*").

2.2.1 Features of the House of Quality

Product Attributes

To determine the desired attributes, list customer requirements for the product. The list can be obtained through surveys, interviews, observations, field contacts, focus groups, employee feedback, publications, sales records, and complaints. Classify customer attributes into specific categories or subdivisions, such as what is primarily important and what is secondary in nature. Weigh each customer attribute by its degree of relative

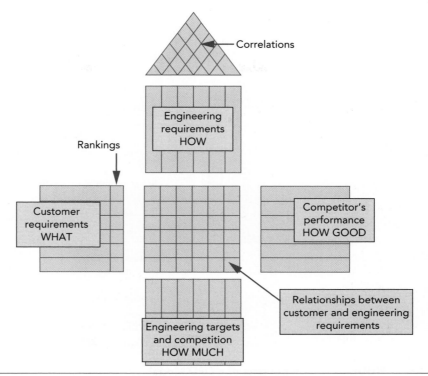

FIGURE 2.3 Typical QFD Matrix for Product Design

importance. To obtain an objective weighing, a brainstorming session amongst product design group members must take place.

Product Control Characteristics
Develop engineering characteristics of the product that reflect customer-desired features. This requires conducting another brainstorming session by the product team. Categorize engineering characteristics into specific items as required and list the product characteristics on the axis across the top of the house.

Interaction Matrix
The product group combines the two axes into a correlation matrix identifying strong/medium/weak relationships. A set of symbols is used to represent correlation (**Figure 2.4**). A matrix that shows a majority of weak relationship signs indicates that certain customer requirements have not been met.

Interaction between Parameters
Develop the roof of the house of quality by building a diagonal matrix above the engineering characteristics. Doing so allows the product team to rate weak–strong relationships among different characteristics. Changing a parameter can influence

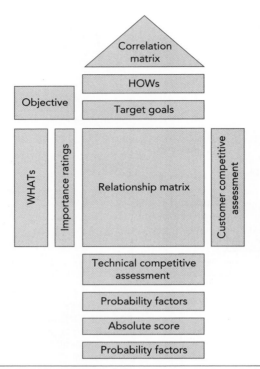

FIGURE 2.4 Components of a QFD Model

other parameters. It is important to know the nature and the strengths of these interactions.

Target Values and Technical Analysis

The basement of the house of quality can be used when objective measurements must be made to compare the specifications of a competitor's products versus those of your product. This comparison provides an insight into the possibilities for improving your product and assists in setting up new target values to be followed.

Full extension of the house of quality concept then allows the *"voice of the customer"* to cascade through the product introduction process via process selection to operational instructions. Each level of the matrix relates to the important elements of *how* and *what*. This concept also ensures that difficult-to-meet requirements are met and the focus on design and fabrication is passed from one matrix to the next.

As illustrated in the following model, the house of quality includes the following components:

- Voice of the customer (VOC)
- Importance ratings
- Assessment of competitor products

- Target goals
- Correlation matrix
- Technical assessment
- Probability factors
- Relationship matrix
- Absolute score
- Relative score

2.2.2 House of Quality—Step by Step

‣ **Step 1: Voice of the Customer**
  - Identify actual customers' needs, wants, and requirements to ensure that the product design decisions are based on them, not just on perceived customer needs.
  - Identify the relative value that customers place on these items.
  - Ensure the involvement of all groups in a company that receive any feedback from customers.

‣ **Step 2: Customer Requirements and Rating**
  - Expand customer requirements to specific points as shown in **Figure 2.5**.
  - Perform this with a "what-to-how" technique.
  - Construct the following chart.

Requirements		
Primary	**Secondary**	**Tertiary**
Dependable	Trouble free	No breakdowns
		Non-critical parts
		Components don't break
	Long life simple fast repairs	
Economical		

FIGURE 2.5 Primary, Secondary, and Tertiary Requirements

- Expand the "*how*" list until each point has a measurable assessment.

▸ **Step 3: Planning Matrix for assessment of competitor product**

▸ **Step 4: Fill out the Correlation Matrix to Determine How Factors Relate to target**
  • Establish positive changes to any one of the control characteristics to the other members.
  • Reconsider the product and process if the matrix indicates more negative than positive effects.

▸ **Step 5: Complete the Relationship Matrix (correlation) and Importance Rating Values**
  • Identify the relationships between the "*whats*" and "*hows*" on the matrix by assigning weights. For example, use 0–9, with 0 indicating very weak and 9 indicating very strong.
  • Total the column values to determine an importance rating. This should result in a few clearly important features and some that are clearly not important.

▸ **Step 6: Evaluate Customer Importance and Technical Assessment**
  • Quantify the opinions of the customer (as collected in Step 1) in terms of importance of the "*what*" requirements and enter numbers in the Customer Importance Rating column.
  • Rank the overall ratings for your product and those of competitive products for each requirement "*what*" from poor to good. These values are derived from information gathered in Step 1 and entered in the Market Evaluation column.
  • Review the product's clearly identified strengths and weaknesses.
  • Reexamine the customer requirements if these values have a reasonable distribution.

▸ **Step 7: Control the Characteristics Competitive Evaluation**
  • Compare competitor products to the internal product. Consider the performance criteria in terms of the Final Product Control Characteristics column.
  • Enter values in the Control Characteristics Competitive Evaluation column of the chart and rank them from good to poor.
  • Compare these numbers to the numbers in the Importance Rating row, which should clearly identify any technical deficiencies in the product and make sure their importance is clear.

▸ **Step 8: Evaluate the Chart**
  • Make some critical evaluations when the chart has adequate information to do so.
  • Identify candidates for change from the Control Characteristics Competitive Evaluation column and the Importance Rating row. Focus on features where the competitor product has higher ratings.
  • Identify how the candidates should be ranked to determine how to correlate the results in the Customer Importance Rating and Market Evaluation columns. If the customer considers a factor as less important, consider it less important as well. (Use the planning matrix to find the effects.)

- If any of the customer requirements are unanswered, the design requirements must be reconsidered.

▸ **Step 9: Develop New Target Values**
- Select new target values using currently implemented design parameters and the relative importance identified in Step 8.
- Use the values determined for the competitor's products to compare to those of the in-house product.
- Use a separate sheet or document for any additional documentations.

▸ **Step 10: Technical Difficulty and Probability Factors**
- Estimate the difficulty of achieving the target value by considering the target values and previous production performance.
- Enter a ranking in the Degree of Technical Difficulty row.

▸ **Step 11: Deployment Selection**
- Identify the cost of quality and consider the cost/benefit trade-off.
- Select one or more factors. If there are not a few clear choices, repeat the process.
- Select the elements with the lowest technical difficulties, but the greatest importance ratings. This decision will be slightly arbitrary, but it should not be very different from what the chart suggests.

▸ **Step 12: Deployment Matrices**
- Develop a deployment matrix for each control characteristic selected in Step 11.
- Develop the top of the matrix discussed in creating the planning matrix.

Measure the control characteristics and position the target values. The values shown in the bottom represent components or products that can be influenced by design.

▸ **Step 13: Design and Test**
- Use the deployment matrix to perform design work, test the results, and compare them to the target values.

The house of quality procedure is inherently a group work approach designed to ensure that everyone works together to give customers what they want. It has changed the way people think and brings quality to products and the manufacturing processes.

Figure 2.6 is a chart that was created for information concerning an electric receptacle (20 amps) to be used along a raceway. In a typical laboratory installation, two standard 20-amp receptacles are installed to allow the user to plug in electrical devices. In the example considered, electricians using commonly available tools can install the receptacle created by the new design.

The receptacle is composed of many different parts such as receptacle face and base, rivet, ground contact subassembly, receptacle yoke, and wire screw terminal assembly. The customer's perception for a receptacle is obtained by technicians or

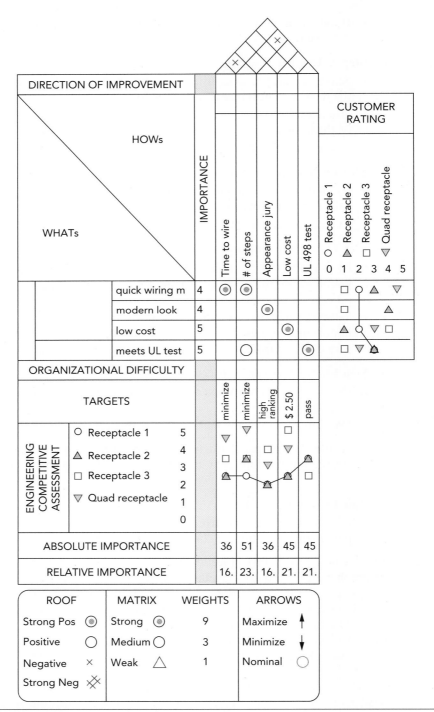

FIGURE 2.6 House of Quality Chart for an Electrical Receptacle

contractors as they have the primary responsibility for installation. As shown in the figure, the number of steps influences the time for installation, which is the most important *customer want*, followed by cost, then its ability to pass the tests required by electrical codes. See **Figure 2.7** for a house of quality chart for a refrigerator, which also shows the relationship between the voice of the customer and design requirements.

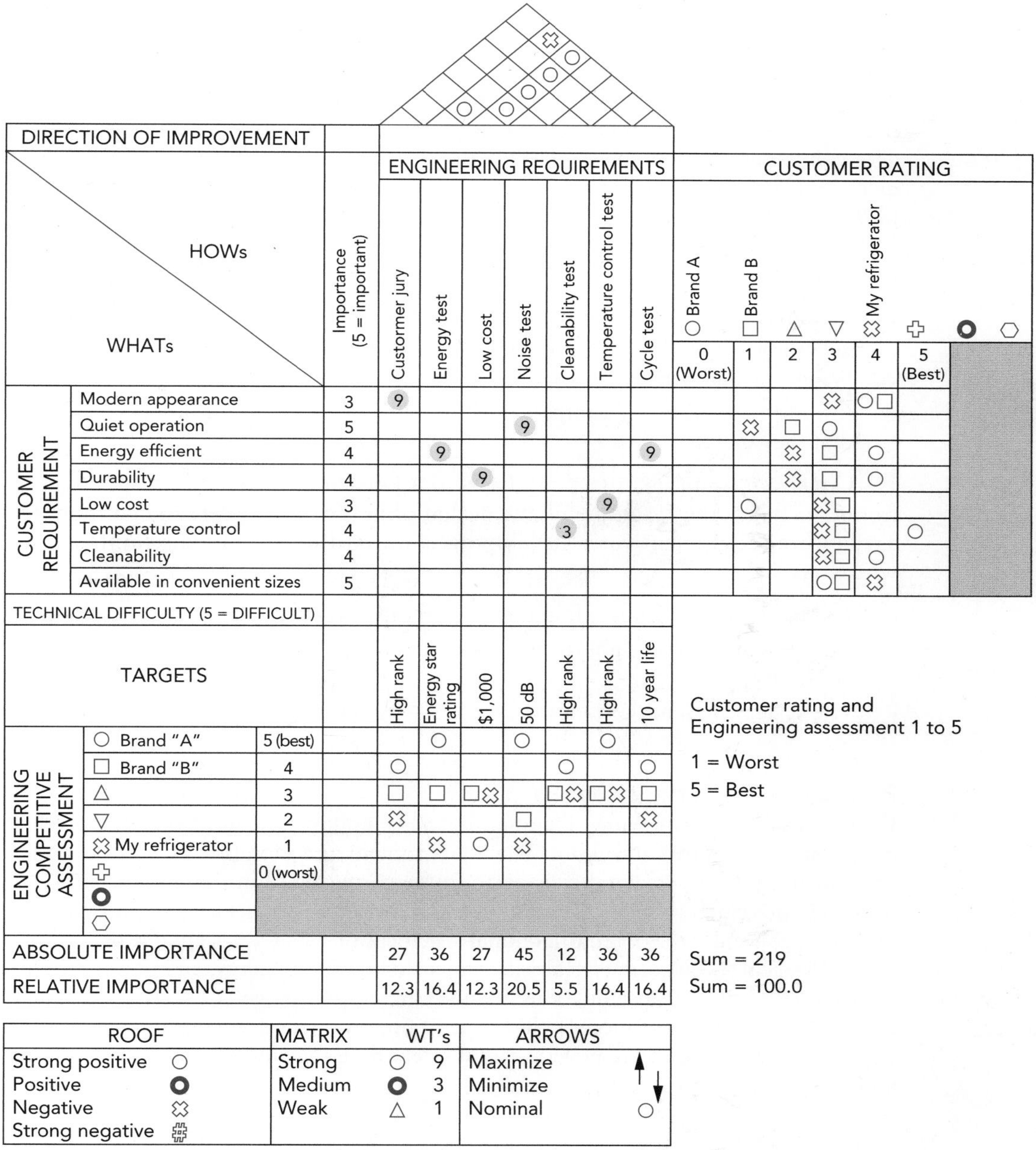

DIRECTION OF IMPROVEMENT

HOWs / WHATs

ENGINEERING REQUIREMENTS: Customer jury, Energy test, Low cost, Noise test, Cleanability test, Temperature control test, Cycle test

CUSTOMER RATING: ○ Brand A, □ Brand B, △, ▽, ⊠ My refrigerator, ✛, ●, ○ — scale 0 (Worst) to 5 (Best)

CUSTOMER REQUIREMENT (WHATs)	Importance (5 = important)	Customer jury	Energy test	Low cost	Noise test	Cleanability test	Temperature control test	Cycle test
Modern appearance	3	9						
Quiet operation	5				9			
Energy efficient	4		9					9
Durability	4			9				
Low cost	3						9	
Temperature control	4						3	
Cleanability	4							
Available in convenient sizes	5							

TECHNICAL DIFFICULTY (5 = DIFFICULT)

TARGETS

		High rank	Energy star rating	$1,000	50 dB	High rank	High rank	10 year life
○ Brand "A"	5 (best)		○		○		○	
□ Brand "B"	4	○				○		○
△	3	□	□	□⊠		□⊠	□⊠	□
▽	2	⊠			□			⊠
⊠ My refrigerator	1		⊠	○	⊠			
✛	0 (worst)							
●								
○								

		High rank	Energy star rating	$1,000	50 dB	High rank	High rank	10 year life
ABSOLUTE IMPORTANCE		27	36	27	45	12	36	36
RELATIVE IMPORTANCE		12.3	16.4	12.3	20.5	5.5	16.4	16.4

Sum = 219
Sum = 100.0

Customer rating and Engineering assessment 1 to 5

1 = Worst
5 = Best

ROOF		MATRIX	WT's	ARROWS	
Strong positive	○	Strong	○ 9	Maximize	↑
Positive	●	Medium	● 3	Minimize	↓
Negative	⊠	Weak	△ 1	Nominal	○
Strong negative	⌗				

FIGURE 2.7 House of Quality Chart Developed for a Refrigerator

2.2.3 **Design Constraints and Product Specification**

Constraints define the permissible range of design specifications and the limits of performance range. If a design does not satisfy the constraints, the product is not considered irrespective of its performance. Constraints basically represent allowable limits placed on a design. An equality constraint specifies that some characteristics of the design must adhere to some specific value, and any variations are not allowed. An inequality constraint may mean certain satisfaction of design parameter to be greater than or less than some certain specified value

For example, let us list some constraints in the case of automotive design. If the overall goal is to design an improved automobile bumper in order to address the damages to bumpers in a low speed collision, the problem can be defined as follows: *Design an inexpensive front bumper so that car withstands 5mph head-on collision with concrete wall without significantly damaging the bumper or other car parts. In addition, at the end of its useful life, the bumper must be easily recyclable.* An example of the equality constraint can be:

- To prevent overriding bumpers in collision between automobiles, the federal government requires all bumpers be installed 18" above ground level.
- Weight of bumpers can not exceed 50 lbs.
- Mounting brackets on bumpers must be between 8" and 12" from the center so that they match.

Constraints can identify limitations to a particular product in a special environment. In another example, if the design objective is to create human powered agricultural equipment for a developing country, the design constraints can be written as,

- Use materials that can be found locally in the country.
- The system should be able to be produced and maintained without the use of electricity.
- The system shall not cost more than $150 to complete.

Product Design Specifications

Design specifications identify the quantitative boundaries within which the product design procedures are carried out. The establishment and evolution of product design specifications are considered as a basic frame of reference in designing a product. Specifications can be of several types, such as, functional, physical, economic, environmental, ergonomic, and legal. Functional specifications look at acceptable operational limits. Physical specifications deal with dimensional needs, energy, and power requirements. Environmental specifications include temperature ranges, dust levels, and noise limits. Economic specifications include operation, production, and maintenance costs. Legal specifications include governmental safety requirements, production standards, and environmental codes.

CASE STUDY 2.1
The Portable Coffeemaker

The objective of any product development process is to design a product that meets the needs of target customers and exceeds their expectations, while allowing the company to decrease manufacture and assembly costs. The best products are those that are easy to be used, maintained, and repaired over time, and then to be disposed of at the end of their lifecycle. These requirements do not make the process of product development an easy task. Because various persons, each with a different background and function within the organization, are involved in the product development process, it is important that the process be carefully coordinated and planned.

Market Study

Market research represents a critical step in the product development process. For the product to be successful, developers must understand the nature of the broad market, its needs, as well as alternatives currently available. Next a specific market segment needs to be chosen as the focus of new product development efforts. An appropriate marketing strategy is critical for product success. Coherency among four "P" factors (product, price, promotion, and place) is essential. This section of the case study examines the current U.S. coffee market and the existing options for brewing coffee "at home." The need for a portable coffeemaker is identified.

The U. S. Coffee Market

The National Coffee Association found in 2000 that 54% of the adult population in the United States drinks coffee daily; each person drinks an average of 3.1 cups of coffee per day (average daily consumption for the entire population is 1.9 cups per day for men, 1.4 for women). Refer to **Figure 2.9** for the coffee market composition in terms of customer segments and their preferences. Three criteria have been identified for market analysis: taste, convenience, and price. The majority of coffee consumers (48%) belong to the Type 4 segment, indicating that the most important element in defining their purchasing behavior is the price. In the order of importance, price is followed by convenience and taste.

Phase	Approach
Customer focus	• Market studies • Voice of the customer • Quality function deployment

FIGURE 2.8 Approaches for Customer Focus

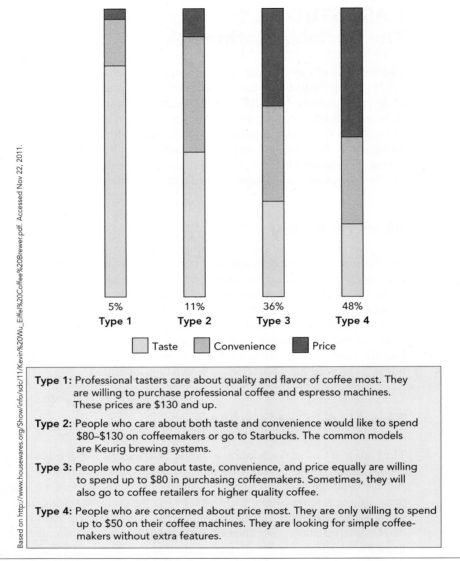

FIGURE 2.9 Coffee Market Segmentation

Existing Products and Procedures

There are currently three main methods for brewing coffee (**Figure 2.10**). Because of the different techniques on which these methods are based, each of them produces a different result in terms of coffee intensity and taste.

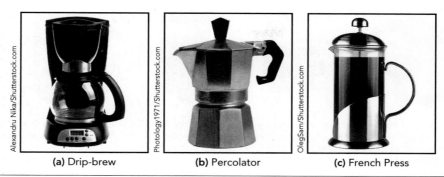

FIGURE 2.10 Methods for Brewing Coffee

Drip-Brew

Water is rapidly heated through a tubular heating element until it boils. The vapor then passes through a one-way valve and disperses once it has cooled again into liquid, over the coffee powder. The water first passes through the coffee grounds and then through a filter, usually made of paper, before dropping into a pot. The pot is usually kept warm by the same heating element that boils the water.

Percolator

The percolator method is less common in United States than in European countries, particularly Italy and Spain. This version uses a heating element on a stove to bring water to the boiling point. After boiling, the water is forced up a tube, typically made of metal and placed in the center of the pot. The vapor then cools to water, passes over the coffee grounds and through a metal filter before returning back into the reservoir of water.

French Press

This method is probably the simplest of the three methods. Coffee grounds are placed into a pot with a plunging element. Boiling water is added and left in the pot for enough time to brew the coffee from the grounds. Finally, the water is separated from the coffee grounds by the plunging element and then poured into a mug.

Figure 2.11 is a picture of the coffee machines most used in the current U.S. market based on their sales on Amazon.com.

Problem Definition

For effective and efficient problem solving, a satisfactory definition of the problem is critical. The coffee market is a major market in the United States. The coffeemaker is one of the most used appliances in everyday life. The problem definition starts by acknowledging the market need for a more convenient yet inexpensive solution for making coffee. Existing coffee machines are not easily stored or carried because of their rigid shapes and bulky structures. The new and redesigned portable coffeemaker combines the

Most popular coffee machines are made by Cuisinart, Keurig, Aerobie, and Black and Decker. The models vary in capability and size. Cuisinart equipment has a range of brewing options. Models under the name Keurig, Aerobie, and Black and Decker have an array of special features such as descale indicators, removable drip trays, disposable paper filters, and automatic shut-off.

FIGURE 2.11 Five Best-Selling CoffeeMakers from Amazon

Source: http://www.t-falusa.com
Source: http://www.mrcoffee.com/
Source: http://www.mrcoffee.com/
Source: http://www.cuisinart.ca/
Source: http://www.keurig.com/

convenience of making coffee at anytime and anywhere with the saving that comes from making your own coffee instead of buying it. The reduced size and limited functionality of a portable coffeemaker allow for a different product that has a special market appeal.

Problem

Based on the previous research and the product definition procedures, the problem statement can be defined as follows:

"High volume and weight of regular coffeemakers make them not easily transportable for use outside the home."

Customer Requirements

To ensure that customer needs are considered and incorporated in the concept design, organizations use QFD or house of quality methodology. The main goal of QFD is to translate customer requirements into product specifications at each stage of the product development process from design to engineering, manufacturing, and marketing/sales and distribution.

The first step in the QFD is to identify customers' requirements and rank each of them based on their importance for the customer. Then customer requirements are translated into engineering specifications. The correlation matrix is completed to determine how each customer specification relates to technical specifications. Finally, target values for each specification are set as shown in **Figure 2.12**.

- **Safety**. The most important requirement identified by customer needs-analysis was safety. The portable coffeemaker should provide maximum safety (the goal is to minimize the number of hazards) when carried around and used in different types of environments.

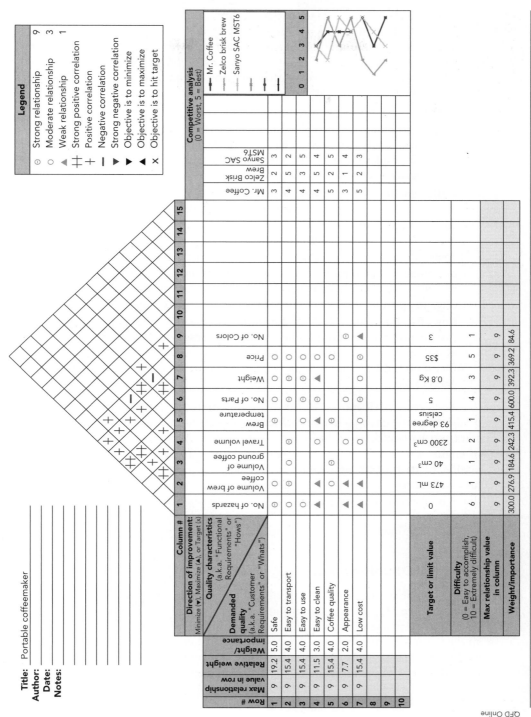

FIGURE 2.12 Quality Function Deployment (QFD) Diagram

- **Convenience for transportation**. Because this requirement is the innovative feature that distinguishes a portable coffeemaker from a traditional one, special attention is given to the convenience of portability. To accomplish this, product volume and weight will be minimized. A sensible target for portable coffeemaker's overall volume is approximately 140 cm^3, for brewing a volume of two cups (around 29 cm^3).
- **Ease of use**. The ease comes from a relatively simple product design.
- **Ease of cleaning**. To maximize this characteristic, the number of parts will be minimized. Parts that need to be cleaned are: filter, ground reservoir, and water reservoir besides the mug. Because all of them are removable, maintenance and cleaning are easy.
- **Coffee quality**. An automatic switch that turns off as the target temperature is reached guarantees the quality of the coffee. The optimal brewing temperature has been identified, through research, to be 93°C. The automatic switch contributes to the ease of use characteristic.
- **Appearance**. Consumers ranked this characteristic as low on desire. However, to be competitive, a minimum aesthetic aspect is incorporated into the design. Also, because coffeemakers usually come in assorted colors, the decision was made to offer the product in four options: black, white, blue, and green.
- **Low cost**. Based on U.S. daily consumption reported earlier, an average American drinks about 3 cups per day. Using an average price of $2 per purchased cup of coffee, each coffee purchaser spends about $6 daily or $30 every work week. Based on these assumptions, a price of $35 for the portable coffeemaker can be a good choice.

Technical specifications/objectives were entered at the bottom of the house of quality. Because changes in one engineering specification affect other technical parameters, a cross-correlation table (the roof of the house) with relationships between pairs of engineering specifications was completed; for example, the cost will increase if the number of parts of the product is lowered (negative relationship). Finally, to determine how well market competitors are performing on the market, a competitive analysis (on the right of the house) was performed. Competitive products were given a score on each characteristic (1 the minimum to 5 the maximum) based on consumers' reviews. The problem is defined based on the identified customer requirements.

Concept Development

After the problem has been identified and the product concept defined, the needs of customers are identified and translated into technical specifications in the concept development phase. The product's subfunctions are identified, and each of them is studied as a separate problem. Alternative product concepts are generated and evaluated until one is selected as best satisfying the customer requirements and meeting the need of functions such as engineering and manufacturing.

REFERENCES

1. Scott Fogler and Steven E. LeBlanc, *Strategies for Creative Problem Solving*, 3rd ed, (Upper Saddle River, NJ: Prentice Hall, 2013).
2. Y. Akao "*QFD: past, present, and future*," *Proceedings of the 3rd International Symposium on Quality Function Deployment*, pp. 19–29. 1997.
3. L.P. Sullivan, "*Quality function deployment*", *Quality Progress*, Vol. 19 No. 6, pp. 39–50, 1986.
4. G. Pahl and W. Beitz, *Engineering Design: A Systematic Approach*, (London: Springer-Verlag, 2008).
5. Nigel Cross, *Engineering Design Methods–Strategies for Product Design*, 4nd ed. (Chichester, UK: John Wiley and Sons, 2008).
6. Karl Ulrich and Steven Eppinger, *Product Design and Development*, 5th ed. (New York: McGraw-Hill, 2011).
7. N.F.M Roozenburg and J. Eekels, *Product Design: Fundamentals and Methods* (Chichester, UK: John Wiley and Sons, 1995).
8. Robert L. Nagel, Kenneth L. Perry, Robert B. Stone, and Daniel A. McAdams "Function-based design process for an intelligent ground vehicle vision system," *Journal of Electron Imaging* 19 (December 2010).
9. Weiqun Cao, Stefan Conrad, Ernst Kruijff, Dirk Langenberg, and Ralph Schultz, "Digital Product Development in a Distributed Virtual Environment," *Proceedings SPIE* 5444 (no. 322) (2004).
10. Brian Mac Cleery and Nipun Mathur, "Right the First Time," *Mechanical Engineering* (June 2008).
11. G. Caruso, S. Polistina, and M. Bordegoni, "Collaborative Mixed-Reality Environment to Support the Industrial Product Development," *Proceedings of ASME World Conference on Innovative Virtual Reality, June 2011, Milan, Italy.*
12. Coffeesearch.org. www.coffeeresearch.org/market/usa.htm.
13. Eiffel Coffee Brewer. http://www.housewares.org/Show/info/sdc/11/Kevin%20Wu_Eiffel%20Coffee%20Brewer.pdf.
14. Dart K., Howard T., McFarlane K., and Plunkett M., "Collapsible coffeemaker" ME 450 Report, December 2007, University of Michigan.

EXERCISES

2.1. Conduct a complete QFD analysis on a household product of your choice such as a dishwasher, washing machine, or refrigerator. The analysis should include the voice of the customer and the product's ranking compared to that of competitors. Briefly explain your reasoning for the selection.

2.2. Select a simple product of your choice (such as a can opener or umbrella) and identify the parts and objectives.

2.3. Create a figure that shows the elements of a QFD chart. Explain how a cascade of these charts can be used to cover the total design and development process.

2.4. Generate a number of concepts for a potato peeler and then create a product design.

2.5. Show a simple QFD-matrix relationship identifying three customer requirements and three design features.

2.6. If you are designing a technologically integrated product, identify the key criteria.

2.7. You are interested in creating a sustainable product. However, the supply chain influences its features. Prepare a survey for the supply chain companies to identify the issues of sustainability.

2.8. Write the design specification for a smart air pump station for filling automobile tires. This example requires the design of a new "Smart Air Pump Station" for filling automobile tires to the right pressure.

The design features include the exact car manufacturer's recommended tire pressure in an internal database without requiring more input from the customer than a basic knowledge of their car's make, model, and year. There will be little possibility of inflating tires to the wrong pressure even in extreme and temperature conditions.

2.9. Write the design specification for creating the next generation of dishwasher.

2.10. The design team decided to implement a full QFD chart for designing a garden implement such as weed cutter. It can be used in maintaining quality gardens. The requirements include (a) durability, (b) adequate power requirement, (c) light weight, (d) consistent starting, and (e) easy maintenance. Develop customer requirements and engineering parameters. Compare your design to a commercially available product.

CHAPTER 3
Creative Concept Generation and Selection

OBJECTIVES

The concept generation stage starts by identifying customer requirements and results in a list of ideas that are candidates for product concepts. The concept generation stage is a creative process. A structured approach by the design team can benefit by identifying ideas in a systematic way. At the concept generation stage, the designer defines the real problem, decomposes the problem into subproblems, gathers the information, and examines patents and published literature. At the concept selection stage, the designer evaluates concepts with respect to customer needs and identifies the relative strengths and weaknesses of proposed concepts and identifies the best candidate for further examination.

3.1 Introduction

Creativity involves the manipulation by the human mind of past experiences with concepts to produce new ideas. It is a mental process in which past experience is reorganized to form a new combination that will satisfy some need. Although there is no single definition for creativity, most of the definitions identify creativity as a combination of experience, intelligence, and motivation (**Table 3.1**).

Adams (1976)	Creativity has sometime been called the combination of seemingly disparate parts into a functioning and useful whole.
DeBono (1970)	Creativity is the operating skill with which intelligence acts upon experience for a purpose.
Koestler (1969)	The creative act is a combination of previously unrelated structures in such a way that you get more out of the emergent whole than you have put in.
Stein and Heinze (1960)	A creative person is an individual who is motivated, curious, self-assertive, aggressive, self-sufficient, not conventional, persistent, self-disciplined, independent and autonomous, constructively critical, widely informed, open to feelings and likes work, uses aesthetic judgment, and has environmental values.

TABLE 3.1 Definitions of Creativity

3.1.1 Creative Concept Generation Techniques

Creative environment: Creativity is important at various stages of product design. It has been said that innovative ideas are not the result of straightforward analytical procedures or complicated algorithms. They come instead from the creativity of designers. Vertical thinking and lateral thinking are two procedures for understanding creativity.

- **Vertical thinking:** Analytical thinking or deductive reasoning is termed *vertical thinking*. In vertical thinking, the individual always moves forward in sequential steps only after a positive decision has been made. Vertical thinking is analytical, judgmental, critical, and selective. If no positive decision can be made, the vertical thinking pattern ends abruptly. Vertical thinking is then used to focus ideas on real work solutions to problems. Vertical thinking moves in a straight line until it is stopped by a positive or negative conclusion.

- **Lateral thinking:** Lateral thinking moves in many different directions, combining various bits of information into new patterns until several possible solutions are exposed. Then all the solution directions are developed until several possible alternatives have been identified. Lateral thinking is random, sporadic, nonjudgmental, and generative. Its basic function is to take experiences and revise them several times to generate new ideas. Lateral thinking moves randomly in fits and starts until the mind is tired without necessarily reaching a conclusion.

Brainstorming environment: Brainstorming is the oldest and best-known technique for group creative thinking. It has also come to mean making a serious effort to think about a problem. The objective of a brainstorming session is to use the disconnected ideas of individuals to trigger new ideas in each participant. The technique relies heavily on group interaction for exchanging ideas and provides an excellent means of building on members' ideas. Triggering ideas in others is the key to successful group brainstorming.

Brainstorming guidelines: A brainstorming environment should nurture and protect creative ideas. It must foster the further development of an idea to interact with other ideas to develop a pattern. A continual exposure to new experiences in the fine arts, sports, industry, sciences, music, and literature provides the format from which new patterns are made.

The guidelines for successful brainstorming sessions are:

- **Get as many ideas as possible.** The more ideas generated, the higher is the probability of reaching a great idea.
- **Do not judge.** No negative comments or judgments are made at this stage. An individual can come up with as many ideas as possible without having the merits of each evaluated. Statements like "too expensive," "not enough time," "this won't work," or "it's against company policy" should be avoided.
- **Do not criticize.** In the group discussions, people tend to criticize ideas as soon as they are expressed. A good facilitator can create a sense of security within the group and can ensure that the ideas are not criticized at this stage.
- **Generate ideas by creative thinking.** An intense creative thinking exercise during brainstorming might generate a number of ideas that should be followed by an incubation period during which ideas are sorted out.
- **Seek to improve ideas.** In addition to contributing ideas of their own, participants should suggest how others' ideas can be turned into better ideas or how two or more ideas can be joined.

The brainstorming group should consist of a small group of individuals who have different perspectives concerning a product and different degrees of knowledge about it. The group leader explains the rules of brainstorming, gives members several practice exercises, and then states the specific problem as accurately as possible. When the session begins, each participant shouts ideas as rapidly as they come to mind. There should be no sense of formality or order to the exchange of ideas. To prevent boredom, the group leader should close the meeting when it becomes obvious that the members have a mental block, then distribute an unedited list of ideas, and call for another group session in a week or two. The most important role of the group leader is to maintain an environment free from criticism to stimulate the flow of ideas.

The common causes of conceptual mental blocks have been explained by Scott Fogler and Steven LeBlanc, 2014, as listed below:

- Definition of the problem is stated narrowly
- Assumption that there is only one answer
- Frustration from not having immediate success
- Consideration of the symptoms rather than the problem
- Attachment to the first answer that comes to mind

Mental walls prevent a creative designer from correctly perceiving the problem. The most common mental blocks are characterized as perceptual, emotional, intellectual, environmental, and expressive (**Table 3.2**).

Perceptual block	Prevents a designer from clearly seeing the problem itself and the information needed to solve it, primarily caused by the combination of information overload and saturation of information
Emotional block	Interferes with a designer's ability to conceptualize because of fear of taking risks, approaching the problem with a negative attitude, and lack of challenge
Intellectual block	Results from a designer not having the necessary background, training, and knowledge to solve the problem
Environmental block	Results from the lack of physical and organizational support to translate creative ideas to practice
Expressive block	Results from a designer's inability to communicate in written form

TABLE 3.2 Examples of Mental Blocks

A number of structured techniques are available to overcome mental blocks; some of them address attitude adjustment by focusing on the positive aspect of the problem and trying out new and bold design alternatives. Some of the accepted techniques for overcoming mental blocks are:

- Random simulation
- Osborn's checklist
- Attribute listing
- Morphological analysis
- Futuring
- Other viewpoints
- Synectics

Random simulation: This is a way to generate ideas that differ totally from those previously considered to end the mental block. The mind looks for similarities in patterns and then groups them. One of the suggested procedures is to use a dictionary to select a random word to act as a trigger and generate other words that can stimulate the flow of ideas.

The dictionary technique is the simplest and offers the most possible combinations of objects. Procedures such as rolling dice or looking at a table of random numbers can be used to locate a page number. All resulting ideas or wordplays should be recorded. One of the important attributes of this technique is the ability to leapfrog from one idea to another. Looking in a journal or newspaper for a starting word, or choosing a picture from a magazine or a catalog, can also be used. These activities can be fun for the participants.

Osborn's checklist: This technique can help a group build on others' ideas. It is based on asking a set of questions to stimulate the mind to change its perspective on the problem.

Adapt?	How could we adapt this product?
Modify?	Can we change the product's shape, color, material, or focus?
Magnify?	Can we make it longer, thicker, and higher?
Substitute?	What else can we use, and where else can we develop it?
Rearrange?	Can we interchange parts, change positive to negative, or use a different pattern?
Combine?	Can we combine different components and ideas or compromise?

The nature of the questions is not important. They are used merely as a mechanism to change viewpoints of a problem. The leader often introduces a checklist of ideas or stimulating questions during a group problem-solving session.

Attribute listing: The first step in this technique is to write down all design attributes of the problem, such as product specifications. The second step is to apply a list of modifiers to each attribute one at a time to generate new alternatives.

- What shape?
- How deep?
- Can it be adjusted?
- Can it be removed?

This technique considers each parameter of the product in isolation but ignores interactions between two attributes that might lead to a different solution. The difficulty with attribute listing is that designers must be familiar with the product and its features.

Morphological analysis: This is an organized method to enable designers to compare the various attributes of a problem and create new forms of design. A morphological chart is used to encourage designers to identify novel combinations of elements and recombine them to derive a solution. It helps designers to generate the complete range of alternative design solutions for a product and to widen the search for potential new solutions.

The steps involved in morphological chart method are as follows:

▸ **Step 1. List the essential features of the product.**

▸ **Step 2. List the means by which each feature can be achieved.**

▸ **Step 3. Prepare a chart that contains subsolutions.**

▸ **Step 4. Identify the possible combinations of subsolutions to make the product.**

Nigel Cross, 2008, discusses the generation of a morphological chart for a forklift truck. The first step is to identify its essential and common features. The main features, such as the methods of support that allow moving the vehicle, steering it, stopping it, lifting the loads, and identifying the operator's location, are among the items in the first column of **Table 3.3**. The means of achieving the identified features are listed in the remaining columns.

FEATURE	METHODS				
Support	Wheels	Track	Air cushion	Slides	Pedipulators
Propulsion	Driven wheels	Air thrust	Moving cable	Linear induction	
Power	Electric	Petrol	Diesel	Bottled gas	Steam
Transmission	Gears and shafts	Belts	Chains	Hydraulic	Flexible cable
Steering	Turning wheels	Air thrust	Rails		
Stopping	Brakes	Reverse thrust	Ratchet		
Lifting	Hydraulic ram	Rack and pinion	Screw	Chain or rope hoist	
Operator	Seated at front	Seated at rear	Standing	Walking	Remote control

Based on *Engineering Design Method*, Cross, Nigel, John Wiley & Sons, pp 118–119, Devdas Shetty, *Design for product success*, SME (Society of Manufacturing Engineers), pp 84., 2002.

TABLE 3.3 Combination of Subsolutions from the Morphological Chart (Nigel Cross)

According to **Table 3.3**, combining different combinations of subsolutions presents distinct design possibilities for different environments. **Table 3.4** shows a morphological chart for a cooking range by combining different combinations of heat, temperature setting, timer, control position, location, and heat circulation.

Heat: Electrical element—chosen for environmental consideration

Temperature setting: Digital push button—selected for accuracy and in keeping with earlier choice of electrical element.

Timer: Digital clock—chosen for accuracy and in keeping with earlier choice of electrical element.

Control position: Back panel—chosen to allow the user to see and change the control settings comfortably.

FEATURE	METHODS			
Heat	Electrical element	Propane	Microwave	Wood
Temperature setting	Knob	Digital push button		
Timer	Mechanical clock	Digital clock	Hourglass	
Control position	Front edge	Top left	Top right	Back panel
Mounting location	On floor	On pedestal	In wall	On countertop
Heat circulation	Nonconvection	Convection fan		

TABLE 3.4 Morphological Chart for a Cooking Range

Mounting location: On floor—selected for the unit to be mounted on the counter for easier movement of pots and of food to be cooked to or from the counter.

Heat circulation: Convection fan—chosen for the unit's increased efficiency.

Futuring: This is another technique to overcome mental blocks by imagining a solution that is currently not feasible but could be in the future. The technique asks questions about an ideal solution and its benefits and how to devise ways to achieve them. As an example, consider designing a special product. During the production process, it is fabricated but causes manufactured waste. Because treating the waste is expensive, the product is not only expensive but also causes environmental problems. As a futuring exercise, group members try to imagine an ideal solution in which the product is not only profitable but also prevents any scrap. When generating solutions, group members imagine processes that have no waste.

Other viewpoints: At times it becomes easier to solve a problem by examining it from different viewpoints. The problem solution becomes different depending on whose viewpoint is selected.

An example of a problem for a space capsule is given by Scott Fogler and Steven LeBlanc (2014):

- *Problem*: A space capsule burns upon entering the atmosphere.
- *Project manager*: The project must be completed on time.
- *NASA accountant*: Solve the problem but keep the cost low.
- *Engineer*: Choose a new material that will not interfere with the capsule's performance.
- *Material scientist*: Find a material that can handle the high temperature on re-entry.
- *Astronaut*: Find a way to return alive, which is more important than saving the capsule.
- *Solution*: Allow the surface of the capsule to be destroyed but protect the astronaut.

Synectics: The term *synectics* means joining of different and apparently irrelevant elements. It is a way to study the problem-solving process and the way in which creative people think. It emphasizes the emotional component and understanding of the irrational element in decision making. Synectics research has shown that creative efficiency in people can be increased if they understand the psychological process by which they operate. The first step in the process is to define a problem in a way that can be understood by the people solving the problem. The designer initially analyzes a problem by looking for bits of it that are familiar. Using small and familiar pieces, the mind rearranges the problem to a situation that can be visualized. This step is known as *problem statement formulation*.

Synectics research recommends stopping once the problem statement has been clearly formulated instead of continuing to analyze it. When it is understood, the designer must twist the problem statement into a totally different form. The main focus of synectics is to distort the problem so that the perspective of the problem changes dramatically. It is like making a familiar situation into a strange situation. By forcing a change of viewpoint of the problem, this technique has the potential to generate very unusual solutions.

Synectics research has generated four mechanisms for idea generation. Each mechanism is intended to distort the problem.

1. Personal analogy
2. Direct analogy
3. Symbolic analogy
4. Fantasy analogy

1. Personal analogy: Individuals place themselves in the position of the product being designed. If the product is an elevator, the designers go inside the elevator's driving mechanism and experience the movement of the carriage going up and down. The designers are at this point inside the problem looking out at the external forces.

2. Direct analogy: Designers compare similar but different technologies. Biological systems allow human biology to be compared to everyday engineering mechanics.

3. Symbolic analogy: Using symbolic analogy, the problem is described by giving examples in other fields through metaphors.

4. Fantasy analogy: Use of fantasy analogy involves imagining an ideal solution while suspending judgment regarding whether the product is feasible or not.

Design groups using the synectics technique require proper training in using the analogy mechanisms.

EXAMPLE 3.1

A bicycle manufacturer is losing sales of its product to competitors who have a new product. This has motivated the company to design a new bicycle that could be appealing to the younger generation as well as to adventurous bicyclists. Using the techniques of brainstorming, random simulation, Osborn's checklist, attribute listing, and a morphological chart (**Table 3.5**), the company begins to produce a standard size bicycle with a new body shape available in three colors.

Brainstorming: A brainstorming session generated ideas for some of the basic features that the bicycle will have: distinctiveness, bright color, sporty shape, durability, extra features to hold water bottles, special grip handles, ease of repair and maintenance, and catchy advertising.

Random simulation: The word selected using the random dictionary method was *release*. The ideas generated for the bicycle are:

- Easy removal of flat tire
- Easy removal of handle bar

- Seat that can be ejected
- Self-contained bicycle lock
- Body shape like letter R
- Easy and quick mounting and dismounting
- Quick release to the market

Osborn's Checklist

Adapt: Use product for road racing, mountain biking, dirt bike riding, and normal transportation; meets needs of everyone for every purpose and for every age.

Modify: Modify (magnify or minify) to appealing shape and color, and add features such as water bottle, clock, timer, and storage compartment.

Substitute: Use durable tires and lightweight material.

Combine: Combine attractiveness with usability.

Attribute List

Color: Provide a choice of three colors to appeal to men, women, and youth.

Tires: Offer standard tires with option of street, race, or dirt track tires.

Seat: Add adjustable replaceable seat for different sizes of bicyclists.

Shape: Form to fit standard shapes and sizes for all riders; necessary modification if needed.

Morphological Chart of a Bicycle

FEATURE	METHODS				
Body material	Steel	Plastic	Carbon fiber	Aluminum	Titanium
Transmission	Belt	Gears	Chain and sprockets	Rope and reel	
Capacity	One rider	Two riders	Three riders		
Power	Feet	Legs	Hands	Arms	
Pedals	Clip	Clipless	Straps	Attached shoes	No pedals
Drive	Front wheel	Rear wheel	All-wheel drive		
Steering	Front wheel	Rear wheel	All-wheel drive		
Stopping	Disk brakes	Drum brakes	Rim brakes	Parachute brakes	
Seat	Padded	Unpadded	Contoured	Flat	

TABLE 3.5 Morphological Chart of a New Bicycle

Defining the Problem

A satisfactory definition of a problem is crucial to its success. Time spent in defining a problem properly and then writing a complete problem statement results in efficient problem solving. It has been said that a problem well defined is as good as half done. The goals of the product design project should be expressed in very broad terms at the beginning. It is a mistake to plunge headlong into a problem solution before setting appropriate goals and identifying the real need.

Product goals generally are derived from the need analysis, which consists of a list of the needs of users who may be customers, contractors, marketing agents, trade associations, government agencies, technicians, and servicing agencies. The product designer must understand and weigh each of those basic needs. For example, the customer wants a product to be functional, aesthetically appealing, durable, and inexpensive. The marketing group wants it to have sales appeal, minimum service factor, ease of transport, and a good profit margin. The manufacturing group wants the product to be easily fabricated with low cost labor with already approved material from an existing supplier. These varying preferences and needs can be difficult to achieve.

All viewpoints should be considered in needs analysis, including technical, time, and cost needs. A study of these factors can take the form of a detailed cost/benefit analysis or a detailed estimate of the cost of manufacturing the design, including profit and marketing costs. Most product design problems have certain boundaries or constraints within which the solution must be found. Legal constraints on engineering design are becoming increasingly important. These include federal and state regulations pertaining to environmental pollution, energy consumption, and public health and safety.

Problem definition is based on identifying customers' real needs and formulating them as goals for the product. The problem statement expresses what is intended to be accomplished to achieve the goals. Design specifications represent a major component of the problem statement. The key role of the problem statement in the design process is shown in **Figure 3.1**.

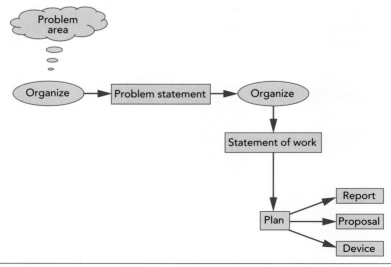

FIGURE 3.1 Problem Statement in the Design Process

3.1.2 **Step-by-Step Design Methodology**

The key factors that distinguish an excellent from a good designer are attitude in approaching a problem, aggressiveness, and depths to which the person uses design methodology. Smart designers believe in the steps of design process and effective use of tools and techniques, heuristics, feedback, model development, and analysis. Designers take enormous effort to understand the various relationships that exist and to describe the problem situation, create a mental picture, ask themselves questions, and break the design problem into several subproblems.

Building blocks of the product design road map consist of the following stages as shown in **Figure 3.2**.

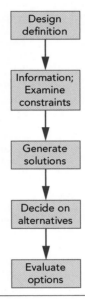

FIGURE 3.2 Design Stages

Definition

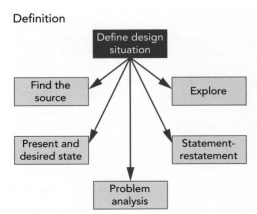

The identification and proper definition of the problem is very important. They express a desire to achieve a transformation from one situation to another. Defining the problem correctly is the first step in which the real problem must be identified. This process involves reporting a significant amount of information, obtaining feedback, and considering value-related issues of the solution. Identifying the problem correctly at this stage saves time and money and makes reaching a satisfactory solution easier.

Find the Source
A few recommended tips are:
- Collect and analyze the information and available data.

(continued)

- Consult people familiar with similar products.
- Inspect the problem personally.
- Verify the collected information.

Information Gathering

Finding whether the problem at hand reflects the real situation is very important. To do this, ask the following questions: Where did the problem originate? Who needs the product? Who initiated it? Can the person responsible explain the reasoning as to how the definition of the need was determined? Are the assumptions valid? Have you considered all viewpoints?

Explore

The five-point strategy (define, explore, plan, act, and reflect) is a technique that helps designers understand and define a project's real problem. Based on that strategy, the following steps can be used to explore the problem:

- Identify pertinent relationships among inputs, outputs, and other variables of importance.
- Recall past experiences.
- Discover the real problems and constraints.
- Consider short-term and long-term implications.
- Collect missing information.
- Hypothesize, visualize, idealize, and generalize.
- If the proposed problem cannot be solved, try to solve part of it or related problems.
- Sketch a pathway that leads to a solution.
- After using some of the activities, write a statement defining the real problem.

Defining the Problem Using the Present State and Desired State Techniques

The present state (PS) and desired state (DS) techniques help designers visualize the starting point and how to proceed so that an appropriate path can be found to the DS that represents solution goals. The designers try to modify the PS or the DS until a satisfactory correlation between them is found. It is important that the PS statement and DS statement should contain solutions that go to the heart of the problem.

(continued)

Information Gathering

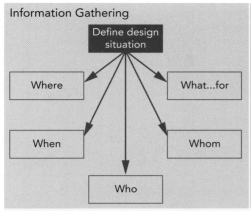

The following questions can function as a guide in defining the product.

- Where does this need arise?
- What is its frame of reference?
- Why is it needed?
- Whom does it benefit?
- By what means can it be best carried out? (How?)
- In what conditions of time and place is the product occurring?
- Who is responsible for the product?

Design Criteria

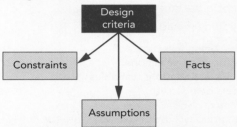

Constraints are factors that affect a project's outcome and cannot be changed. They can reflect customer values.

Assumptions are statements that are accepted as true; there is no doubt. The first step is to clarify the assumptions in regard to the problem. Assumptions can be modified to simplify the problem.

Facts are listed to help clarify what is known and what should be found before proceeding with the project.

Sources of Information Needed

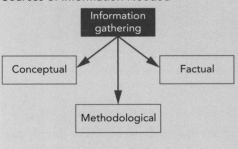

Conceptual information covers fundamental principles and laws such as principles of conversion of mass, energy, and momentum.

Factual information, such as the properties of substances to be used, can be found in reference handbooks. This type of information keeps growing and changing as new substances are invented and new products are developed.

Methodological information represents a link between conceptual and factual information. It identifies the methods and ways by which conceptual information can be applied to generate factual information or more conceptual information.

Generating Options and Solutions

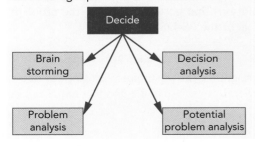

The goal in this step is to generate as many potentially useful concepts and solutions to the problem as possible. Designers need to be innovative and maintain an open, receptive mind to new ideas. An appreciation for the unusual or extraordinary is also important.

(continued)

Evaluation of Options	The evaluation of a particular concept and solution should be based on how it relates to certain design parameters and ecological, social, and cultural factors: • Does it satisfy the basic objective? • Is it theoretically feasible? • Is it practical? • Is its cost within the means? • Is it safe to operate? • Is it the optimum solution? • Does it satisfy the constraints? • Does it satisfy all the human, social, and ecological factors involved? • Is it aesthetically acceptable? • Is it legal? • Does the solution allow the project to be completed in the time allotted?
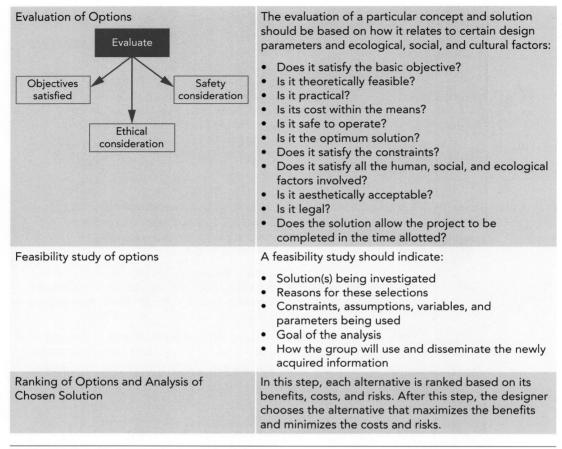	
Feasibility study of options	A feasibility study should indicate: • Solution(s) being investigated • Reasons for these selections • Constraints, assumptions, variables, and parameters being used • Goal of the analysis • How the group will use and disseminate the newly acquired information
Ranking of Options and Analysis of Chosen Solution	In this step, each alternative is ranked based on its benefits, costs, and risks. After this step, the designer chooses the alternative that maximizes the benefits and minimizes the costs and risks.

TABLE 3.6 Step-by-Step Design Methodology

Additional Methods of Product Definitions

Defining the Problem Using the Present State and Desired State Techniques
The present state (PS) and desired state (DS) techniques help designers visualize the starting point and how to proceed so that an appropriate path can be found to the DS that represents solution goals. The designers try to modify the PS statement or the DS until a satisfactory correlation between them is found. It is important that both the PS and DS statements should contain solutions that go to the crux of the problem. Consider an example that involves the usage of the PS-DS technique.

EXAMPLE 3.2

In colleges with challenging engineering programs, the freshman year dropout rate is high. This problem can be addressed in several ways. Use of PS-DS technique can be helpful.

PS: Freshman engineering dropout rate is high.

DS: The freshman retention rate should be increased in engineering programs.

There is no one-to-one match, and the PS does not have anything common with the DS. The statements need to be modified.

PS: We need to reduce the freshman dropout rate.

DS: We can increase the freshman retention rate in engineering programs by making freshman courses more interesting.

A difference between the PS and DS still exists. It is necessary to revise the statements so that gap between them is narrowed.

PS: The freshman dropout rate is high because the students are only exposed to nonengineering courses such as mathematics and physics. The students do not see the connection with the engineering profession.

DS: We want to expose the first-year students to engineering principles in combination with physics and mathematics so that they see that the engineering profession is interesting.

We can see a relationship between these two statements and can clarify their differences.

PS: The reasons for the high freshman dropout rate are that the students in the engineering program do not see the connection between the courses in engineering and those in the basic sciences such as physics and mathematics.

DS: We will provide first-year students integrated engineering courses that teach engineering principles in combination with physics and mathematics as integrative learning blocks.

The Duncker Diagram

Duncker diagrams help designers to examine possible paths from the starting point to the desired state. By going through various paths, it can also lead to solutions. The unique feature of the Duncker Diagram is that it points out ways to solve the problem by making it OK not to reach the desired solution.

There are two types of solutions:

1. The first type examines the path to be followed and actions to be taken to achieve the DS.
2. The second type points out ways to solve the problem by making it acceptable not to reach the desired solution. The DS is transformed until it matches PS as shown in **Figure 3.3**. This procedure eliminates the need to achieve the DS.

Figure 3.3 shows the three stages of the Duncker diagram. The first stage represents a general solution, and the second stage represents functional solutions. Functional solutions are possible paths to the desired state but need not necessarily examine the feasibility of the solution. They give solutions by considering "what-if" situations. The third stage represents specific solutions to implement the functional solutions.

Figure 3.4 is an example related to coffeemakers. Some coffee drinkers want to make their own coffee while they are outside of their homes. The size and weight of regular coffeemakers make them not easily transportable. In this case, the Duncker diagram analyzes two types of situations: one that addresses people who want to make

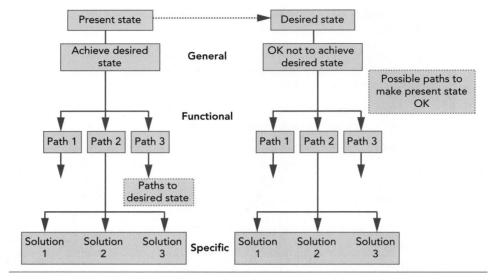

FIGURE 3.3 Three-Stage Duncker Diagram

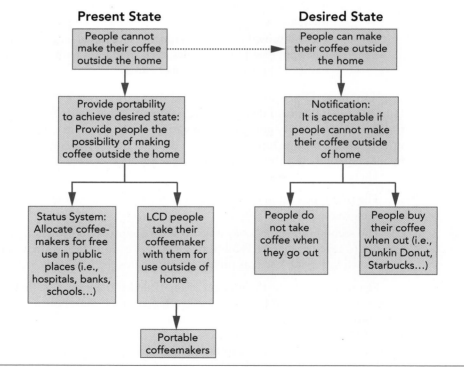

FIGURE 3.4 Duncker Diagram for a Coffeemaker

their coffee outside the home and one that analyses how to modify the DS so that it corresponds to what designers want. As shown in the figure, the solution that goes from the PS to the DS is to provide people with a portable coffeemaker.

Trigger Problem Statement-Restatement

A. F. Osborn developed a technique that uses words and questions to trigger a different thought. The triggers focus on possible changes in the problem statement by rewording it. The technique involves the following:

1. Vary the stress pattern by trying to emphasize different phrases and words.
2. Choose a word that has an explicit definition and substitute the explicit definition each time that the word appears.
3. Make an opposite statement by changing positives to negatives and vice versa.
4. Change every to *some, always* to *sometimes, sometimes* to *never*, and vice versa. Replace persuasive and implied words in the problem statement such as *obviously, clearly*, and *certainly* with the use of such phrases as: "Is the reasoning valid?"
5. Express words in the form of an equation or a picture and vice versa.

Statement-Restatement Method

The statement-restatement method seeks to achieve stated objectives by rephrasing the problem in a number of ways. It can be applied to various problem statement triggers such as varying the stress pattern on certain words, changing positive terms to negative terms, substituting explicit definitions of certain terms in the statement, and so on. For example, consider the case of an industrial staple gun. It is a handheld device that pushes heavy-duty metal staples into different materials such as wood or concrete. An easily disassembled staple gun is easier to take a part to repair or rework parts. This allows the company to reduce scrap and add a department for consumer repairs. The problem can be redefined in the following format.

To start with the redesign, use a statement-restatement technique.

1. Develop a *removable cartridge* that holds and advances the staple.
2. Develop a loadable cartridge that *orientates and advances* staples.
3. Develop an *easily removable loadable* cartridge for staples.
4. Develop an *easily removable* cartridge that is readily loadable and advances staples.
5. Develop an *easily removable*, easily loadable cartridge to orient and advance staples.

EXAMPLE 3.3

Consider an example from the aerospace industry. Jet engines may fail due to numerous reasons that can cause an airplane to crash. One possible reason could be due to system malfunction of the engine turbine. The failure of the engine turbines could be due to defects in the engine turbine blades. A designer perceives the situation as a "need for a methodology to reduce surface defects in turbine blades." Based on this information, the initial problem statement is: "Surface irregularities of turbine blades of a jet engine causes blade failure that causes danger to the aircraft."

Vary the Stress Pattern

Trigger 1

This step helps to determine whether the focus of the problem itself has changed. Consider these statements and notice the different stress patterns in the following identical sentences:

> The *turbine blade* failure is because of surface irregularities in jet engines for commercial aircraft.
>
> The turbine blade failure is because of *surface irregularities* in jet engines for commercial aircraft.
>
> The turbine blade failure is due to the surface irregularities in the *jet engines* for commercial aircraft.

Trigger 2

Substitute a word that has a similar definition each time that the original word appears.

> The *physical surface characteristics* of an *engine component* cause it to fail in an aircraft. (Think about the physical characteristics of the component and how it can be designed to last, not fail.)

Trigger 3

Change positive to negative and vice versa.

> Ask about the worst-case surface characteristics that could always cause the engine failure. (Think how to maintain and measure surface characteristics and how to control them.)

Trigger 4

Replace persuasive words such as *every* with *some*, *always* with *sometimes*, and *sometimes* with *never*. (Challenge the fundamental assumption in the problem definition.)

> The physical surface characteristic of a turbine blade is always maintained to prevent the engine of an aircraft to fail. (Open new areas of discussion, such as why each blade should be inspected and measured all the time.)

Trigger 5

Express words in the form of an equation or a picture and vice versa. Surface roughness (R_a) is a function of cutting tool geometry, vibration of the machine, and machining process.

R_a = F (cutting tool geometry, vibration of the machine, machining process).

EXAMPLE 3.4

In the power generation industry, there are many reasons why the overall design affects boiler efficiency. The variables such as boiler pressure, temperature of feed water, water flow rate, and calorific value of the fuel influence the efficiency. In some current designs, the feed water inlet components are not able to generate temperatures high enough to provide proper steam flow for maximum efficiency.

Based on this information, the initial problem statement is: "Improper design temperatures of the feed water inlet component in a power generation boiler are the causes of poor efficiency in boiler performance."

Apply the Statement-Restatement Method

Trigger 1

The *boiler efficiency performance* reduction results from improper temperature designs in the main feed water pipeline. (Vary the structure pattern and focus on efficiency.)

The boiler efficiency performance reduction is the *result of improper temperature* designs in the main feed water pipeline. (Vary the structure pattern and focus on the temperature.)

The boiler efficiency performance reduction is the result of improper temperature *designs in the main feed water pipeline*. (Vary the structure pattern and focus on the feed water pipeline.)

Trigger 2

Ask how design engineers can determine temperatures that will always create poor performance resulting in reduced efficiency. (Make an opposing statement.)

Trigger 3

The temperature of main feed water pipeline is always maintained and monitored to prevent reduced boiler efficiency in a power generation plant.

$T = f(x_1, x_2, ...x_n)$

where $x_1, x_2 .... x_n$ are the variables such as geometry, flow, and velocity.

Figure 3.5 shows the Duncker Diagram method for solving the problem of power generation. **Table 3.7** shows the morphological chart for the hot water recirculation system for power generation.

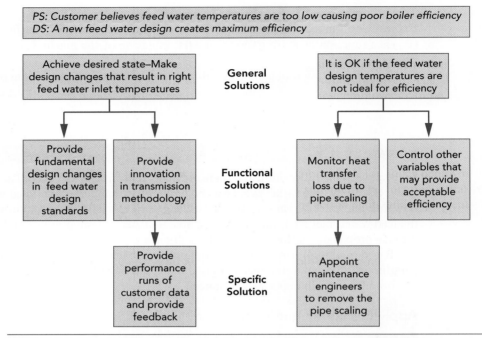

FIGURE 3.5 Duncker Diagram for Solving the Problem of Power Generation

FEATURE		METHODS	
Piping materials	Carbon fiber	Steel	Stainless steel
Valves	Control	Stop/Check	Isolation
Supports	Lugs	Structural	Weld straps

TABLE 3.7 Morphological Chart for the Hot Water Recirculation System

3.1.3 Design Concept Development Methodology

The concept development phase of product design requires increased coordination among many functions. Concept development is the front-end of the product development process in which customers' needs are identified and translated into technical terms, target specifications are established, alternative product concepts are generated and evaluated, and one or more concepts are selected. At this stage of the design concept development methodology, the product team explores various product possibilities to meet customers' requirements, including external as well as internal search, brainstorming, and exploration of various ideas. This phase also involves most of the company's functions, customers, outside suppliers, and government agencies.

Figure 3.6 is a graphic representation of a four-step concept generation methodology starting with clarifying the problem by breaking it into subproblems and focusing

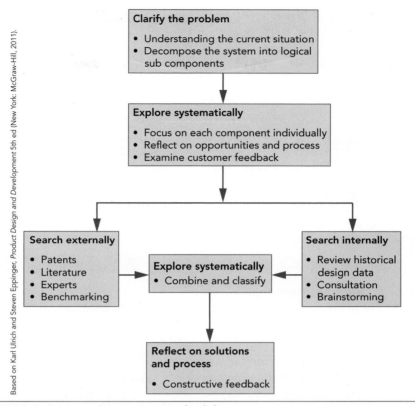

Based on Karl Ulrich and Steven Eppinger, *Product Design and Development* 5th ed (New York: McGraw-Hill, 2011).

FIGURE 3.6 Concept Generation Methodology

on critical subproblems. The second step involves searching for web-based and other information, literature, patents, and benchmarking reports as well as investigating customer feedback and consulting others. The third step is a systematic exploration of ideas, morphological classification, and combination charts for identifying the best possible plan. The fourth stage reflects on the problem statement's solutions, feedback, and revision.

As an example, consider a problem statement for an electrical receptacle to be used in a raceway as shown in **Figure 3.7**. Two standard 15/20 amp receptacles typically are installed in the raceway to allow the user to plug in an electrical device. Electricians using common tools must be able to install the new receptacle design easily. The device must meet all code requirements for its type.

Figure 3.7 includes the functional diagram in which the input is the electrical energy and the output is the generation of power to switch the machine on/off. The diagram shows how to decompose the problem into subproblems. The function of the product is subdivided into two subfunctions: (1) localizing the energy access and (2) allowing a connection to the outside source.

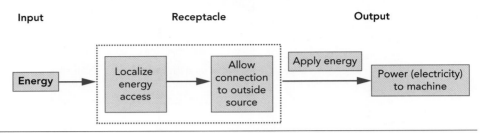

FIGURE 3.7 **Function Diagram of an Electrical Receptacle**

EXAMPLE 3.5

Problem Statement

The problem of receptacle design discussed in the last section can be stated and restated in many ways. Using statement-restatement techniques, the problem can be expressed in a number of ways by varying the stress pattern:

The *receptacle* is too time consuming to install. (To make the designer think about other products that are faster to install.)

The receptacle is too *time consuming* to install. (To focus on faster installation.)

The receptacle is too time consuming to *install*. (To focus on the installation process.)

Develop a receptacle that is *easy* to wire and install. (To emphasize design features.)

Develop a receptacle that is easily *moved from place to place*. (To emphasize portability.)

The methods for achieving these subfunctions are shown in **Table 3.8**; it also includes a combination chart outlining features such as mounting, electrical contact, wiring, and total appearance. By examining each of these solutions, the designer has the correct combination of subfeatures and components.

Feature	Methods		
Mounting	Screws	Snaps	Glue
Electrical contact	Terminals	Contacts	Welding
Wiring	Soldering	Terminals	Insulation displacement
Total appearance	Flat	Rounded	Obtrusive

TABLE 3.8 **Methods for Achieving Subfunctions**

EXAMPLE 3.6

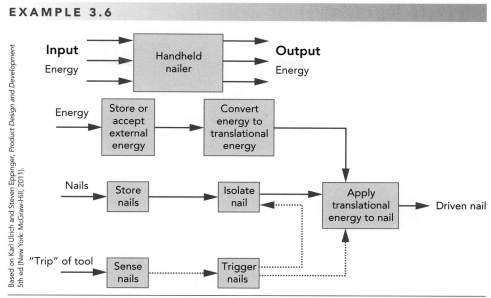

Based on Karl Ulrich and Steven Eppinger, *Product Design and Development* 5th ed (New York: McGraw-Hill, 2011).

FIGURE 3.8 Handheld Nailing Device and its Subfunctions

As Ulrich and Eppinger (2011) explained, a handheld nailer, as shown in **Figure 3.8**, operates using a solenoid, which compresses a spring and then releases it repeatedly to drive each nail with multiple impacts. The motor winds a spring and accumulates potential energy, which is then delivered to the nail in a single blow.

The motor repeatedly winds and releases the spring, storing and delivering energy over several blows. Multiple solutions arise from the combination of a motor with transmission, a spring, and a single impact. The combination is explained in **Figure 3.9**.

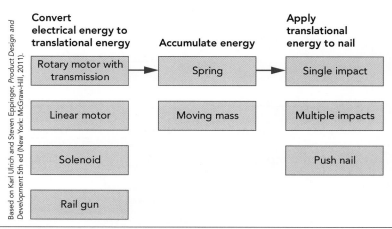

Based on Karl Ulrich and Steven Eppinger, *Product Design and Development* 5th ed (New York: McGraw-Hill, 2011).

FIGURE 3.9 Combination Table for Handheld Nailer

EXAMPLE 3.7

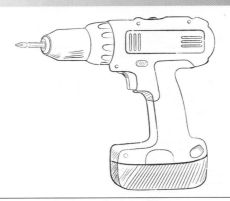

FIGURE 3.10 Conceptual Sketch of Cordless Drill/Driver

The development of a cordless drill/driver unit for the home repair market is illustrated in **Figure 3.10**. After establishing a set of customer and target requirements, the product specifications are established. The heavy-duty cordless drill/driver unit will have a 12-volt electrical supply and will be provided in two speed ranges of 0–600 rpm and 0–1500 rpm; its maximum drill size is three-eighths of an inch. The unit will operate with batteries and have an adjustable clutch and electric brake for disabling it.

The problem is represented in the form of a function diagram (**Figure 3.11**) with inputs and outputs represented around a black box. The black box is operated on input

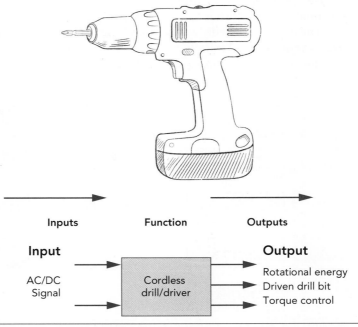

FIGURE 3.11 Representation of Cordless Drill/Driver Function

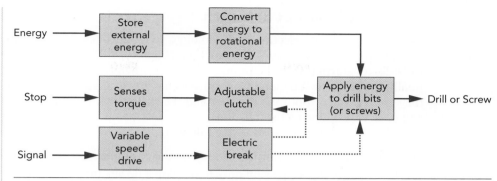

FIGURE 3.12 Functional Decomposition of Cordless Drill/Driver

represented by AC/DC input and a start/stop signal. The output is represented by rotational energy, inserted drill bits, and adjustable torque control. **Figure 3.12** depicts the functional decomposition of the cordless drill/driver.

A concept classification chart as shown in **Figure 3.13** is used to identify alternative energy sources, which could be electric, chemical, hydraulic, pneumatic, or nuclear. The choice of energy is narrowed to three or four subconcepts. Multiple solutions are available including the combination of rechargeable batteries, power pack, fuel cell with torque sensors, piezo-resistive sensors, and variable speed drives. In this solution, rechargeable batteries are used with a combination of torque sensors and a variable clutch.

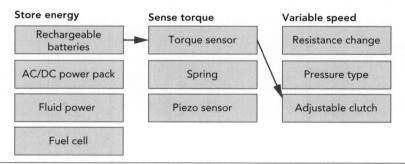

FIGURE 3.13 Multiple Solutions from Combination of Ideas Regarding Alternative Solutions to Energy Source

3.2 **Concept Selection**

At the concept generation stage, we can develop as many concepts as possible. All the generated concepts may not be manufacturable. We must eliminate those concepts that are not feasible.

3.2.1 **Pugh Concept Selection**

In the Pugh concept, selection methodology and product design criteria are written in the form of customer features and functions (Pugh 1981, 1991). Product functions are to be listed, starting with the primary functions, then the secondary functions, and, finally, the aesthetic functions. Concept sketches are arranged simply in the order that they are generated. Both the designer and the customer should understand the brief and open description of the function.

The methodology consists of the following:

1. Define the product's functions based on the customer's needs and desires.
2. Use creative thinking techniques to develop a dozen different design concepts that might satisfy the functions. All concept designs must be developed to the same degree of detail as their sketches. Concepts need not be in final form.
3. Establish a concept and evaluation matrix to compare each concept to corresponding datum.

Refer to **Table 3.9** for the basic Pugh concept selection matrix. In the first step, concepts are evaluated relative to the original product design reference datum. If this were a conceptual design stage and a new product creation, a new design could be chosen as a datum. Each concept is compared function by function to the datum design. The concept designs are not compared to each other but are individually and separately compared with the datum. Evaluation can be made using the following criteria:

+   This concept is clearly better than the datum.

−   This concept is not as good as the datum.

0   This concept is (about) the same as the datum.

If no clear concept selection can be made after the first evaluation, a second selection process should be performed comparing three or four of the most promising design concepts. One good redesign concept is selected as the datum, and then each redesign concept is compared function by function with the new datum. The concept with the highest score should be used for redesign.

	Concepts							
Functions	1	2	3	4	5	6	7	8
A								
B								
C								
D								
E								
F								
Score								

Source: Based on Pugh, S., "Concept Selection–A Method that Works", Proc. I.C.E.D. Rome (March 1981), WDK 5 Paper M3/16, pp 497–506.

TABLE 3.9 Selection Matrix

EXAMPLE 3.8

This example examines the joining method of a header to the heat exchanger. The heat exchanger is a device in which two or more fluids exchange thermal energy through a heat transfer surface. Heat exchangers are found in applications ranging from space heating, air-conditioning, power production, waste heat recovery, and chemical processing. The purpose of this case study is to analyze and determine the best method of attaching low-pressure headers to the core while considering customer requirements, ultimately resulting in a reduced product cost.

The heat exchanger typically has four headers welded to the brazed core (**Figure 3.14**). Hot and cold fluids enter and leave through the headers. The core is made of thin fins that are brazed to channels for support. The two end plates hold the assembly together. The core bands, which provide structural support to the assembly, are welded at four places to the end plates. The headers are welded to the core bands. Because of the frequent need to inspect and repair the fins, the headers must be removable and replaceable.

After brainstorming, 15 alternatives were generated for the header attachment. They were made in accordance with customer requirements and target specifications.

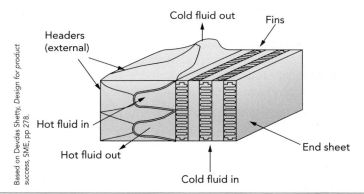

Based on Devdas Shetty, *Design for product success*, SME, pp 278.

FIGURE 3.14 Schematic of a Heat Exchanger

Design Solutions/Alternatives

1. Snap fit the header to the core band and end plates.
2. Rivet the header with solid rivets.
3. Clinch the header to the core band with a flanged header.
4. Rivet the header with blind rivets.
5. Clinch the header to the core band and end plate without a flange.
6. Snap fit the header in the machine groove.
7. Rivet two sides of the header and snap fit two sides of the header in a groove.
8. Bond the header.
9. Bolt the header.
10. Rivet or clinch the header with a U-channel.
11. Clamp the header with a full plate band.

(*continued*)

Design Solutions/Alternatives
12. Snap on the header with a finger seal.
13. Tab the header in the slots.
14. Replace the core band with a finger seal.
15. Join the header to the core with screws.

TABLE 3.10 Heat Exchanger Concept Generation

Identifying the Best Alternative

The key factors to be considered in the new design, as assessed from the customer requirements, were:

- Header should be easy to assemble.
- Assembly set-up time should be reduced.
- Header joining method should not damage the core.
- Header joint design should meet leak criteria.
- Header joint should be free from welding.
- Header should be easily removable.
- Header should withstand operating conditions (pressure/temperature).

After generating the concepts, one or more concepts need to be selected, which will proceed in product development. More than one concept is sometimes chosen, in cases where manufacturing the prototype and testing it is not expensive.

The concepts are selected based on the following:

- Interchangeability
- Easy-to-repair heat exchanger
- Cost-effectiveness
- Process capability
- Durability
- Weight
- Ease of manufacturing

The concept is evaluated using the Pugh table (**Table 3.9**). In the Pugh Matrix, the concepts generated in **Table 3.10** are rated on the basis of the design criteria with (+) indicating a relative score of better than, (0) same as, or (−) worse than in each cell of the matrix.

The design option with a highest score in **Table 3.11** is chosen and further investigated for product characteristics and functional characteristics that are evaluated based on the following factors:

- Thickness
- Weight
- Material
- Joint sealing
- Joint strength

Design Criteria	1	2	3	4	5	6	7	8	9	10	11	12	13	14	15
Performance	--	−	--	−	--	−	−	0	0	−	−	0	--	−	−
Durability	--	0	--	0	--	−	−	0	−	−	--	0	−	−	0
Assembly	0	0	--	−	--	−	--	+	0	−	+	+	--	--	0
Manufactur-ability	−	0	−	−	−	--	−	+	0	--	0	0	0	0	−
Complexity	−	−	−	--	−	--	−	0	0	−	+	0	−	--	−
Weight	0	--	0	−	0	0	−	+	--	−	−	−	−	0	0
Repairability	−	0	--	·	--	−	−	+	0	−	0	+	−	+	−
Cost	−	−	−	--	−	--	--	+	0	−	0	−	−	−	−
Total	−8	−5	−11	−9	−11	−10	−10	5	−3	−9	−2	0	−9	−6	−5

TABLE 3.11 Pugh Matrix

3.3 Structured Innovation Using TRIZ

A number of design methodologies are used to create new products and processes. Some of these methods overlap and use different definitions of key concepts such as function. Although these methodologies emphasize the importance of the knowledge of physical effects to solve technical problems, they were developed independently and have different backgrounds. TRIZ is the acronym for a Russian term that translates to theory of inventive problem solving (TIPS) developed by Genrich Altshuller in 1946. He began with the hypothesis that universal principles of invention serve as the basis for creative innovation across all scientific fields. Codifying and teaching these principles make innovation more predictable. To test this theory, Altshuller reviewed about 200,000 patents submitted in the Soviet Union. The analysis showed that most patents suggested the *means for eliminating system conflicts* in a system.

For a problem to be considered inventive, it had to include at least one contradiction, which can occur when a certain parameter cannot be improved without causing another parameter to deteriorate. A contradiction between speed and sturdiness is one example. Designing an automobile to be sturdy means that it will have more weight. More weight generally results in less speed. How do we design the same vehicle to run faster?

Furthermore, TRIZ researchers found about 39 parameters, each of which could contradict another. The initial step in using TRIZ is to find which design parameters are contradictory. Another technique, Systematic Approach to Engineering Design (SAPB) developed by Pahl and Beitz, has a European origin. SAPB states that solving design problems is a variant of general problem solving. A designer usually follows a path consisting of certain fundamental activities, such as problem and requirements formulation, search for alternative solutions, evaluation and documentation, and communication of the results. Design methodologies support this process by providing specific design methods and design knowledge.

Step-by-Step TRIZ Process

TRIZ methodology systematically investigates a problem to identify an innovative solution by applying a series of steps to generate solution alternatives as shown in **Figure 3.15**. It improves the product parameters while maximizing product changes and costs. Many organizations have adopted TRIZ as an effective concept-generating tool. In addition to solving technological issues, it has the capability to affect key functions in management.

‣ **Step 1. Identify the Problem**
   The operating environment, resource requirements, primary useful functions, harmful effects, and ideal result in the engineering system under consideration should be identified.

‣ **Step 2. Formulate the Problem in Terms TRIZ and Contradiction**
   Restate the problem in terms of physical contradictions. Identify problems that can occur and identify technical conflicts in the problem that might force a compromised solution.

‣ **Step 3. Search for a Previously Solved Problem**
   According to TRIZ research, 39 standard technical characteristics can cause conflicts. Find the contradictions.

‣ **Step 4. Examine TRIZ Inventive Principles**
   First find the principle that needs to be changed and the principle that has an undesirable secondary effect. Then identify any technical conflict. Apply the necessary algorithms to solve the contradictions.

‣ **Step 5. Examine the Effects**
   Determine whether any harmful actions need to be eliminated. If there are harmful actions, define the best way to reduce them. Is it necessary to maximize useful actions?

‣ **Step 6. Determine a Specific Solution**
   Reduce the designed system in such a way to fulfill the objective satisfactorily while the system is reduced or eliminated.

Technical Contradictions and Resolution

Physical contradictions arise when a certain parameter cannot be improved without causing another to deteriorate. A contradiction arises from mutually exclusive demands that may be placed on the same system. Improvement of one of the system parameters

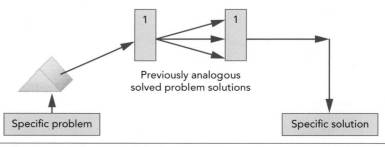

FIGURE 3.15 TRIZ Approach to Problem Solving

can lead to deterioration of others. TRIZ helps to find the physical contradictions at the hidden root of the problem. The physical contradictions and principles are combined in a matrix, the rows and columns of which contain 39 generalized parameters and correspond to the most common parameters that engineers try to improve. The contradiction idea does not reflect the whole design problem but deals with conflicting elements of the weak spots of the project. See **Table 3.12** for a list of technical contradictions. After reviewing it, a designer can have some idea about where to start looking for solutions.

Table 3.13 is a part of a matrix representation of ideas of contradiction provided along two axes.

1. Weight of moving object	21. Energy spent by nonmoving object
2. Weight of nonmoving object	22. Power
3. Length of moving object	23. Waste of energy
4. Length of nonmoving object	24. Waste of substance
5. Area of moving object	25. Loss of information
6. Area of nonmoving object	26. Waste of time
7. Volume of moving object	27. Amount of substance
8. Volume of nonmoving object	28. Reliability
9. Speed	29. Accuracy of measurement
10. Force	30. Accuracy of manufacturing
11. Tension, pressure	31. Harmful factors acting on object
12. Shape	32. Harmful side effects
13. Stability of object	33. Manufacturability
14. Strength	34. Convenience of use
15. Durability of moving object	35. Repairability
16. Durability of nonmoving object	36. Adaptability
17. Temperature	37. Complexity of device
18. Brightness	38. Complexity of control
19. Energy spent by moving object	39. Level of automation
20. Productivity	

TABLE 3.12 Table of Technical Contradictions

Worsening Features

Features	1	2	3	4	5	6	7
1: Weight of moving object	*	-	15 8 29 34	-	29 17 38 34	-	29 2 40 28
2: Weight of stationary	-	*	-	10 1 29 35	-	35 30 13 2	-
3: Length of moving object	8 15 29 34	-	*	-	15 17 4	-	7 17 4 35
4: Length of stationary	-	35 28 40 29	-	*	-	17 7 10 40	-
5: Area of moving object	2 17 29 4	-	14 15 18 4	-	*	-	7 14 17 4
6: Area of stationary	-	30 2 14 18	-	26 7 9 39	-	*	-

Improving Features

Features	1	2	3	4	5	6	7
7: Volume of moving object	2 26 29 40	-	1 7 4 35	-	1 7 4 17	-	*
8: Volume of stationary	-	35 10 19 14	19 14	35 8 2 14	-	-	-
9: Speed	2 28 13 38	-	13 14 8	-	29 30 34	-	7 29 34
10: Force (Intensity)	8 1 37 18	18 13 1 28	17 19 9 36	28 10	19 10 15	1 18 36 37	15 9 12 37
11: Stress or pressure	10 36 37 40	13 29 10 18	35 10 36	35 1 14 16	10 15 36 28	10 15 36 37	6 35 10
12: Shape	8 10 29 40	15 10 26 3	29 34 5 4	13 14 10 7	5 34 4 10	-	14 4 15 22
13: Stability of the object	21 35 2 39	26 39 1 40	13 15 1 28	37	2 11 13	39	28 10 19 39
14: Strength	1 8 40 15	**40 26 27 1**	1 15 8 35	15 14 28 26	3 34 40 29	9 40 28	10 15 14 7

TABLE 3.13 Contradiction Matrix from TRIZ

See more at: http://www.triz40.com/aff_Matrix_TRIZ.php#sthash.sw3I50Qf.dpuf

3.3.1 **Concept Generation Using TRIZ**

EXAMPLE 3.9

Identify a new structural material to be used for a bulletproof vest that must be strong but not heavy.

▸ **Step 1. Identify the Contradictions.**
In this case, the material strength (*improves*) and at the same time weight (*worsens*). (Ref. http://www.triz40.com/)

▸ **Step 2.** In **Table 3.13**, look at the list of features and identify those important to the contradictions: strength: #14; weight: #2.

▸ **Step 3.** In **Table 3.13**, identify which are improving features and which are worsening features: strength (#14) improves and weight (#2) worsens.

▸ **Step 4.** By referring to **Table 3.13**, we observe the relevant TRIZ principles to use are 40, 26, 27, and 1.

Row 14 (strength) and column 2 (weight) of the contradiction matrix indicate that the following principles may be useful:

40. Composite materials
26. Copying

Principle	TRIZ Explanation
40. Composite materials	Change from uniform to composite (multiple) materials such as reinforced polymers.
	For lightweight and stronger vests, research on composites is an active area. Polymers reinforced with carbon nano fibers are currently being investigated as a strong, lightweight alternative to steel for structural materials.
26. Copying	Instead of an unavailable, expensive, fragile object, use simpler and inexpensive copies of objects used elsewhere.
	Copy the design of suits used in scuba diving to use as a bulletproof garment.
27. Cheap short living	Replace an inexpensive object with multiple inexpensive objects that have certain qualities such as service life.
	This principle does not appear to be readily applicable to this problem.
1. Segmentation	Divide an object into independent parts; make it easy to disassemble and increase the degree of fragmentation or segmentation.
	This idea should lead to consideration of several different coverings for different parts of the body (pants, vest, etc.) rather than a one-piece suit and different materials to cover the critical areas such as chest and head, each of which takes advantage of the specific properties that would be customized for different applications.

TABLE 3.14 TRIZ Principles Selection

27. Cheap short life

1. Segmentation

These principles are only general suggestions to help focus on areas that have proven fruitful in previous problems. A brainstorming session could follow to determine how to use these four principles to solve the problem.

By identifying problem contradictions, TRIZ elements can help reach a solution. Note that the principles do not constitute a final solution to the problem but high-level strategies for finding ideas. Forty inventive principles of TRIZ, which are essentially strategies, should assist designers in finding a highly inventive solution. These principles force designers to preformulate the problem in terms of standard engineering parameters.

One TRIZ design principle involves segmentation or division. It involves dividing an object that is usually seen as a whole. As an example, consider a car radio that is detachable. The component's removable nature prevents theft of the radio because it can be detached when the driver leaves. Contradicting is also a design principle that is reliable and has the least harmful side effect. In this case, reliable usage can be interpreted as being able to use a car radio whenever and wherever desired.

EXAMPLE 3.10

Design an automobile component and consider different ideas. Start with the assumption that "an automobile should be strong/sturdy for crash protection but lightweight for better fuel economy."

▸ **Step 1.** Applying TRIZ principles, identify the contradictions. In this case, *improving* the material could cause some parameters to worsen. By referring to **Table 3.13**, we observe the relevant TRIZ principles to use are 14, 15, 19, and 1.
  Data in **Table 3.13** indicate that (14) strength and (15) durability of the moving object improves, but 19 energy spent by moving object worsens.

▸ **Step 2.** We use the contradiction matrix to obtain the principles list: Using the combination principle of improvement/worsening results in a set of principles.
  - 14/1 − 1, 8, 40, 15
  - 14/19 − 19, 35, 10
  - 15/1 − 19, 5, 34, 31
  - 15/19 − 28, 6, 35, 18

▸ **Step 3.** Generate detailed ideas from the identified principles.

Principles List

1. Segmentation
   - Use crumple zones to absorb energy during a crash.
   - Create a high-strength safety "cage" around the passenger compartment to protect passengers.
   - Use lighter materials (aluminum, plastics) where less strength is required, such as body skins.

8. Antiweight
   - Improve aerodynamics to reduce drag.
   - Reduce exterior items that create drag; for example, use cameras rather than side view mirrors.
   - Use spoilers to create lift and reduce rolling resistance.

40. Composite materials
   - Use composite materials in the safety cage frame and/or the skin.

15. Dynamics
   - Use adaptive spoilers that respond to speed, steering, braking, and environmental inputs to improve handling and braking, reduce rolling resistance, and so on based on the current situation.
   - Use crumple zones to absorb energy during crash.

- Use hydraulic dampers as part of an improved bumper system to absorb energy during front and rear impacts.
- Use a hybrid system with regenerative braking.
- Use a flywheel for energy storage.

19. Periodic action
    - Shut off engine when stopped or coasting.

35. Parameter changes
    - Nothing was generated for this category.

10. Preliminary action
    - Use adaptive spoilers that respond to speed, steering, braking, and environmental inputs to improve handling and braking, reduce rolling resistance, and so on based on the current situation.

5. Merging
    - Use a hybrid system with regenerative braking.

34. Discarding and recovering
    - Make passenger seats and certain amenities removable to reduce weight when not needed.

31. Porous materials
    - Make safety cage frame from high-strength porous materials to act as the bones in a skeleton.
    - Remove weight in safety cage frame by adding holes near neutral axes to maintain stiffness.
    - Make plastic skins and panels thicker for higher stiffness but with porosity to reduce weight.

28. Mechanics substitution
    - Reduce exterior items that create drag; for example, use cameras rather than side view mirrors.

6. Universality
    - Use a flywheel to store energy during braking, which is then used to help accelerate.

18. Mechanical vibration
    - Nothing was generated for this category.

EXAMPLE 3.11

An e-mail spam filter should be efficient enough to screen unwanted e-mail but discriminating enough not to block wanted ones.

▸ **Step 1.** Data indicate improvement in (25) loss of time, (30) object-affected harmful, and (31) object-generated harmful and worsening in (24) loss of information.

▸ **Step 2.** Use the contradiction matrix to obtain principles list: (Improves/Worsens – Principles)
  - 25/24 − 24, 26, 28, 32
  - 30/24 − 22, 10, 2
  - 31/24 − 10, 21, 29.

▸ **Step 3.** Starting with the principles identified, detailed ideas can be generated:

Principles List

24. Intermediary
  - Segregate questionable e-mails into a separate folder for the user to screen.

26. Copying
  - Nothing was generated for this category.

28. Mechanics substitution
  - Nothing was generated for this category.

32. Color changes
  - Use different colors to flag e-mails based on spam score to sort e-mails either in spam folder or inbox.

22. "Blessing in disguise" or "turn lemons into lemonade"
  - Use heuristics to learn and develop filters.
  - Allow user to whitelist/blacklist senders.
  - Allow user to help define spam and allowed terms such as subjects, keywords, sender, and so on.

10. Preliminary action
  - Allow user to approve or disapprove senders.
  - Allow user to help define spam and allowed terms such as subjects, keywords, sender, and so on.
  - Combine with antivirus and antimalware/spyware program.
  - Arrange spam folder based on spam score or make multiple spam folders that are sorted by spam score.

2. Taking out
   - Segregate questionable e-mails into a separate folder for user to screen.

21. Skipping
   - No ideas were generated in this category.

29. Pneumatics and hydraulics
   - Nothing was generated for this category.

3.4 **Axiomatic Thinking in Product Design**

Many times people can identify a distinguishing piece of art or music but find it difficult to explain why a particular combination of elements in a work causes it to be excellent. In other words, these results lack an absolute frame of reference, which often leads to different opinions. Much depends on intuition and experience when people compose music or design a product or a process. It is difficult to reduce these facts and observations into a consistent set of statements and descriptions. Nam Suh (1990) proposed the use of axioms to represent design. Their use is based on the assumption that a fundamental set of principles that represents a good design practice exists.

The design methods of diverse fields such as industrial design, architecture, mechanical design, software engineering, and the development of management policies have many similarities. In other words, it can be said that they represent a set of common factors in a good design. These common factors can be applied to other design situations just as natural laws are applied to natural science problems. The goal of axiomatic design philosophy are to:

- Make human designers more creative.
- Reduce the random search process.
- Minimize the iterative trial and error process.
- Determine the best designs among those proposed.
- Generate a scientific base for the design field.

Nam Suh (1990) developed a set of axioms and corollaries to represent design.

- *Axioms* are fundamental truths that are always expected to be true.
- *Corollaries* are propositions that follow from the axioms.
- *Functional requirements (FRs)* are characteristics of perceived needs for a product or a process. In addition, there is a minimum set of independent requirements that characterize the design objectives for a specific need.
- *Design parameters (DPs)* are variables that characterize the physical entity created by the design process to fulfill the FRs.

Axiomatic Principles in the Design Process

Design begins with defining the problem from an array of facts into a coherent statement of questions. The objective of design is stated in the functional domain, whereas the physical solution is generated in the physical domain. Design involves continuous interaction between the objectives of what we want to achieve and how we want to do it with a physical solution. The design process links these two domains, which are independent of each other.

The next step in the design process is to determine the objectives in terms of specific FRs. To satisfy these functional requirements, a physical embodiment is developed in terms of DPs. Design process relates FRs of the functional domain to the DPs of the physical domain. This mapping feature between FRs and DPs is illustrated in **Figure 3.16**. The design axioms provide principles that aid the creative design process by enabling good designs to be identified from an infinite number of designs.

Two main axioms are:

Axiom 1—The Independence Axiom: Maintain the independence of functional requirements.

Axiom 2—The Information Axiom: Minimize the information content of the design.

The axioms provide insight into issues such as how to make design decisions and why a particular design is better than others. Axiom 1 is related to the process of translation from the functional to the physical domain. Axiom 2 states that after Axiom 1 has been satisfied, the complexity of the design should be reduced. Axiom 2 explores questions such as whether something is a rational decision and how many design parameters are needed to satisfy the functional requirements. The same principles are used in all design situations regardless of whether the design is related to a product, process, or organization.

Mathematical Representation

The independence axiom can be represented as follows:

$$[FR] = [DM][DP] \tag{3.1}$$

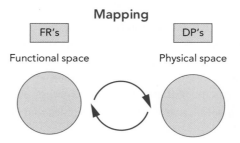

FIGURE 3.16 Mapping Functional Space to the Physical Space

where

[FR] = vector of the functional requirement

[DP] = vector of design parameters

[DM] = relationship matrix between functional and physical domain

$$[DM] = \begin{bmatrix} X_{11} & X_{12} & X_{13} & .... & X_{1n} \\ \\ X_{m1} & X_{n2} & X_{n3} & .... & X_{mn} \end{bmatrix} \tag{3.2}$$

An element of [DM], X_{ij} represents the relationship between each FR_i and DP_j. If the FR_i is affected by DP_j, then X_{ij} has a finite value. If FR_i is not affected by DP_j, then X_{ij} is zero. We can write a design equation and design matrix for each possible solution.

The implementation of the independence design axiom results in the case in which every functional requirement is associated with a single design parameter. This is called the *uncoupled design* and is represented by the diagonal matrix.

$$\begin{bmatrix} FR_1 \\ FR_2 \\ FR_3 \\ .. \\ FR_N \end{bmatrix} = \begin{bmatrix} X & 0 & 0 & .. & 0 \\ 0 & X & 0 & .. & 0 \\ 0 & 0 & 0 & .. & 0 \\ .. & .. & .. & & .. \\ 0 & 0 & 0 & & X \end{bmatrix} \begin{bmatrix} DP_1 \\ DP_2 \\ DP_3 \\ .. \\ DP_N \end{bmatrix}$$

Functional Requirements   Diagonal Matrix   Design Parameters

It can be observed from the first axiom that for a design to be uncoupled, it requires the number of FRs and DPs to be the same.

When the matrix is triangular (e.g., $X_{nm} = 0$ when $n \neq m$ and $m > n$), the design is decoupled. Both uncoupled and decoupled designs satisfy the independence axiom. All other matrices, which do not satisfy Axiom 1, are called coupled designs.

Mathematical Relationships: Functional requirements are broken down into sub-functions *(FR$_1$ ... FR$_n$)*. Design parameters can also be subdivided into (DP$_1$, DP$_2$ ... DP$_n$).

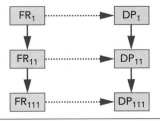

FIGURE 3.17 Decomposition and Mapping of FRs and DPs

Three types of design equations are used to represent the FR and DP relationships in **Figure 3.17**:

1. Uncoupled: satisfies Axiom 1(**Figure 3.18**)

$$\begin{bmatrix} FR_1 \\ FR_2 \\ FR_3 \end{bmatrix} = \begin{bmatrix} a_{11} & 0 & 0 \\ 0 & a_{22} & 0 \\ 0 & 0 & a_{33} \end{bmatrix} \begin{bmatrix} DP_1 \\ DP_2 \\ DP_3 \end{bmatrix} \tag{3.3}$$

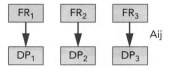

FIGURE 3.18 Graphical representation of Uncoupled Design

2. Coupled: always violates Axiom 1 (**Figure 3.19**)

$$\begin{bmatrix} FR_1 \\ FR_2 \\ FR_3 \end{bmatrix} = \begin{bmatrix} a_{11} & a_{12} & a_{13} \\ a_{21} & a_{22} & a_{23} \\ a_{31} & a_{32} & a_{33} \end{bmatrix} \begin{bmatrix} DP_1 \\ DP_2 \\ DP_3 \end{bmatrix} \tag{3.4}$$

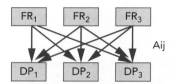

FIGURE 3.19 Graphical Representation of Coupled Design

3. Decoupled: satisfies Axiom 1, independence of FRs assured if DPs are arranged in a certain order (**Figure 3.20**)

$$\begin{bmatrix} FR_1 \\ FR_2 \\ FR_3 \end{bmatrix} = \begin{bmatrix} a_{11} & 0 & 0 \\ a_{21} & a_{22} & 0 \\ a_{31} & a_{32} & a_{33} \end{bmatrix} \begin{bmatrix} DP_1 \\ DP_2 \\ DP_3 \end{bmatrix} \tag{3.5}$$

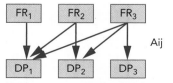

FIGURE 3.20 Graphical Representation of Decoupled Design

Corollaries: These design rules are derived from basic axioms to facilitate the applications.

Corollary 1 (decoupling of coupled design): Decouple or separate parts or aspects of a solution if FRs are coupled or become interdependent in the designs proposed. Decoupling does not mean that a part must be broken into parts or new element must be added to the design.

Corollary 2 (minimization of FRs): Minimize the number of FRs and constraints. Increasing these elements of design increases information content. Designers should not create a design that does more than what is intended. Unnecessary designs tend to be more expensive and less reliable.

Corollary 3 (integration of physical parts): Integrate design features in a single physical part if FRs can be independently satisfied with the proposed solution.

Corollary 4 (use of standardization): Use standardized or interchangeable parts if they are consistent with the FRs and constraints.

Corollary 5 (use of symmetry): Use symmetrical shapes and/or arrangements if they are consistent with the FRs and constraints. Symmetrical parts require less information to manufacture and assemble.

Corollary 6 (largest tolerance): Specify the largest allowable tolerance in stating FRs.

Corollary 7 (uncoupled design with less information): Seek an uncoupled design that requires less information than coupled designs in satisfying a set of FRs. This corollary implies that if designers propose an uncoupled design that has more information than a coupled design, they should begin the design again because a better design could be available.

When applying axiomatic design, Axiom 1 must be satisfied at all stages from functional domain to physical domain. Therefore, the matrix [DM] should be either triangular or diagonal.

Axiom 2 is stated in terms of information and complexity. If a product is complex, additional information is needed to describe it. The major objective of product design is to determine the right combination of product and process parameters as well as the material selection for the most economical solution for a product with the required quality. The axiomatic approach is intended to help designers to choose the matrix [DM] combination of information content to maximize the probability of achieving the FRs.

EXAMPLE 3.12

Consider the example of a two-knob water faucet. The basic objective is to provide continuous water at a desired flow rate and temperature. Hot and cold water are supplied separately.

Conventional Design

There are two functional requirements:

FR_1 = Obtain water flow rate
FR_2 = Obtain water temperature
DP_1 = Means to adjust cold water flow
DP_2 = Means to adjust hot water flow

$$\begin{bmatrix} FR_1 \\ FR_2 \end{bmatrix} = \begin{bmatrix} X & X \\ X & X \end{bmatrix} \begin{bmatrix} DP_1 \\ DP_2 \end{bmatrix} \tag{3.6}$$

This equation illustrates that the needs of the two-knob faucet can be represented as a coupled system. This is because the hot and cold water flow amounts must be adjusted at the same time to provide the needed flow rate and temperature. In this representation, water flow temperature and water flow are linked.

Alternate Design

To obtain a different solution in which we can maintain water temperature independent of water flow, we can reformulate DPs without saying how to do it. By uncoupling the DPs with original FRs, we might be able to make alternate designs. One possible uncoupled design is

$$\begin{bmatrix} FR_1 \\ FR_2 \end{bmatrix} = \begin{bmatrix} X & X \\ X & X \end{bmatrix} \begin{bmatrix} DP_1 \\ DP_2 \end{bmatrix} \tag{3.7}$$

DP_1 = Water flow regulating device
DP_2 = Water temperature regulating device

This design shows a solution representing independence of water flow from water temperature requirement.

EXAMPLE 3.13

Consider a refrigerator design using axiomatic design procedures. The examination of functional requirements and design parameters can provide insight into different design possibilities. The main requirement for the refrigerator is to preserve (freeze) food for long-term use and to keep some food at a cold temperature for short-term use. (**Figures 3.21** and **3.22**).

The needs of the refrigerator can be summarized in the form of functional domains and physical domains (first level):

First Level: Functional Domain

FR_1 = Freeze food for long-term preservation
FR_2 = Maintain food at a cold temperature for short-term preservation

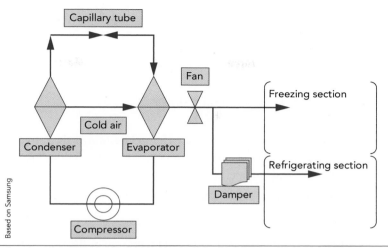

FIGURE 3.21 Conventional Refrigerator (One Cooling Fan Type)

A conventional refrigerator has a compressor, condenser, and evaporator with one fan to circulate the cold air. The FR requirements are satisfied by designing a refrigerator with two physical domain compartments.

DP_1 = Freezer section

DP_2 = Chiller section

To address FR_1 and FR_2, the freezer section should affect only the freezer area, and the chiller section should affect only the food to be chilled. The design matrix to satisfy this is a diagonal matrix.

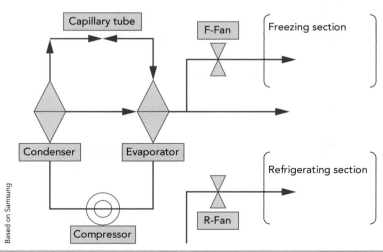

FIGURE 3.22 New Cooling System Refrigerator (Two Cooling Fan Type)

$$\begin{bmatrix} FR_1 \\ FR_2 \end{bmatrix} = \begin{bmatrix} X & 0 \\ 0 & X \end{bmatrix} \begin{bmatrix} DP_1 \\ DP_2 \end{bmatrix} \tag{3.8}$$

We can now proceed to the second level and decompose the functional domain.

Second Level: Functional Domain

The decomposition of FR_1 follows:

FR_{11} = Temperature control of freezer section in the range $18-25°F$

FR_{12} = Maintain the uniform temperature throughout the freezer section at preset temperature

FR_{13} = Control relative humidity to 50%

Decomposition of FR_2

FR_{21} = Control chilled section temperature in the range of $45-55°F$

FR_{22} = Maintain uniform temperature within the chilled section within $2°F$

To satisfy the second level FR_{11}, we must design DP in such a way to satisfy FRs at the level of decomposition. We must make sure that FRs at this level are independent.

In other words, DP_{11}, DP_{12}, and DP_{13} satisfy FR_{11}, FR_{12}, and FR_{13} and are independent. The freezer section's functions are met by pumping the chilled air into that section (FR_{11}) and by circulating it uniformly throughout the freezer area (DP_{11}).

DP_{11} = Turn on/off the compressor when the air temperature is higher or lower than set values

DP_{12} = Blow the air into freezer section and circulate it uniformly

DP_{13} = Condense the return air moisture when the dew point is exceeded

The equations can be represented as

$$\begin{Bmatrix} FR_{12} \\ FR_{11} \\ FR_{13} \end{Bmatrix} = \begin{bmatrix} X & 0 & 0 \\ X & X & 0 \\ X & 0 & X \end{bmatrix} \begin{Bmatrix} DP_{12} \\ DP_{11} \\ DP_{13} \end{Bmatrix} \tag{3.9}$$

This equation indicates that the design is *decoupled*.

We can further decompose FR_2 as

FR_{21} = Control temperature of chilled section in the range of $3-5°F$

FR_{22} = Maintain a uniform temperature throughout the chilled section at the preset temperature

$$\begin{Bmatrix} FR_{21} \\ FR_{22} \end{Bmatrix} = \begin{bmatrix} X & 0 \\ 0 & X \end{bmatrix} \begin{Bmatrix} DP_{21} \\ DP_{22} \end{Bmatrix} \tag{3.10}$$

The corresponding DP_2 is

DP_{21} = Circulate the chilled air by a fan

DP_{22} = Vents

Features of Conventional Refrigerator: Dependent Control

- Fan − Freezing room (thermal sensor)
- Fan + Damper − Refrigerating room
- Damper required (produces pressure loss)

Features of New Cooling System: Independent Control

- F-fan Freezing room (with sensor)
- R-fan Refrigerating room (with sensor)
- No damper required

Manufacturing Domain

Axiomatic procedures can be extended to the manufacturing domain. See **Figure 3.23** for the mapping between the product's FRs (defined in the functional space) and the DPs (defined in physical space). If the variables in the manufacturing space are defined as process variables (PVs), there can be additional mapping between the DPs and PVs of the manufacturing space.

Axiom 1 applies to both of these mapping situations as,

$$\{FRs\} = [A]\{DPs\} \tag{3.11}$$

$$\{DPs\} = [B]\{PVs\} \tag{3.12}$$

where the matrices [A] and [B] must be either uncoupled or decoupled to satisfy the independence axiom. The functional space is then related to the manufacturing space as

$$\{FRs\} = [A][B]\{PVs\} = [C]\{PVs\} \tag{3.13}$$

Therefore, for the product to be able to be manufactured, the matrix [C] must also be an uncoupled or decoupled type. That means that the product design cannot

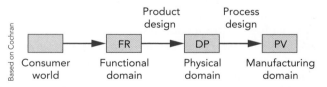

FIGURE 3.23 Domains of the Design for Manufacturing

be manufactured within the design specifications unless both the product and process design are either uncoupled or decoupled designs. In searching for a design solution to satisfy a given set of FRs, designers know that the design matrix must be diagonal or triangular and that the number of DPs equals the number of FRs in an ideal design. The uncoupled design equation is expressed as

$$\begin{Bmatrix} FR_1 \\ FR_2 \\ FR_3 \end{Bmatrix} = \begin{bmatrix} X & 0 & 0 \\ 0 & X & 0 \\ 0 & 0 & X \end{bmatrix} \begin{Bmatrix} DP_1 \\ DP_2 \\ DP_3 \end{Bmatrix} \tag{3.14}$$

The only unknowns in the equation are the DPs. Thus, we can proceed to conceptualize a design solution that consists of at least three DPs that satisfy equation (**3-14**). To achieve good quality products, the manufacturing process and system must satisfy the functional requirements specified by the designer in terms of geometry, hardness, and so on. When the manufacturing process can meet the design specifications, the probability of producing high quality parts is 1, and the information required to achieve the task is 0 because the manufacturing process can produce good parts each time. When the probability is less than 1, additional information must be supplied by the operator or by some other means to meet the functional specifications.

3.4.1 Axiomatic Approach for Production System Design

A manufacturing system is a complex arrangement of physical elements characterized by measurable parameters. A manufacturing system design consists of the design of physical elements and operations required to produce a product. A production system provides the functions to support the manufacturing system and defines its performance measures. The production system consists of the design of all elements and functions that support the manufacturing system.

Designing production systems for a product is crucial for understanding the relationship between system design objectives and physical design implementation. This process can provide a framework for explaining why low-level decisions tend to affect the viability of an entire production system design. Designing a system requires understanding what is important and the need to redesign or change those variables that had previously impacted the manufacturing system's operation negatively.

A lean (Toyota) production system represents a new production design and therefore requires a new set of performance and cost measurement criteria, which are inherently part of the system design. Most production systems today are measured in a way that causes their design to move in the opposite direction of meeting lean production design objectives. Axiomatic design helps in defining *what* the design system objectives are and *how* they are to be accomplished and implemented from a system design perspective.

Mapping Conventional Production to Mass Production

The two elements of axiomatic design that are significant in the development of the production system design are the idea of *zigzagging design* and *axioms*. The idea of zigzagging means that any design, no matter how complex, can be decomposed into its constituent levels. The production system design decomposition provides an organized method for designing production systems. The process of decomposition includes the functional and physical domains of design. The functional requirements or objectives defined by the functional domain are the measurable parameters of the production system design.

Benefits of the Production System Design Decomposition

- Ability to concretely describe and distinguish between various production system design concepts
- Adaptability to create different products and manufacturing environments
- Ability to design or create new system designs to meet new environments (for example, what happens when the FRs or DPs change, as in lean production versus mass production)
- Portability of a production system design methodology across industries
- Indication of the impact of lower-level design decisions on total system performance
- Provision of the foundation for developing a new set of manufacturing performance measures from a system design perspective
- Ability to make the connection between machine design requirements and manufacturing system objectives

Figure 3.24 provides a graphic explanation of the difference between mass and lean production. The difference is the result of a change in the design parameters affecting the functional requirements of sales revenue, production costs, and production investment. During the mass production period, increasing sales revenue simply meant making more products.

Figures 3.24 (a) and **(b)** show how the axiomatic approach can be applied to clarify the differences between mass production and lean manufacturing. As shown in **Figure 3.24 (a)**, if the business objective is to *increase the return on revenue*, the FRs can be shown as *sales revenue, production cost*, and *production investment*. One component of lean manufacturing is increasing sales revenue. **Figure 3.25** illustrates how increase in sales revenue can be mapped to maximizing customer satisfaction.

$$\begin{Bmatrix} FR_{111} \\ FR_{112} \\ FR_{113} \end{Bmatrix} = \begin{bmatrix} X & 0 & 0 \\ X & X & 0 \\ X & X & X \end{bmatrix} \begin{Bmatrix} DP_{111} \\ DP_{112} \\ DP_{113} \end{Bmatrix} \qquad (3.15)$$

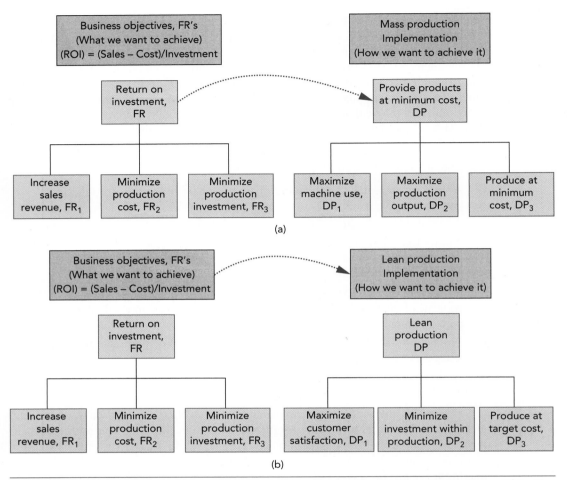

FIGURE 3.24 (a) Mapping Business Objective to Mass Production (b) Mapping Business Objective to Lean Production

For mass production, the main aim is to provide *products at minimum* cost by maximizing production output and maximizing machine utilization. In lean manufacturing, the emphasis is placed on reducing defective products, increasing customer satisfaction, and providing on-time delivery. In mass production, the response to *increase sales revenue (FR)* leads to *maximizing production output (DP)*. In lean production, the response to *increase sales revenue (FR)* comes from *satisfying the customer (DP)*.

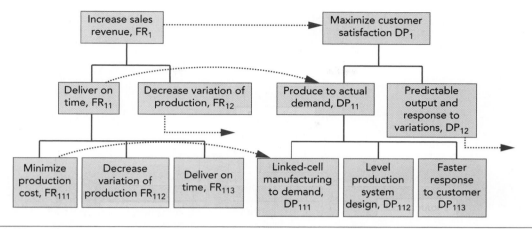

FIGURE 3-25 Axiomatic Thinking Applied to Increasing Sales Revenue

CASE STUDY 3.1
The Portable Coffeemaker (continued from Chapter 2)

Coffee has a major market in the United States, involving the majority of the U.S. population. The coffeemaker is one of the regularly used appliances in everyday life. The product definition starts by identifying customers' needs as the requirement to design a more convenient, yet inexpensive solution for their coffee-making process. A portable coffeemaker provides the convenience of being able to have coffee at any time. The reduced size and limited functionality of a portable coffeemaker allow for a different product with a special market appeal.

Concept Development

After the problem has been identified and the product concept defined, the concept development phase identifies customers' needs that are translated into technical specifications. The product is broken down into its subfunctions, each of which is studied as a separate problem. Alternative product concepts are generated and evaluated until one is selected over the others as the concept that better satisfies customer requirements and meets the function needs such as engineering and manufacturing.

Problem Definition Using Present State (PS) and Desired State (DS) Technique

By defining the problem, the PS-DS technique helps product designers to identify the starting and desired final points. Following that, an appropriate path for reaching the target can be found easily. Refer to **Figure 3.4** for a Duncker diagram representation of concept development for a coffeemaker.

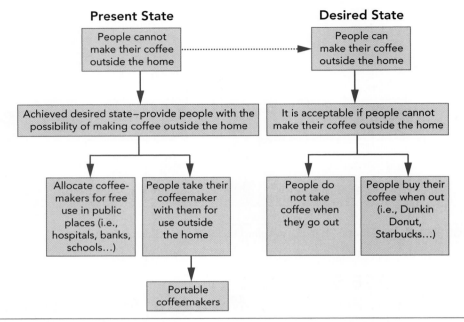

FIGURE 3.26 Duncker Diagram

As shown in **Figure 3.26**, the solution that goes from the PS to the DS is to provide customers the possibility of making their coffee outside the home by using a portable coffeemaker.

Functional Decomposition

To develop a product concept, the main product function is identified and broken into subfunctions.

Figure 3.27 illustrates a functional diagram for the portable coffeemaker. Its main function is to make coffee; also it has four main subfunctions: (1) containing water, (2) boiling water, (3) filtering coffee/water, and (4) storing coffee. Each subfunction is then broken down until the product part/physical component providing a specific function is identified.

In the present case, the water reservoir contains the water. The heating element then boils the water. Water then pours into the ground coffee reservoir and is filtered. The brewed coffee is stored in the coffee reservoir.

Morphological Analysis

Morphological analysis enables designers to compare various attributes of a product to evaluate each different attribute's possibilities in isolation, and then to consider them as a whole to derive a solution.

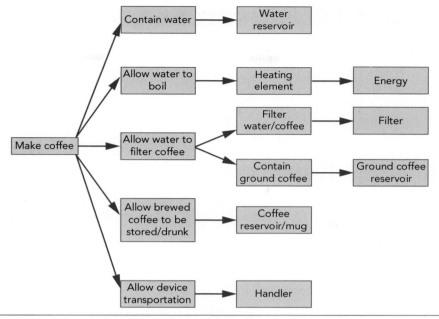

FIGURE 3.27 Functional Design of Portable Coffeemaker

A morphological chart (**Figure 3.28**) is based on the subfunctions identified in the functional analysis. Each subfunction is listed, and three solutions by which each can be achieved are identified.

The following is a description of the features listed for each subfunction.

- *Boil water.* There are three options for boiling water. (1) In its reservoir, water can be heated by a hot plate through conduction. (2) A tubular caldron can be used to hold the water that is fed externally to the water reservoir. (3) A wire heating element can be placed directly into the water reservoir.

- *Direct water.* The three methods—drip-brew, percolator, and French press—of directing water through coffee beans are considered.

- *Filter water.* Three types of filters can be used. Paper and mesh filters are easy to maintain because they can be disposed after use. A removable permanent metal filter is less convenient because the coffee grounds need to removed after use but is more convenient than a permanent nonremovable filter.

- *Store coffee.* Coffee can be contained in (1) the same or (2) a separate reservoir of water or (3) in the coffee mug. This last option provides easier maintenance because the mug can be removed and cleaned in a dishwasher.

- *Handle device.* Three handle devices are considered: a handle, a grip, and a string. The first two provide a similar function; the string also allows the coffeemaker to be placed on a hanger. However, the string is less convenient for carrying the coffeemaker around.

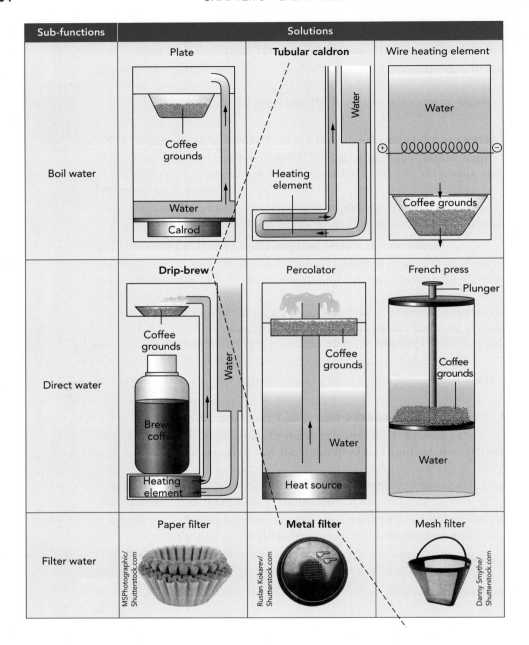

Sub-functions	Solutions		
	Plate	**Tubular caldron**	Wire heating element
Boil water	Coffee grounds / Water / Calrod	Water / Heating element	Water / Coffee grounds
	Drip-brew	Percolator	French press
Direct water	Coffee grounds / Water / Brewed coffee / Heating element	Coffee grounds / Water / Heat source	Plunger / Coffee grounds / Water
Filter water	Paper filter / MSPhotographic/ Shutterstock.com	**Metal filter** / Ruslan Kokarev/ Shutterstock.com	Mesh filter / Danny Smythe/ Shutterstock.com

FIGURE 3.28 Morphological Chart with the Selected Combination of Subsolution

A percolator's drip-brew mechanism has two separate reservoirs, one for water and one for coffee. The French press mechanism uses just one reservoir where water and coffee are temporarily mixed together before being eventually separated by a plunging element.

Among several combinations of features, the optimal one for the task has been selected. The solution chosen is bolded in **Figure 3.28**.

Based on Dart, Kyle. Howard, Tyler. McFarlane, Kurt. Plunkett, Matt. Collapsible Coffee Maker. Final Report, ME 450, Section 002. Dec. 2007.

CASE STUDY 3.2
Product Design of a Bottle Unscrambler (Sorting) Machine

Problem Definition: Design a Bottle Sorting Machine

The problem is to design a simple 12 oz. plastic soda bottle sorter capable of sorting no less than 100 bottles per minute.

Current bottle-sorting equipment are capable of sorting bottles at a rate of 80–100 bottles per minute. The new product is designed with a target of 100 bottles per minute (bpm). The overall bottle dimensions are 3" round, 7" tall with a 1" neck (typical plastic soda bottle size).

The unscrambling equipment will have the following features:

- Simple design
- Minimal number of components
- Competitively priced
- Ease of service
- Robustness

▸ **Step 1. Concept Generation**

The following conceptual sketches reveal some of the options available for the bottle-sorting equipment. These options are created to inspire originality in design and to help the designer consider new design alternatives. **Figure 3.29** illustrates the initial concept explore.

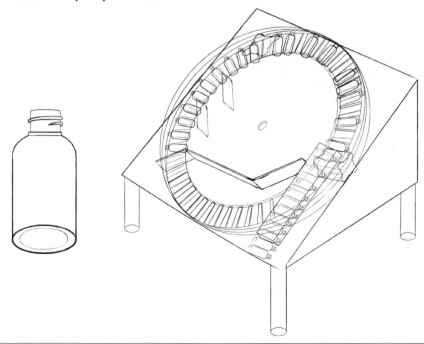

FIGURE 3.29 CAD model of Bottle Unscrambler

Concept 1
Bottle Movement.

The sorting mechanism on a conveyor can carry bottles out of a hopper for delivery as shown in **Figure 3.30 (a)**.

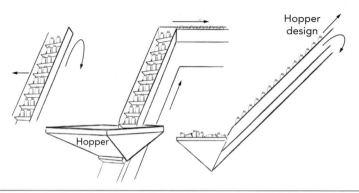

FIGURE 3.30 (a) Sketch of Concept 1

Bottles migrate toward the side of a barrel, or a paddle wheel picks them up when they are oriented correctly, and then they are transferred to a conveying system as shown in **Figure 3.30 (b)**.

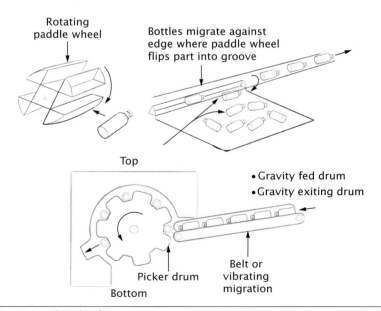

FIGURE 3.30 (b) Sketch of Sorting Concept

Concept 2
Bottle Feeding Mechanism

This mechanism can be used to separate bottles facing opposite one another so they can later be recombined to an upright orientation as shown in **Figure 3.30 (c)**.

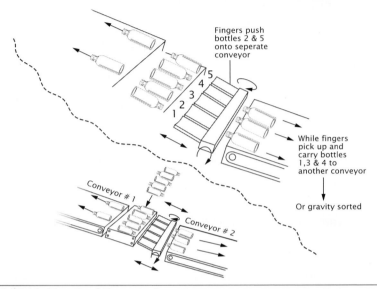

FIGURE 3.30 (c) Sketch of Concept 2

Concept 3
Bottle Orientation

A slotted vibratory table or drum can deliver bottles in a standing position so that upside-down bottles can be sorted as shown in **Figure 3.30 (d)**.

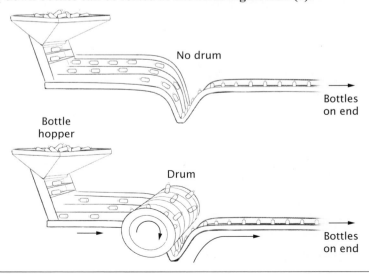

FIGURE 3.30 (d) Sketch of Concept 3

Concept 4
Bottle Reorientation

Rails can be used so that bottles in different orientations sit at different heights; this could prove advantageous because upright bottles are easier to topple, while upside-down bottles will continue along the path, making sorting much easier.

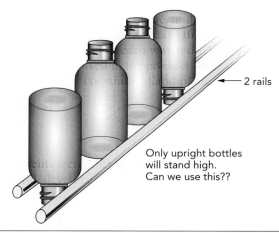

FIGURE 3.30 (e) Sketch of Concept 4

Concept 5
Bottle Size Recognition

A slotted vibratory table can easily orient bottles lengthwise. Square grooves might allow stray bottles (of the wrong orientation) to migrate over the aligned bottles without hampering the aligned bottles progress.

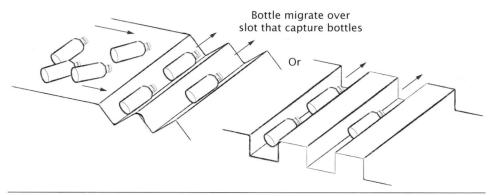

FIGURE 3.30 (f) Sketch of Concept 5

Concept 6
Alternate Concepts on Bottle Orientation

The bottles fall gently onto a rotating cone-shaped carousel; then as they orient themselves lengthwise, they fall through a slot on the side of the carousel.

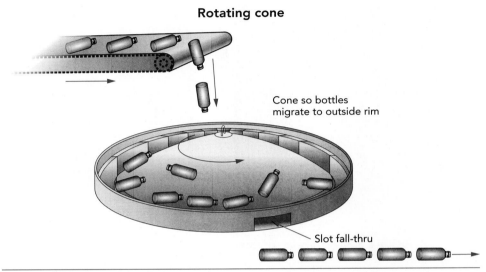

FIGURE 3.30 (g) Sketch of Concept 6

> **Step 2. Benchmarking and Market Evaluation**
> See **Table 3.15** for a list of websites showing sorting machines in the market.

Manufacturer	Model	Link	Simplistic	Bottles/Min.	Footprint
Nalbach	NS-54	http://www.nalbach.com/ fr_index.html?/Products/ NECOSORT.htm	10	Moderate–High	23
Hoppman	FW-2500	http://www.hoppmann.com/ photolib/fw2500.htm	10	None	N/A
Kaps-All	AU-6	http://www.kaps-all.com/ 5.htm	10	Moderate	N/A
SeungJin Co.	DUSC-0200	http://www.packingsj.co.kr/ Etc/BOTTLE%20UNSCRAM-BLER.htm	10	Moderate	N/A
AVE Industrial	Auto-Unscrambler	http://www.pack.it/ave/ p03.htm	10	Moderate–High	N/A

(continued)

Manufacturer	Model	Link	Simplistic	Bottles/Min.	Footprint
Hoppman	FT-15-RD	http://www.hoppmann.com/photolib/ft15ac.htm	10	Moderate–High	4
Hoppman	FS-50-RD Scallop	http://www.hoppmann.com/photolib/fs50rd.htm	9	Moderate–High	42
Omega	RP3 Rotary Pocket	http://www.omegadesign.com/unscramblers.htm	9	Moderate–High	14
Pace	Omni Line	http://www.pacepkg.com/Unscramblers.htm	8	Moderate–High	N/A
Hoppman	FC-4500	http://www.hoppmann.com/photolib/fc4500.htm	8	Need robot	16
Hoppman	FR-20 Centrifugal	http://www.hoppmann.com/photolib/fr20.htm	8	Moderate–High	5
Hoppman	FT-30-RD Centrifugal	http://www.hoppmann.com/photolib/ft30rd01.htm	8	Low–High	11
Hoppman	FR-30 Centrifugal	http://www.hoppmann.com/photolib/fr30.htm	8	Moderate–High	11
New England	NEHAE	http://www.neminc.com/	8	Moderate–High	N/A
Ramco	MultiSort 70	http://www.romaco.com/bosspak/products/-1-Unscrambling/	7	Slow–Moderate	N/A
Perl	Perl Rotary Orienter	http://www.djsent.com/Product.asp?ProductId=757	7	Slow–Moderate	9
Hoppman	FT-40-RD	http://www.hoppmann.com/Centr40.htm	7	Moderate–High	24
Hoppman	Linear Bottle Feeder	http://www.hoppmann.com/photolib/linbot01.htm	6	Moderate	N/A
New England	NEHHLPE	http://www.neminc.com/UNSC/puHHLP.htm	6	High	N/A
Halcord	Item# BOTT2344S	http://www.processplant.com/image.asp?ID=2344&pID=997354.jpg	6	Moderate–High	14
Hoppman	FL-110	http://www.hoppmann.com/Fl100.htm	4	Slow–Moderate	128
Hoppman	FL-200	http://www.hoppmann.com/Fl200.htm	4	Slow–Moderate	144

TABLE 3.15 Websites Showing Sorting Machines on the Market

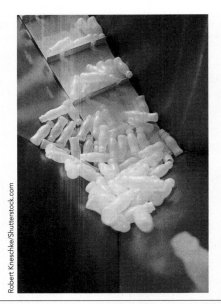

Robert Kneschke/Shutterstock.com

FIGURE 3.31 Unsorted bottles in Omni-Line M500 SSD on display at the Pace Packaging booth at the ANUGA FoodTec industry trade show in Cologne, Germany on March 27, 2012

▸ **Step 3. Concept Selection Based on Customer Needs**
Customer requirements focus on minimizing components, compact design, and maximizing product life. The house of quality identifies the customer requirements for the descrambler to be compact, modular, low cost, and energy efficient.

From the information obtained, it was decided to use barrel or drum design pitched at an angle to take advantage of gravity. This minimizes the mechanical requirements and facilitates the use of gravity for loading, orienting, sorting, and discharge. To provide flexibility to the customer, the front-end device such as hopper and a rear-end device such as a conveyor are not considered as a part of the product shown in **Figure 3.32**. This enables the customer to easily adapt the unscrambler to an existing process. **Table 3.16** represents the morphological chart of a bottle unscrambler machine. The highlighted components are used for further design.

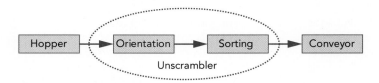

FIGURE 3.32 Overall Function Diagram

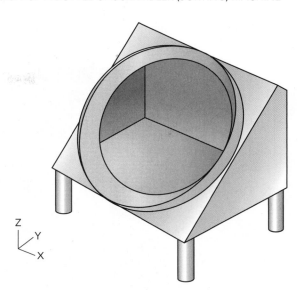

Function		Solution (component)	
Hopper integration	Mechanical	Sensor feedback	Manual speed adjustment
Bottle orientation	Vertical	Lateral (horizontal)	Radial (horizontal)
Table pitch	30 degrees	45 degrees	60 degrees
Deflectors	Adjustable from above	Fixed to rotary table	
Table drive	Fixed speed	Variable speed	
Sorting	Redirect by cam	Flip by mechanical arm	Wheel ejection
Picker	Belt	Wheel ejection	Fall through
Chute	Enclosed slide	Open slide	Boxed rails
Conveyor integration	Mechanical	Sensor feedback	Manual speed adjustment

TABLE 3.16 Morphological Bottle Unscrambler (Sorting) Machine

Component	Specifics		
Rotary table	Stainless steel (SS)	Brass	Plastic
Frame/Housing	SS weld	Cast iron	Cast resin
Deflectors	Steel	Brass	Plastic
Table drive	Direct	Belt	Gear
Sorting wheel drive	Direct	Belt	Gear
Picker wheel drive	Direct	Belt	Gear
Chute	SS	Brass	Plastic

‣ **Step 4. Preliminary Design of Body Structure for the Bottle Unscrambler**

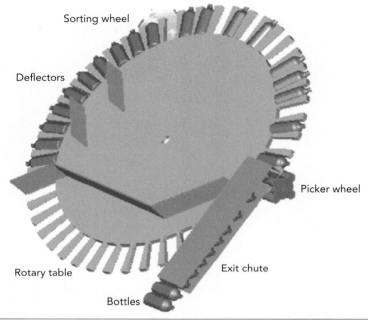

FIGURE 3.33   Key Components of Bottle Unscrambler

A detailed design follows.

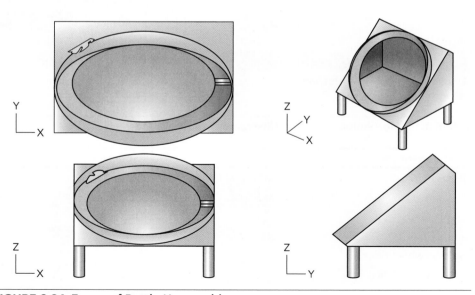

FIGURE 3.34 Frame of Bottle Unscrambler

▸ **Step 5. Identification of Critical Components for the Unscrambling Machine**
The conceptual design of the sorting wheel is shown in the next figure.

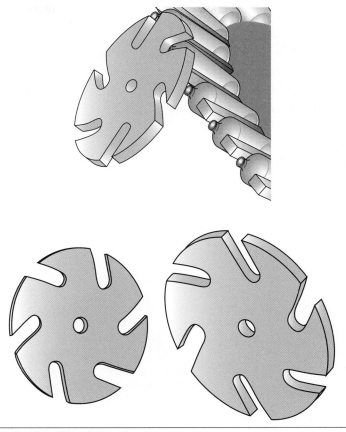

FIGURE 3.35 Sorting Wheel Details

A rotary table with a 72" diameter can hold 45 bottles radially. Knowing this, we can calculate the required gear ratio.

Rotary table $= 72$ inches

Inside diameter of sorting wheel for neck

$$\frac{72 \text{ inches}}{42 \text{ bottles}} = \frac{\text{inside diameter}}{6 \text{ slots}}$$
inside diameter $= 9.6$ inches

$$\text{Gear Ratio} = \frac{72}{9.6} = 7.5$$

$\therefore$ 7.5 to 1

The picker wheel details follow.

The picker wheel that lifts the bottles from the rotary table into the exit chute can be of any diameter; however, to keep the part small, the picker wheel is designed with only four fingers.

Rotary table = 72 inches

Inside diameter of picker wheel for exiting botles

$$\frac{72 \text{ inches}}{45 \text{ bottles}} = \frac{\text{inside diameter}}{4 \text{ fingers}}$$

inside diameter = 6.4 inches

$$\text{Gear Ratio} = \frac{72}{6.4} = 11.25$$

$\therefore$ 12 to 1

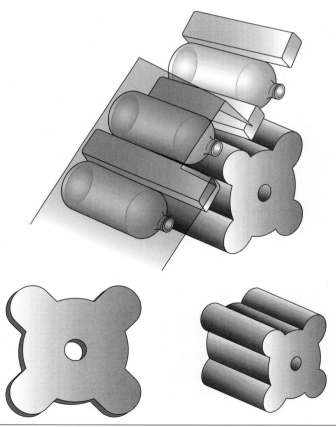

FIGURE 3.36 Picker Wheel Details

The gear ratio of 11.25 is most likely nonstandard. To accommodate standard gear reducers, we can lower the picker wheel and compensate with longer fingers.

▸ **Step 6. Graphical Representation of the Component Assembly**

The design assembly for bottle unscramble layouts are shown in **Figure 3.37**.

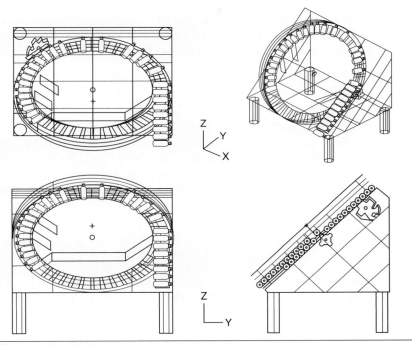

FIGURE 3.37 3D Solid Model

▸ **Step 7. Design Ideas for Equipment Performance and Synchronization**

Tooling: An interchangeable rotary table for different bottle sizes is desirable. To accommodate this, the picker and sorting wheel also need to be interchangeable. These three components must be designed concurrently as a set to accommodate any fixed gear ratios.

Sorting rate: The number of bottles sorted is inversely proportional to the bottle diameter; more thin bottles can fit along the circumference of the rotary table. Because there is a terminal speed for sorting capabilities, wider bottles lessen the sorting rate. For the device to sort 100 soda bottles per minute, the estimated variable table speed must be 5 to 10 RMP, assuming that 33–50% of the slots are occupied by bottles of the proper orientation.

Table Conveyor Synchronization

The tangential velocity of the rotary table is synchronized with the conveyor for a smooth transfer of bottles. This can be done by using a feedback sensor or a mechanical device. A feedback sensor would allow the system to be more flexible, making it easier to adapt the unscrambler to a conveyor already on hand. However, the mechanical synchronization of the unscrambler with the conveyor is less flexible.

Design without air: Many unscramblers on the market use air diverters to push bottles to the outside of a sorting bowl. This design completely avoids the use of air diverters. This is beneficial because some users do not have compressed air at their disposal.

Hopper: Many commercially available hoppers can be easily adapted. Frequently used designs consist of a hopper with a variable speed motor driven by a controller using sensory feedback from both the exit chute and the rotary table.

Orientation: Radial orientation was used for this design because bottles pack better radially; however, a tangential orientation in which bottles lie along the circumference of the table might be tried. The present design lends itself well to experimenting with different table patterns. Because the table can be made of a plastic such as Delrin©, it would be easy to drill holes to enable further experimentation with pins inserted into the table. The pins might help flip, rotate, or guide bottles to maximize the number of slots occupied per revolution.

Sorting: Although a wheel design was chosen, it is simple enough to allow sorting by ramps, fingers, and air.

Conveyor: Similar to the hopper, the conveyor speed should be synchronized with the hopper and unscrambler so that bottles are constantly fed to the next operation. It is best to do this with a sensor-controlled gate at the end of an exit chute long enough to act as a buffer for bottles being transferred from chute to conveyor.

‣ **Step 8. Typical Product Costing as shown in Table 3.17.**

Motor	½ HP 220 Voltage AC	$1,500
Frame	Fabricated stainless steel	4,000
Plastic tooling	Table	1,200
	Sorting wheel	200
	Picker	400
	Deflectors	500
Controls	Conveyor sensor	120
	Exit chute sensor	120
	Table sensor	120
	Main motor drive	2,200
	Total estimate	$10,360

TABLE 3.17   Product Cost

Appendix

3.5.1 **Additional TRIZ Tools**

Problems of a more difficult nature are solved by using the following precise TRIZ tools:

1. ARIZ (algorithm for inventive problem solving)
2. Su-field analysis (substance-field analysis)
3. DPE (directed product evolution)
4. AFD (anticipatory failure determination)
5. Technical systems
6. Laws of ideality
7. Psychological inertia

ARIZ: ARIZ (algorithm approach for inventive problem solving) is an analytical tool of TRIZ. It provides specific sequential steps for developing a solution to complex problems without apparent contradictions. Depending on the nature of the problem, as many as 5 to 60 steps may be involved. From an unclear technical problem, the underlying technical problem can be revealed through ARIZ. Following is a brief description of eight of the steps.

▸ **Step 1. Analysis of the problem.**
Begin by making the transition from vaguely defined statements of the problem to a simply stated mini problem. At this step, an analysis of conflicting situations (technical contradictions) can be performed once a decision is made as to which contradiction will be considered for further resolution. Once a decision is made, a model of the problem is formulated.

▸ **Step 2. Analysis of the problem's model**
A simplified diagram modeling the conflict in the operating zone (a specified narrow area of conflict) is drawn. Then an assessment of all available resources is made.

▸ **Step 3. Formulation of the Ideal Final Result (IFR)**
Usually the IFR statement reveals contradictory requirements to the critical component of the system in the operating zone, which is called the physical contradiction.

▸ **Step 4. Utilization of outside substances and field resources**
Use an informational data bank. Consider solving the problem by applying standards in conjunction with a database of physical effects.

▸ **Step 5. Reformulation of the problem.**
If the problem still remains unsolved at this stage, ARIZ recommends returning to the starting point and reformulating the problem in respect to the supersystem. This looping process can be repeated several times.

▸ **Step 6. Analysis of the method that removed the physical contradiction**
The main goal of this step is to check the quality of the solution after the physical contradiction has been removed.

▸ **Step 7. Use of the current solution**
This step guides designers through an analysis of effects that the new system could have on adjacent systems. It also forces the search for applications to other technical problems.

▸ **Step 8. Analysis of the steps that led to the solution.**
This is a checkpoint where the real process used to solve a problem is compared with the process suggested by ARIZ. Deviations are analyzed for possible further use.

Su-field analysis (substance-field): This is a tool for expressing function statements in terms of one subject acting on another subject. The objects are called *substances* and the action is a *field*. Su-field analysis is helpful in identifying functional failures. By looking at actions as fields, undesirable or insufficient actions can be countered by applying opposite or intensified fields. There are 76 standard solutions that permit the quick modeling of simple structures for Su-field analysis.

There are essentially five steps to follow in making the Su-field model:

▸ **Step 1. Identify the elements.**

▸ **Step 2. Construct the model.**

▸ **Step 3. After completing the first two steps, stop to evaluate the completeness and effectiveness of the system. If an element is missing, identify it.**

▸ **Step 4. Consider solutions from the 76 standard solutions.**

▸ **Step 5. Develop a concept to support the solution.**

DPE (Directed Product Evolution): Traditional techniques forecasting tries to predict the "future characteristics of machines, procedures or techniques." It relies on surveys, simulations, and trends to create a probabilistic model of future developments. It gives a forecast but does not precisely detail the technology being forecasted.

DPE is essentially a prediction with a level of confidence of a technological achievement in a given time frame with a specified level of support. Most of the innovations of the next 20 years will be based on scientific and technological knowledge that exists now. However, it is difficult to identify what knowledge will be of real significance. The role of DPE is to evaluate today's knowledge systematically, thereby identifying what is achievable and how one technological advance—perhaps in conjunction with another—could fulfill a human need.

The principle of DPE basically works on the following eight patterns of evolution:

1. Stages of evolution
2. Evolution to increased ideality
3. Non-uniform development of system elements
4. Evolution to increased dynamism and controllability
5. Evolution to increased complexity and then simplification
6. Evolution with matching and mismatching elements

7. Evolution toward the micro level and increased use of fields
8. Evolution toward decreased human involvement

By analyzing the current technology level and contradictions in products, this TRIZ tool enables designers to analyze evolutionary processes and create the future. Using the eight patterns of evolution, the DPE method considers past and present scenes and recommends directions of innovation.

Anticipatory Failure Determination (AFD):   AFD is an efficient and effective method for analyzing, predicting, and eliminating failures in systems, products, and processes. AFD guides **designers** in documenting situations, formulating related problems, developing hypotheses, verifying potential failure scenarios, and finding solutions to eliminate problems. It is a distinctive and powerful approach that favorably impacts costs associated with quality, safety, reliability, recalls, and warranty claims because prevention of unanticipated failures is important in new product development. In effect, AFD invents failure mechanisms and then examines the possibilities that they would actually occur; as a result, factors contributing to the failures can be eliminated with this highly proactive technique.

The AFD system consists of two modules:

- Failure analysis of previous failures
- Failure to predict failures that might occur in the future

The AFD system supports these applications by providing a disciplined, rigorous process by which the designer can:

- Analyze failure mechanisms thoroughly
- Obtain an exhaustive set of potential failure scenarios
- Develop inventive solutions to prevent, counteract, or minimize the impact of failure scenarios

More Principles of TRIZ

Technical Systems: In TRIZ methodology, anything that performs a function is a technical system and consists of one or more subsystems. The hierarchy of a technical system spans from the least complex with only two elements to the most complex with many interacting elements. When a technical system produces inadequate functions, it might need to be improved. This requires imaginatively reducing the system to its simplest state. In TRIZ, the simplest technical system consists of two elements with energy passing from one element to another.

People face essentially two groups of problems: those with generally known solutions and those with unknown solutions. In the case of known solutions, people follow a general pattern of problem solving as shown in **Figure 3.38**. The specific problem is considered as a standard one of a similar or analogous nature. A standard solution that

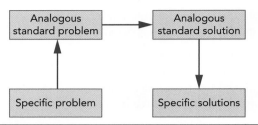

FIGURE 3.38 General Model of Technical Systems

results in a specific solution is known. The functions of technical systems are realized by using physical, chemical, and geometrical effects. It follows that knowledge of such effects is crucial in inventive situations. **Figure 3.38** shows a general model of technical systems.

Law of Increasing Ideality: The most fundamental law is that of the *ideal system* in which the given function is realized but no resources are consumed. This ideal solution may never be found, but the ratio function divided by resources is likely to increase over time. The ideal system law states that any technical system throughout its lifetime tends to become more reliable and effective. Ideality always reflects the maximum utilization of existing resources, both internal and external to the system. If more readily available resources are utilized, the system becomes more ideal. The further the inventive work is from its ideal state, the more complex the system will be. The *law of increasing ideality* means that technical systems evolve toward increasing degrees of ideality.

Psychological Inertia: If the problem being explored has no solution forthcoming or lies beyond the designer's experience, the number of trials could vary depending on the designer's intuition, field of expertise, and creativity. But the drawback is that extrapolating these psychological tools to other people in the organization could be difficult. This leads to *psychological inertia* when the solutions being considered are only within the designer's domain of expertise omitting other alternative solutions. The ideal solution may lie outside the designer's experience and could occur randomly within a solution space.

Combination of TRIZ and QFD: Because TRIZ can help engineers and developers solve technical contradictions and invent new technologies, its use in new product development is very important. With QFD, an industry should be able to identify important customer requirements and then solve any technical bottlenecks that occur. TRIZ can also help identify new functions and performance levels to achieve truly exciting levels of quality. **Table 3.18** shows areas where QFD and TRIZ can complement each other.

QFD identifies and prioritizes the voice of the customer and the capabilities of the company technologies; then it helps prioritize new concepts for design and production. TRIZ helps to create new concepts.

Development Phase	Benefit of Combining QFD and TRIZ
Market research	Use directed product evolution (DPE) with concept methods to show customers what new products will be like
Preliminary research	Solve engineering bottlenecks and contradictions Eliminate contradictions discovered by the roof of the house of quality Help determine target values in the quality planning table
Design	Use Su-field analysis and DPE to identify new functions to attract customers Use AFD to identify and prevent failure modes in new products Use TRIZ to develop new concepts with DPE patterns Use TRIZ to lower costs without resorting to trade-offs
Manufacturing	Remove design constraints caused by equipment and manufacturability limitations
Production	Remove design constraints caused by limitations of processes and people
After service	Help design for serviceability and to remove bottlenecks

TABLE 3.18 Combining QFD and TRIZ

TRIZ Inventive Principles

Segmentation
- Divide an object into independent parts.
- Make an object sectional.
- Increase the degree of an object's segmentation.

Extraction
- Extract (remove or separate) a "disturbing" part or property from an object.
- Extract only the necessary part or property.

Local Quality
- Change an object's structure from uniform to nonuniform; change an external environment (or external influence) from uniform to nonuniform.
- Have different parts of the body perform different functions.
- Make each part of an object function in the way most suitable for its operation.

Asymmetry
- Replace a symmetrical form with an asymmetrical form.
- If an object is asymmetrical, increase its degree of symmetry.

Combination
- Combine homogeneous objects in space; assemble identical/similar parts to perform parallel operations.
- Make operations contiguous; bring them together in time.

Universality
- Make part or object perform multiple functions; eliminate the need for other parts.

Nesting
- Contain the object inside another that in turn is placed inside another abject.
- Pass an object through a cavity of another object.

Anti-weight
- To compensate for the weight of an object, merge it with other objects that provide lift.
- To compensate for the weight of an object, make it interact with the environment.

Prior Anti-action.
- If it is necessary to perform an action with both harmful and useful effects, the action should be replaced with anti-actions to control harmful effects.
- If the object is under tension, provide anti-tension in advance.

Prior Action
- Perform the required change of an object (either fully or partially) in advance.
- Arrange objects beforehand so that they can come into action from the most convenient place and without losing time for their delivery.
- Compensate for the relatively low reliability of an object by countermeasures taken in advance.

Equipotentiality
- Change the working conditions so that an object need not be raised or lowered; limit position changes.

Inversion
- Implement an opposite action instead of taking one dictated by the problem's specifications.
- Make movable parts fixed and fixed parts movable.
- Turn the object upside-down.

Spheroidality
- Replace rectilinear parts and surfaces with curved ones; replace cubical shapes with spherical ones.
- Use spirals, rollers, or balls.
- Go from linear to rotary motion; use centrifugal forces.

Dynamics
- Allow the characteristics of an object, external environment, or process to change to be optimal or find an optimal operating condition.
- Divide an object into elements that can change the position relative to each other.
- If an object (or process) is rigid or inflexible, make it movable or adaptive.

Partial or Excessive Actions
- If it is difficult to obtain 100% of a desired effect, achieve something more or less to simplify the problem.

Movement to a New Dimension
- Move an object in two- or three-dimensional space.
- Use a multistory arrangement of objects instead of a single-story arrangement.
- Tilt or reorient the object; lay it on its side.
- Use another side of a given area.

Mechanical Vibration
- Set an object into oscillation.
- If oscillation exists, increase its frequency even as far as ultrasonic.
- Use the resonant frequency.
- Use piezo-vibrators instead of mechanical vibrators.
- Use ultrasonic vibrations in conjunction with an electromagnetic field.

Periodic Action
- Instead of continuous action, use periodic or pulsating action.
- If an action is already periodic, change its periodic magnitude or frequency.
- Use pauses between impulses to perform a different action.

Continuity of Useful Action
- Carry out an action continuously with all parts of an object operating at full capacity.
- Remove idle and intermediate motions.
- Skip some steps.
- Conduct a processor or certain stages at high speed.

Conversion of Harm to Benefit
- Use harmful factors to achieve a positive effect.
- Eliminate the primary harmful action by adding to it another harmful action to resolve the problem.

Feedback
- Introduce feedback to improve a process or action.
- If feedback already exists, reverse it.

Intermediary Item
- Use an intermediary carrier article or intermediary process.
- Temporarily connect one object with another that is easy to remove.

Self-Service
- Make an object serve itself by performing auxiliary helpful functions.
- Use waste resources, energy, or substances.

Copying
- Use a simple and inexpensive copy instead of an object that is complex, expensive, fragile, or convenient to operate.
- Replace an object by its optical copy or image. A scale can be used to reduce or enlarge the image.
- If visible optical copies are used, replace them with infrared or UV copies.

Cheap, Short-Lived Objects
- Replace an expensive object with a multiple of inexpensive objects, comprising certain qualities.

Replacement of a Mechanical System
- Replace a mechanical means with a sensory (optical, acoustic) means.
- Use an electrical, magnetic, and electromagnetic field to interact with the object.
- Change from static to movable fields or from unstructured fields to those having structures.
- Use fields in conjunction with field-activated particles (e.g., ferromagnetic).

Pneumatics and Hydraulics
- Replace solid parts of an object by gas or liquid. These parts can use air or water for inflation or use air or hydrostatic conditions.

Flexible Membranes or Thin Film
- Replace traditional construction with that made from flexible membranes or thin film.
- Isolate the object from the external environment using flexible shells and thin films.

Porosity
- Make an object porous or add porous elements (inserts, coatings, etc.).
- If an object is already porous, use the pores to introduce a useful substance or function.

Color Change
- Change the color of an object or its surroundings.
- Change the degree of translucency of an object or processes that are difficult to see.
- Use colored additives to observe objects or processes that are difficult to see.
- If such additives have already been used, employ luminescent traces or tracer elements.

Homogeneity
- Make objects interact with another of the same material or a material with identical properties.

Rejection of Regenerating Parts
- After it has completed its function or become useless, reject or modify an element of an object.
- Immediately restore any part of an object that is exhausted or depleted.

Parameter Changes
- Change an object's physical state.
- Change the concentration or consistency.
- Change the degree of flexibility.
- Change the temperature.

Phase Transformation
- Implement an effect developed during the phase transition of a substance.

Thermal Expansion
- Use thermal expansion (or contraction) of materials.
- If thermal expansion is being used, use multiple materials with different coefficients of thermal expansion.

Use of Strong Oxidizers
- Replace normal air with enriched air.
- Replace air with oxygen.
- Treat an object in air or in oxygen with ionizing radiation.
- Use ionized oxygen.

Inert Atmosphere
- Replace a normal environment with an inert one.
- Add neutral parts or inert additives to an object.

Composite Materials
- Replace a homogeneous material with a composite one.

REFERENCES

1. James L. Adams, *Conceptual Blockbusting: A Guide to Better Ideas,* 4th ed (Cambridge, Massachusetts: Perseus Publishing,2001).

2. Edward de Bono, *Lateral Thinking: Creativity Step by Step* (New York: Harper Colophon, 1990).

3. M.I. Stein and S.J. Heinze, *Creativity and the Individual* (New York: Free Press, 1960).

4. G.I. Nierenberg, *The Art of Creativity Thinking* (New York: Simon & Schuster Inc, 1986).

5. Scott Fogler and Steven E. LeBlanc, *Strategies for Creative Problem Solving,* 3rd editon, (Upper Saddle River, NJ: Prentice Hall, 2013).

6. G. Phal, W. Beitz, J. Feldhusen, and K.H. Grote, *"Engineering Design: A Systematic Approach",* 3rd Edition (London: Springer-Verlag, 2007).

7. Nigel Cross, *Engineering Design Methods: Strategies for product Design,* 4th ed (Chichester, UK: Wiley and Sons, 2008).

8. Karl Ulrich and Steven Eppinger, *Product Design and Development* 5th ed (New York: McGraw-Hill, 2011).

9. Lee B. Tuttle, *Creative Thinking: The Touchstone of DFMA",* International Forum on DFMA, Newport, RI, May 14-15, 1990.

10. M.I. Stein, *Creativity in Handbook of Personality Theory and Research* (Chicago, Rand Mcnally: 1968).

11. Lawrence D. Miles, *Techniques of Value Analysis and Engineering* 2nd ed (New York, McGraw-Hill: 1972).

12. S. Pugh, "Concept Selection: A Method that Works," *Proceedings International Conference on Engineering Design, (ICED 81), March, 1981.*

13. K. Bemowski, "The Benchmarking Bandwagon," *Quality Progress,* Vol. 24, no. 1 (January 1991), 19–24.

14. L.S. Pryor, "Benchmarking: A Self-improvement Strategy," *The Journal of Business Strategy* (Nov./Dec. 1989), 28–32.

15. B.S. Dhillon, "Engineering Design -A Modern Approach" Irwin, a Times Mirror Higher Education Group, Inc. Company, 1996.

16. Barry Hyman, *Fundamentals of Engineering Design* (Upper Saddle River, NJ: Prentice Hall, 2002).

17. S. Pugh, *Total Design- Integrated Methods for Successful Product Engineering* (Reading, MA: Addison-Wesley Publishing Co., 1991).

18. N.F.M Roozenburg and J. Eekels, *Product Design: Fundamentals and Methods* (Chichester, UK: John Wiley and Sons, 1995).

19. N.P. Suh, "Design Axioms and Quality Control", MIT Industrial Liaison Program, Report 6-22-89, 1989.

20. El Wakil Sherif, *Processes and Design for Manufacturing,* 2nd ed (Boston: PWS, 1998).

21. J. R. Hartley, *Concurrent Engineering* (Cambridge, MA: Productivity Press, 1992).

22. S. G. Shina, *Successful Implementation of Concurrent Engineering Products and Processes* (New York: Van Nostrand Reinhold, 1994).

23. S. G. Shina, *Concurrent Engineering and Design for Manufacture of Electronics Products* (US: Springer, 2012).

24. N. P. Suh. *The Principles of Design* (New York: Oxford University Press, 1990).

25. George E. Dieter and Linda C. Schmidt, *Engineering Design* 5th ed (New York: McGraw-Hill, 2012).

26. Edward B. Magrab, Satyandra K. Gupta, F. Patrick McCluskey, and Peter Sandborn, *Integrated Product and Process and Development* (Boca Raton, FL: CRC Press, 2010).

27. J. Usher, U. Roy, and H. Parsae, *Integrated Product and Process Development* (Hoboken, NJ: John Wiley and Sons, 1998).

28. TRIZ, www.triz-journal.com.

29. Axiomatic Design, www.axiomaticdesign.com.

30. G.S. Altschuller, *Creativity as an Exact Science: The Theory of the Solution of Inventive Problems* (New York: Gordon and Breach Science Publishers, 1984).

31. Mason Brian, "The Aquaduct, a water filtration bicycle" Dartmount Engineer, Winter 2010.

EXERCISES

3.1. Pick any product of your choice and construct a morphological chart for it. Indicate one of its subsolutions and explain a brief reasoning for it.

3.2. A customer has requested a weedcutter that can function for prolonged cycles with little maintenance. It must be lightweight in construction and durable. Go through the steps of attribute listing and create a morphological chart for this product.

3.3. Develop some new design concepts for a *"comfort bicycle"* and contrast them to the design of a mountain bicycle.

3.4. Construct a three-stage Duncker diagram for the following problem statement: "Finding time for sports is becoming difficult."

3.5. Select a hearing aid with styles varying from behind the ear to implantable, compatibility with blue tooth, automated volume control and a variety of maintenance tools. Construct a morphological chart and function diagrams for it.

3.6. TRIZ methodology can help designers to formulate design solutions. Explain the commonality and differences between TRIZ and lateral thinking.

3.7. Patients suffering from cerebral palsy, traumatic brain injury, and muscle diseases benefit greatly by the availability of improved design tools. Identify how the customer requirements in each situation could be a part of a new product design process.

3.8. Assistive technology products for elderly people present challenges to product designers who need to understand the general assessment process and issues connected with the basic customer needs. Identify a product that improves quality of life for elderly people, and generate concepts for the product architecture.

3.9. A new product must be defined based on customer needs. The customer need is for a device (a) that recognizes that an automobile is locked from a distance, and (b) that can send a signal to lock the automobile if it is unlocked. Examine the above scenario and generate a problem statement for a new product.

3.10. In establishing the necessity for rehabilitation technology products, clients' immediate and anticipated medical and functional needs are always considered. Some rehabilitation-related devices include individually configured manual and powered wheelchair systems that have adaptive seating and alternative positioning and require evaluation, fitting, configuration, and adjustment. Using the axiomatic approach and TRIZ, develop concepts and design parameters for an intelligent wheel chair.

CHAPTER 4
Product Configuration and Design for Function

OBJECTIVES

This chapter addresses the issues of product function. A product's functional design, which addresses the upstream phases of design, involves the generation and synthesis of ideas and performance. Functional design steps consist of defining a set of subfunctions and their interrelationships to determine the general product's architecture. Product architecture evolves during concept development, resulting in an arrangement of physical elements into building blocks. The realistic constraints, components, standardization, and design for manufacturability are linked to product performance and have broad implications for the product architecture. Because the creation of product architecture is a dominant activity, the team responsible for it must have input from the design, manufacturing, marketing, and supplier areas.

4.1 Introduction: Design for Function Techniques

Design for function is of interest not only to product designers but also to the people involved in several business aspects, including product, process, and quality improvements. Designers should understand each product's function and parts within the product before performing any analysis with the objective to improve the assembly of the product's existing parts or manufacturability. The functional description of a product is at an abstract level. A product's function can be described in normal language, as a mathematical expression, or as a black box.

Value engineering and function analysis system technique (FAST) are well represented in industry. Both techniques focus more on the design of individual parts, especially in terms of the costs of material and manufacturing, than on redesigning the entire product for improved manufacturability.

Value engineering techniques: Value engineering techniques have been in use for many years. Using these techniques one can identify the individual parts and their contribution to the final product. In addition, they define the function/value/worth and cost of a product.

Function analysis system technique (FAST): This technique identifies the functional relationship between parts at the individual part level. It focuses on the part's design at the individual level. In the concept design stage, two things are known: *function and form*.

Function: The designer specifies the product's function after studying the customer's needs and desires.

Form: The designer generates several different concept designs (forms) that can satisfy the customer's needs to a higher or lesser degree.

4.2 Function Analysis for Product Design

Function analysis provides a clear picture of the objectives of design. It is a loosely structured methodology derived from previous ideas and is based on customer-derived functions rather than engineering-conceived forms. The methodology translates the customer's functional needs into product functions, evaluates manufacturability, and creates alternate products. The function analysis methodology comprises several major steps:

- Determine customer needs and views through a customer/competitor analysis.
- Establish a need for each product function from the customer's perspective.
- Translate the customer's needs, desires, and views into a functional product.
- Develop a symbolic image of the product by constructing a function family tree (FFT).
- Perform design for assembly analysis to identify manufacturability difficulties.
- Use creative thinking techniques to generate new concepts based on function.
- Use the concept selection process to select a product design based on function and form.

Function types: Each function type is defined in the following list and shown in **Figure 4.1**.

1. **Use functions** relate to the purpose or goals.
2. **Basic functions** are the characteristics of a product or a part to fulfill a user need.
3. **Secondary functions** are required to allow designers to choose methods for accomplishing the basic functions. They are based on identifying a part's function that is not absolutely needed for the basic product and that satisfies only user desires.

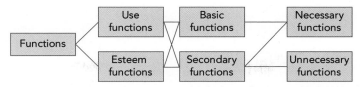

FIGURE 4.1 Function Types

4. **Esteem functions** provide only aesthetic and occasional customer preferences; they are intended to provide sensory satisfaction.
5. **Necessary functions** are those demanded by the customer.
6. **Unnecessary functions** satisfy neither the needs nor the desires of a customer in today's market. As designers consider the customer's perspective, it becomes apparent that a category was needed to identify product features that may have been useful at one time but no longer have value to the customer.

An overall product's function is first determined and is then followed by the definition of the subfunctional group of parts. The next step is to identify the detailed functions of each part. A subfunctional group often utilizes a portion of the parts in several subassemblies. Parts could work together to perform a function even though they are not assembled into the same subassembly. Subassemblies indicate only the order in which parts are joined on the assembly line, not the functional relationships between them.

Function Family Tree: The FFT is a block diagram indicating only the functional relationship between the product and its environment or among various product segments. It is particularly important in the product concept phase because it allows designers to break major products into smaller, more manageable design problems. It also provides more design freedom by removing any visual form from the function, allowing designers to see it from a new perspective.

The development of an FFT begins with the definition of the system boundaries at the level of interest. These boundaries are easy to distinguish. The input of the user on the product represents the *left boundary*. The output of the product to its environment represents the *right boundary*. At the product level, the system boundaries are the interface between the outside world and the product.

4.2.1 Examples of Function Family Tree

Pocket Flashlight

The design features of the pocket flash light in **Figure 4.2** includes a mechanism to provide illumination using a system to deliver power to the bulb in a reliable body support system.

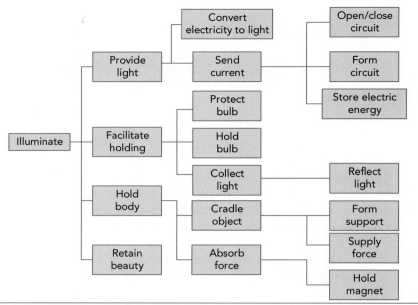

FIGURE 4.2 Function Family Tree for a Pocket Flashlight

The objective of providing light is achieved by doing the following:

1. Convert electricity to light.
2. Send current.
3. Store energy.
4. Form circuit to create open/close situation.

Cellular Phone

A cellular phone plays an important role in a customer's life. **Table 4.1** represents the part names and objectives. Data in **Figure 4.3** are used to analyze the functional components of a personal cellular phone.

Parts	Objective
Antenna	Wireless signal
Mic/Speaker	Voice signal
Vibration/Display	Human interface
Button	Human interface
Rugged frame	Portability

TABLE 4.1 Product: Cellular Phone

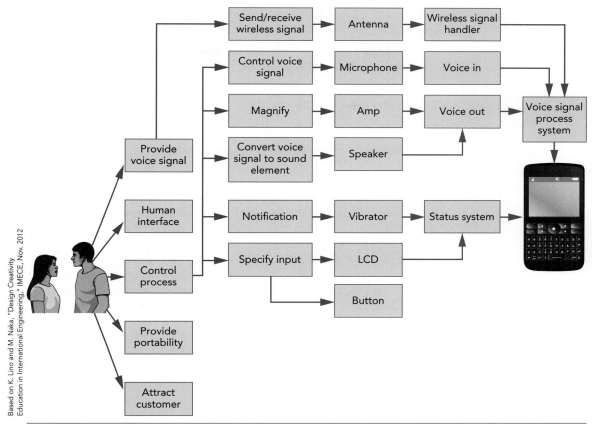

Based on K. Lino and M. Naka, "Design Creativity Education in International Engineering." IMECE, Nov. 2012

FIGURE 4.3 Functional Cell Phone Design

4.3 **Establishment of Product Architecture**

A product's functional design includes upstream phases that involve the generation and synthesis of ideas that can be translated into a product. This phase also includes defining a set of subfunctions and establishing their interrelationships to achieve the general product architecture. Defining the product architecture results in arranging physical elements into building blocks. Realistic constraints, component standardization, and design for manufacturability are linked to product performance and have broad implications to the product architecture. Because the creation of product architecture is a dominant activity, the product development team is required to obtain input from groups involved in design, manufacturing, marketing, and supply chain.

Product architecture can be established in the following stages.

- Developing the concept
- Creating the schematic diagram of the product
- Subgrouping the elements of the schematic
- Generating geometric layout
- Establishing interactions/algorithms
- Prototyping

4.3.1 **Product Architecture for Portable Surface Roughness Probe**

Concept Development Stage

The demand for high product quality requires continuously monitoring various processes and product conditions. One of the essential aspects of quality control in many manufacturing operations is to measure the surface finish quality of the machined parts during the manufacturing process. Variation in the texture of a critical surface of a part influences its ability to resist wear and fatigue. Many techniques have been developed to measure surface roughness, which vary from the conventional profilometer to the recently developed laser diffraction technique. This example of product creation strategy shows steps in developing a simple, robust roughness measurement probe.

Functional Needs in Surface Texture

A traditionally machined surface consists of many components from different sources that are generated during the manufacturing process. The combination of these components composes surface texture. **Figure 4.4** illustrates the requirements of surface inspection strategy.

Roughness

As the material is cut, a spiral or helix pattern is formed on the part surface. This commonly known tool mark is called the *roughness component* of the surface. The roughness height is called *amplitude*; it is caused by the depth of the cut. The roughness

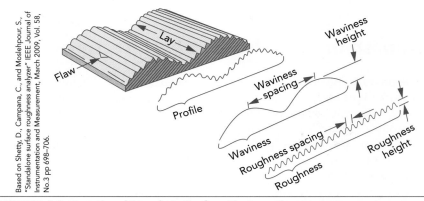

Based on Shetty, D., Campana, C., and Moslehpour, S., "Standalone surface roughness analyzer" IEEE Journal of Instrumentation and Measurement, March 2009, Vol. 58, No.3 pp 698–706.

FIGURE 4.4 Functional Needs in Surface Texture Components

spacing represents the *wavelength*; it results from the rotation speed and feed rate of the cutting tool. The most common representation of roughness uses the notation Ra, which represents arithmetic average roughness value. *Lay* is the direction of the predominant surface pattern, generally determined by the production method used. In general, surface texture can provide information about the manufacturing process.

Development of Measurement Principle

Stylus technique: Surface roughness has been conventionally measured using contact-type instruments that traverse along the lay of the surface of measurement. The instrument's contact with the surface is made by a stylus that records the undulations of the movement to indicate the surface roughness value. However, the contact method is time consuming. With this technique, surface statistics can then be calculated from the profile record. To quantify the average topography, many statistical parameters and functions have been developed. The most common one is the arithmetic average roughness (Ra), given by the following equation.

$$Ra = \frac{1}{L}\left|y(x)\right|dx$$

(4-1)

where L is the sampling length.

The accuracy obtained by the mechanical profilometer is quite high with Ra values in the order of micrometers. However, this technique is limited by its low speed and the need to contact the surface being inspected.

Optical light-scattering techniques: Light scattering has significant promise as a practical tool for measuring surface finish. It is quick and sensitive and provides accurate data on surface finish. The necessary link between scattering light and surface topography can be made using either empirical correlation or by scattering theory. **Figure 4.5** illustrates the basic light-scattering principle. A beam of light called a *wavelength* is projected onto a surface at an incident angle θ. According to the law of reflection, if the

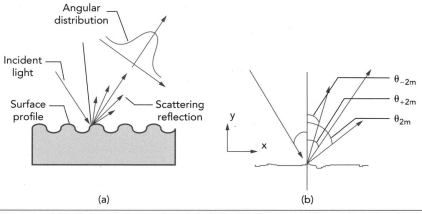

(a)

(b)

FIGURE 4.5 (a) Basic Light-Scattering Principle (b) Scattering Intensity Distribution

surface is perfectly smooth, the light will be reflected at the same angle as θ. However, if the surface is rough, the reflection will be scattered around the direction of specular reflection. The intensity of the diffused light decays exponentially. The diffused light intensity has an almost linear relationship with surface roughness. An object and its far-field diffraction pattern have a Fourier transformation relationship with each other. If the object distribution is represented by $f(x, y)$, its Fourier transformation $F(x, y)$ is given by

$$F(x,y) = f(x,y)e^{-2\pi(xu+uv)}\,dy\,dx \tag{4-2}$$

where x and y are spatial coordinates, and u and v are spatial frequency variables.

Machined surfaces tend to exhibit a grating structure from tool marks made during the machining process. Shetty and Neault have developed a noncontact method of measuring the specular and scattered reflection of a laser beam from the surface of a workpiece. The method takes advantage of the fact that a light source reflected off the surface of a workpiece provides a signature pattern based on the roughness of the surface. For example, an electromagnetic wave of known wavelength is incident on the rough surface at an angle θ_1. The scattering surface where the light is projected can have either one- or two-dimensional roughness. The angle of diffused scatter, θ_{2m}, is related to the period of the roughness.

$$\theta_{2m} = \sin^{-1}(\sin\theta_1 + m\lambda/T) \tag{4-3}$$

where, $m = 0, \pm1, \pm2$, and $T =$ surface period.

Because surfaces produced by various processes exhibit distinct differences in texture, the specimens of machined surfaces can be easily identified by looking at the diffraction pattern as found on ground, shaped, milled, or turned workpieces. A sensor is applied to collect the intensity of specular and scattered reflectance. The output of the sensor with respect to surface roughness has also been described by Beckman and Spizzichino as follows:

$$I = I(r,d,s,m) \tag{4-4}$$

where

$I =$ intensity of reflected light

$r =$ surface roughness

$d =$ distance from the surface to the sensor

$s =$ coefficient related to the sensitivity of the detector

$m =$ material constant

The objective of classifying the surface roughness of a machined workpiece is achieved by using a laser and microcomputer-based sensing system. The intensity of the diffracted light is measured as a function of the output voltages produced by the sensors and is processed by the digitizing circuit. The method also includes a microcomputer-based procedure, which provides operator control in the form of a menu-driven graphical interface.

Measurement and procedure: A computer-based vision system measures the intensity of the collimated, monochromatic light source diffracted in the spectral direction. It is captured by a video system that provides an analog signal to a digitizing system for conversion to digital information, which is subsequently modified to display the surface roughness value. The intensity is measured as a function of the image's gray level, is processed by the digitizing circuit, and is compared to a previously defined calibration standard. The method also includes a procedure that provides operator interaction in the form of menu-driven steps, which guide the operator through the requirements of each phase of the process. The three phases of the methodology during its operation are:

1. Calibration
2. Measurement
3. Analysis

The instrument includes a **calibration** procedure built in as a part of the surface roughness evaluation. The calibration specimens vary depending on the methods of manufacturing process used to create the samples. Calibration must be repeated each time any changes are made irrespective of whether these changes are related to the machining process or the apparatus setup and reorganization. **Measurement** and **analysis** are performed after the calibration. The data acquired in the measurement procedure are mathematically averaged and compared to those data in the calibration procedure. Results are calculated from specific mathematical formulas that have been specifically derived for each machining process and material combination selected via the computer program menu option for that particular combination.

Creation of schematic diagram of the surface probe: The probe in **Figure 4.6** is composed of the following components. A solid state laser is used as a light source. Light from the laser diode shines down on the work surface. The detectors read the reflected light coming back into the probe, and internal voltage divider circuits convert it to a proportional voltage. The tube, which is internally coated with reflective material, is used to collect the diffused reflection and send it to an array of highly sensitive

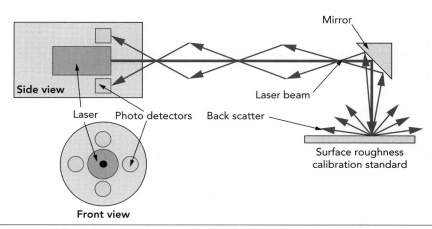

FIGURE 4.6 Schematic Representation of Probe Components

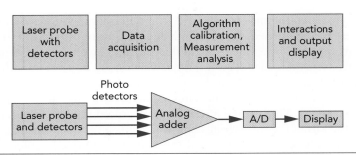

FIGURE 4.7 Functional Representation of Probe

photoelectric sensors. The sensors convert scattered light intensity into electric voltage signals. The amount of scattered light received by the photoelectric sensors is proportional to the degree of roughness of the surface. **Figure 4.7** represents the functional representation of the probe.

Clustering the elements of the instrument: The surface roughness probe is further divided into five subsections: probe, power supply, adder circuit, A/D circuit, and display circuit (including MOSFET decimal point driver circuit). The design ideas come together at the schematic stage; then the physical components come together during the clustering stage as product module.

Algorithm Development and Interactions

Calibration algorithm: The first step in the algorithm development is to derive the relationship between the surface roughness template standard values and the voltages read

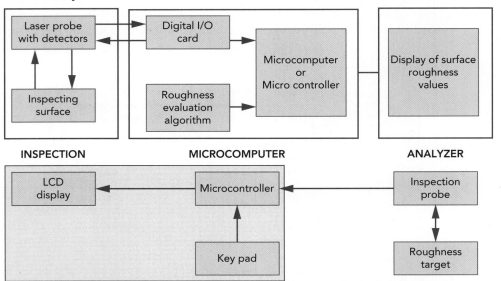

FIGURE 4.8 Layout Representation of the Probe

Index	Roughness (µm)	Trial 1 Voltage (bit count)
1	0.0508	3151.5
2	0.1016	3333.0
3	0.2032	3767.0
4	0.4060	4177.0
5	0.8130	4366.0
6	1.6000	4733.5

TABLE 4.2 Sample Trial Run Results of Measured Voltage against Roughness

from the measurement probe as shown in **Figure 4.8**. The second step is to determine whether this relationship is repeatable and reproducible. The data in **Table 4.2** show the results from a sample trial run.

The goal of any statistical evaluation is to obtain the experimental data to fit a known model. To calibrate the measurement system, sample data are collected from the known surface roughness target. These data are used to fit to a line (**Figure 4.9**), and the equation of the line is stored for later use.

The goal is to use the measured value to predict the surface roughness. This could be done by feeding the measured value into the calibrated line equation and identifying the index value. The index value is then transformed back into a surface roughness value.

Prototyping: Product prototyping can be performed using two methods.

a. PC-based hardware-in-the-loop simulation uses graphic user interface coupled with standard data acquisition and a control interface card (**Figure 4.10**).

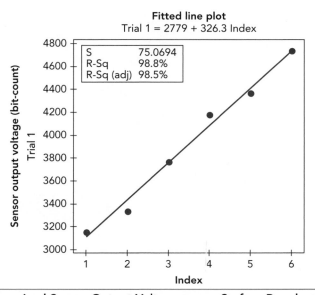

FIGURE 4.9 Linearized Sensor Output Voltage versus Surface Roughness Index

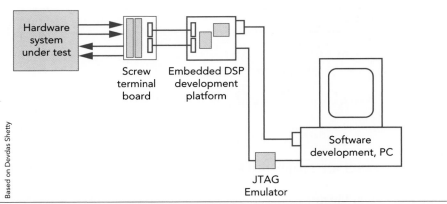

Based on Devdas Shetty

FIGURE 4.10 Embedded Processor Platform for Surface Roughness Measurement

In the PC-based method, the interface card is inserted in one of the expansion slots of the chassis. One major drawback of PC-based simulation systems is the inability to work in systems that need loop responses to be fast (less than 100 microns). This shortcoming is the result of a nonoptimized software code running through an interpreter that then interfaces to an operating system not designed for "real-time" processing.

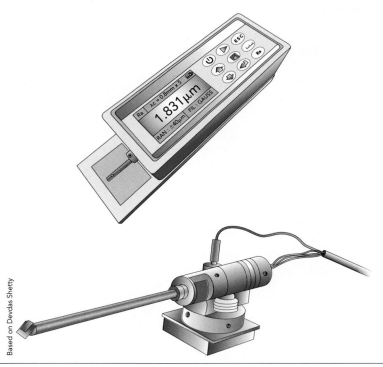

Based on Devdas Shetty

FIGURE 4.11 Surface Roughness Probe and Display

b. The other method for accomplishing hardware-in-the-loop testing involves cross compiling the control algorithm to target an embedded real-time processor platform. The embedded processor platform (**Figure 4.11**) is often a digital signal processor (DSP) with input/output (I/O) tailored for embedded system products. This I/O can be in the form of analog inputs and outputs as well as digital inputs and outputs. The cross-compiled code is downloaded to the embedded processor, sensors are connected to the embedded processor board inputs, and actuators are connected to its outputs. Embedded processor platforms are designed for reduced cost and increased speed; as such, they generally have neither video displays nor standard desktop inputs such as full-function keyboards and mouse interfaces.

Results and display of surface roughness: Different types of machining operations such as grinding, milling, lapping, and reaming provide standard roughness samples. The surface roughness probe is initially fine-tuned and calibrated using standard roughness specimens. The optical diffraction pattern produced by these standards is analogous to ideal values of surface roughness. Calibration is initially performed for the range of readings to be taken. Readings are taken on the ground samples from 0.1 to 1.6 microns. See the results of the test readings using the stand-alone surface roughness analyzer in **Figure 4.12**.

Results from the test measurements show a good correlation with the surface roughness value of the ground specimen. Measurements show the accuracy of the instrument to be within ±0.1μm, which was the accuracy reached by the PC-based data acquisition measurement system. Refer to **Figure 4.12** for the results of experimental evaluation and comparison between a contact-type profilometer and the stand-alone analyzer unit.

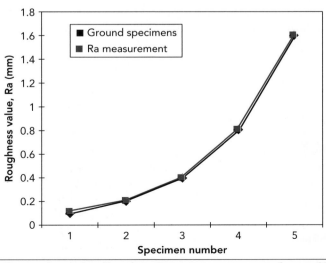

FIGURE 4.12 Comparison of surface roughness measurements for profilometer and noncontact stand-alone laser probe

Techniques Used	Resolution	Roughness Range	Orientation
Stylus-type contact techniques	± 1 micron	0–3.25 micron	Always perpendicular to the lay (machine marks)
Stand alone	± 0.1 micron	Intervals of 0–0.2; 0.2–1.6; 1.6–3.25 micron	Any orientation

TABLE 4.3 Comparison of Quality Parameters

Three ground samples of 0.4, 0.8, and 1.6 microns are used in the experiment.

The above example focused on developing a stand-alone surface roughness inspection instrument that is compatible with shop floor environments using inexpensive components. It is a portable, low-cost, and noncontact measurement system solution achieving the same reading accuracy compared to the existing PC-based data acquisition measurement system. The experimental results of surface roughness values are compared to a contact-type profile meter that is normally used on a shop floor as in **Table 4.3**. The experimental results validate the reliability of the stand-alone instrument.

CASE STUDY 4.1
Function Family Tree: Toaster

While creating an FFT for a toaster, the components as well as their objective functions (**Table 4.4**) are identified in sequence. The FFT for the toaster as shown in **Figure 4.13** at the concept phase allows us to break down the product into smaller components such as electrical cord, body of the product, bread trays, actuating lever, thermostat, and heating elements.

Parts	Objective
Electrical cord	Deliver power
Body	Protect mechanisms
Bread trays	Hold bread
Thermostat	Control temperature
Lever	Open circuit
Heating elements	Create heat

TABLE 4.4 Toaster—Components and Objectives

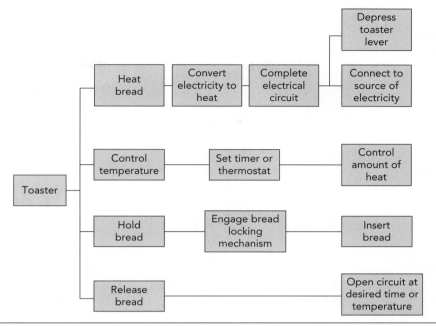

FIGURE 4.13 Function Family Tree for a Toaster

CASE STUDY 4.2
Function Family Tree: Pencil Sharpener

The FFT for a pencil sharpener represents a set of blocks indicating the relationship between the steps involved in the sharpening of a pencil. The FFT allows us to break down the components of the pencil sharpener into smaller subcomponents such as hand crank, switch, body and holding mechanism, and bin as shown in **Table 4.5**.

Figure 4.14 represents the FFT of a pencil sharpener.

Parts	Objective
Hand crank	Rotate and cut wood
Switch	Generate power
Body and holding mechanism	Add stability and support for the pencil
Bin	Provide container for shavings disposal

TABLE 4.5 Pencil Sharpener—Components and Objectives

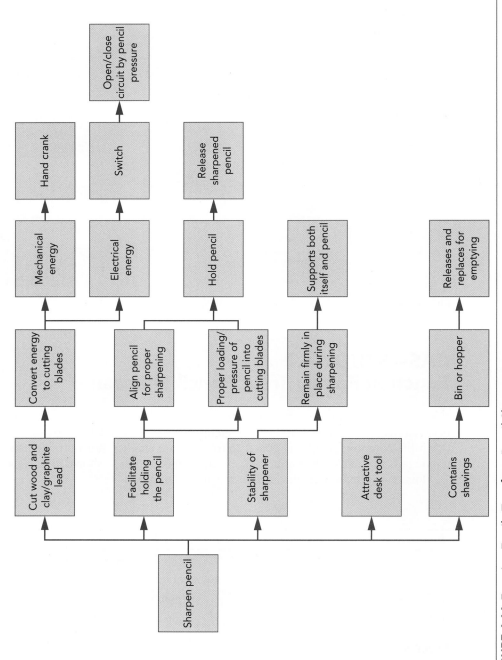

FIGURE 4.14 Function Family Tree for a Pencil Sharpener

CASE STUDY 4.3
Design for Portable Harness Ambulatory System

An ambulatory suspension system is a device that therapists use during physical therapy for patients who suffer from musculoskeletal disabilities, injuries, diseases, muscle weakness, and surgical procedures. This system helps patients to recover more quickly. An ambulatory system supports body weight and keeps patients from falling when exercising. Because this system prevents patients from falling and lifts weight from their bodies, it helps them to gain confidence when exercising to strengthen their muscles sooner than therapy without this system.

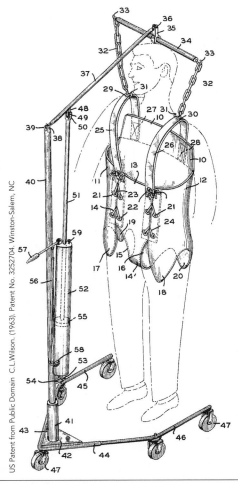

US Patent from Public Domain C.L.Wilson. (1963). Patent No. 3252704. Winston-Salem, NC

FIGURE 4.15 Harness Ambulatory System

Currently hospitals and physical therapy clinics use many devices that are usually complex, sophisticated, large, and difficult to move and use without supervision. Some devices are mounted to ceilings; others use pulley systems and safety features, making them very expensive to purchase, install, and operate and are therefore almost impossible for people to use in their own homes.

A new system to be used by patients in the home environment is needed to enhance the rehabilitation process and help them to return more quickly to their previous lifestyle. This device must be safe and reliable as well as easy to use without additional supervision.

Background: C.L. Wilson invented one of the first ambulatory suspension systems known as a *lifting and walking jacket*, patent number 3252704 as shown in **Figure 4.15**. This device was intended for use by people with a physical condition who needed help and support to relearn to walk. This early invention was able to lift a person from a seated position to a standing position and supported that individual throughout an exercise process. A simple piston with pressure regulators and locking handle was used as a lifting mechanism. The piston was connected to the base of the device, and a rod was connected to the support jacket.

Current inventions are much more sophisticated and require additional features to obtain approval according to quality, safety, and ergonomic standards. Because patients' health and safety are the most important aspects of such a device, inventors must pay

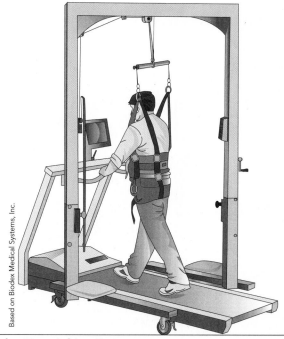

Based on Biodex Medical Systems, Inc.

FIGURE 4.16 Biodex Unweighing System

special attention to these features while designing a new product. Some examples of ambulatory suspension systems are provided in **Figures 4.16, 4.17, 4.18,** and **4.19**.

Biodex Unweighing System: This device incorporates a dynamic suspension system that accommodates the vertical displacement of the center of gravity that occurs in a person's normal gait. The off-loading mechanism maintains constant force by dialing the amount to off-load. This device does not accommodate vertical movements, and it requires another person to assist the patient.

Biodex Supported Ambulation System: The FreeStep SAS model is an overhead track and harness system designed for patient and therapist safety. It supports patients who are weak or have a balance problem without using a large amount of floor space or requiring additional staff. This system is mounted to the ceiling and allows movement only on tracks. Another person is required to assist the patient in moving from a seated to a standing position.

The AquaGaiter by Hudson: This underwater treadmill exercise system combines traditional treadmill training with the natural properties of water (buoyancy, resistance, and heat) for a rigorous cardiovascular workout with reduced impact and stress on joints. The water level can be adjusted according to the amount by which the patient is to be unweighted during the exercise. This method is difficult to use outside the hospital or

Based on Biodex Medical Systems, Inc.

FIGURE 4.17 Biodex Supported Ambulation System

Based on Hudson Acquatic Systems.

FIGURE 4.18 AquaGaiter

clinic because of the complexity of getting in and out of the unit without assistance and is most likely too expensive for an average person.

NaviGATOR: This is an ambulatory suspension and rehabilitation apparatus developed jointly by the author and Albert Einstein College of Medicine/Montefiore Medical Center. It is a new device for clinical and research applications. Patients with neurological or musculoskeletal injuries, diseases, or muscle weakness can use it for physical therapy and exercise. The NaviGATOR enables exercise and movement training in all three planes of motion without the risk of falls and injuries. Operation of the entire system can be directed manually or automatically.

Problem Definition

Physical therapy gait training can be very lengthy and expensive. Patients must use an ambulatory suspension system that is helpful in the rehabilitation process and reduces the risk of injury to them. Such devices are very effective in hospital and physical therapy clinics that have limited room for the equipment and that must have another person to assist patients while using them. Normally, such units are too expensive for home use.

This case study examines the design of a device that can be useful for domestic applications and has functions similar to those of ambulatory systems used in hospitals and clinics. These units will allow patients to continue rehabilitation at home, enhancing the process and helping them to return to their normal lifestyles sooner. If

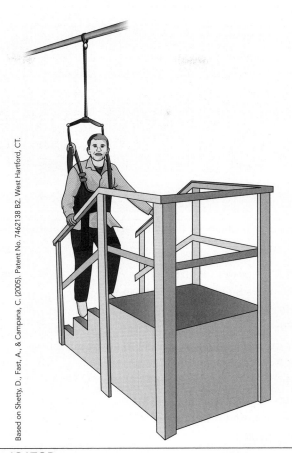

Based on Shetty, D., Fast, A., & Campana, C. (2005). Patent No. 7462138 B2. West Hartford, CT.

FIGURE 4.19 NaviGATOR

the price is relatively low, this device would be very beneficial for people who do not have insurance coverage for therapy or have high co-payments.

The system must be safe, reliable, and easy to use without additional supervision. It should help people to stand up and walk while supporting their weight when they move around the house. Additional features such as safety locks to catch a user who falls during exercise must also be included in the package.

Concept Development

A list of basic requirements for an ambulatory system follows.

- Is light weight in construction
- Is easy to maneuver

Feature	Means				
Body material	**Aluminum**	Steel	Plastic	Carbon fiber	Titanium
Capacity	**1 person**	2 persons	3 persons		
Lifting mechanism	Electric winch	Pneumaticity	Screw	**Pulley**	**Crank**
Transmission	Gears	Belts	Chains	Cables	**Rope**
Steering	Rails	**Turning wheels**	Skids		
Unweighting mechanism	**Spring**	Dampers	Pneumatic/ Hydraulic cylinders		
Support	**Harness**	Self-support	Underarm	Overhead lift	
Stopping	Hand brakes	Ratchet	N/A	**Wheel brakes**	
Power	Electric	Legs	**Arms**		

TABLE 4.6 Morphological Chart for Portable Harness Ambulatory Device

- Has no power requirements
- Requires no assistance when in use
- Has acceptable safety and quality characteristics
- Fits within a home's layout and door frames
- Is durable and requires little to no maintenance
- Lifts up to a 400-pound person
- Has unweighting capability
- Has wheel brakes
- Has rotating brakes
- Is reliable
- Makes change from seated to standing easy
- Makes change from standing to seated easy

While concept generation, the morphological chart **Table 4.6** and the functional diagram **Figure 4.20** was created based on the above listed basic requirements.

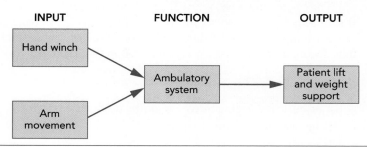

FIGURE 4.20 Functional Diagram

Concept Design

The main idea for the new ambulatory system is to lift a person from a seated position, possibly from a wheelchair, to a standing position. A harness around the chest and waist supports the user by increasing balance and spreading the lift force evenly across the body. The harness connects to a lifting mechanism with a rope or belt that is especially helpful for a patient with weak muscles. Simpler solutions could use a pulley system, gearbox, or crank. It is also possible to combine these systems to obtain the best ratio. Once lifted to a standing position, the person's body needs to be supported when exercising and walking.

A locking device is required for the lifting mechanism to support the body at the desired position. This locking device could be a ratchet feature in the crank to prevent the crank from turning in the opposite direction, causing the rope to unwind from the winch. This crank is released to lower a lifted body to a seated position. An emergency mechanism must be included; it works as a secondary support should the first safety device fail or the user have a problem while using the device. If possible, this emergency mechanism should lower the user slowly to prevent unnecessary injuries.

To be easily moved, this ambulatory system must be light in weight and easy to steer. It must fit in the home design, especially through door frames. A wheelchair should be able to fit inside the system's frame. The base therefore must be wide, and the center of gravity should be low enough to prevent the device from flipping over during use. For transportation, this system must be easy to assemble and disassemble; it will preferably separate into few subassemblies that can be put together with standard fasteners.

The cost of this device must be affordable for average consumers. Standard parts, materials, and operations must be used to minimize the cost. All items described in this case study must be accurately evaluated when the final conceptual design is created. When the concept has been completed and the design meets all requirements, a prototype can be built for further testing and evaluation.

Schematic Diagram Creation

See **Figure 4.21** for the conceptual prototype of an ambulatory suspension system. This system consists of two main components, the frame and the lifting mechanism.

The **frame** is constructed from aluminum extrusion, which is light and strong. The advantage of using this material is that it requires very little machining to mount its parts and subassemblies to the frame. This unit has four quiet glide casters that are rated for 700 pounds each, can rotate 360 degrees along the axis, and are equipped with a simple locking mechanism. Two handlebars are mounted to the frame. This design does not have easily adjustable handles, but later revisions could incorporate this change. Handlebars are located in the position that is comfortable for a person of average height when standing.

The **lifting mechanism** as shown in **Figure 4.23** is built of five main subassemblies:

1. Pulley rigging system
2. Rope

3. Hand winch
4. Counterbalance
5. Harness

1. A **pulley system** is fixed to the top of the frame. A plate secures two top bearings to the aluminum extrusion. The top portion of the pulley system is captured between two plates to eliminate the slope in the assembly and prevent the rope from falling off the tracks. The bottom portion of the pulley system is connected to the top portion with rope. This pulley system is capable of giving a 4:1 mechanical advantage when lifting objects.
2. The **rope** is fixed at one end to the top plate of the pulley, is fed through the pulley links four times, and then goes through the bearing assembly located in the top corner of the frame. From that point, the rope is fed down to the hand winch assembly,

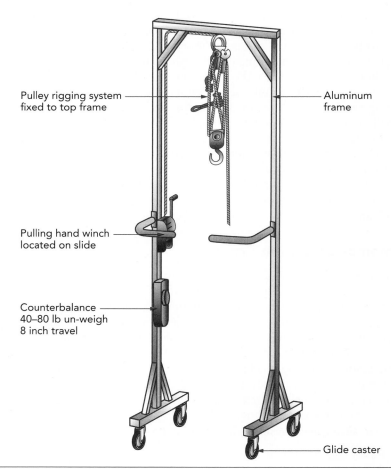

Pulley rigging system fixed to top frame

Aluminum frame

Pulling hand winch located on slide

Counterbalance 40–80 lb un-weigh 8 inch travel

Glide caster

FIGURE 4.21 Schematic of the Ambulatory Suspension System

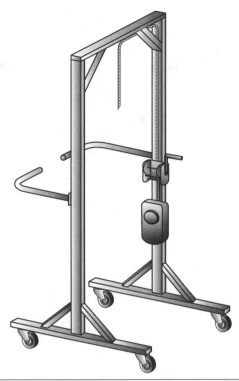

FIGURE 4.22 Prototype of Ambulatory Suspension System

connecting both pulley and winch. When the winch is rotated, the rope is curled on the shaft, vertically displacing the object attached to the end of pulley system.

3. The **hand winch** is mounted to the slide located on the side frame. This slide has two fixed stops that allow only 5 inches of total travel distance in the vertical direction. The center of gravity of a person walking can on average accommodate up to 5 inches of vertical displacement while maintaining a consistent level of unweighing provided by counterbalance. The hand winch assembly gives a 4:1 mechanical advantage when pulling an object. The pulley and winch combination provides a 16:1 advantage for lifting a person. For the system to support the body at a desired position, a locking device is incorporated in the hand winch assembly. This locking device is a ratchet mechanism in the crank that will prevent cranking in the direction that could cause the rope to unwind from the winch. The ratcheting device would be released to lower a lifted person from standing position to seated position.

4. The **counterbalance** is another major lifting mechanism. A counterbalance assembly is fixed to the frame structure and applies a constant vertical pulling force on the hand winch and slide assembly. The force that this counterbalance exerts on the user is adjustable.

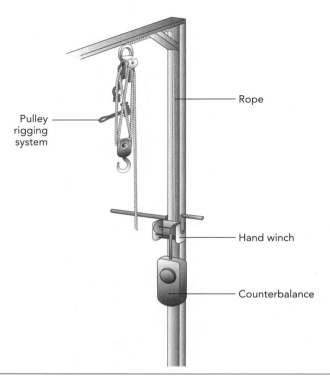

FIGURE 4.23 Lifting Mechanism Assembly

5. A **harness** is required to complete this ambulatory suspension system assembly. To support the user, the harness goes around the chest and waist area to increase balance and spread the lifting force evenly across the body. The harness is connected to a lifting mechanism with a rope or belt with a quickly disconnecting part. **Figures 4.22** and **4.23** represent the final prototype of Ambulatory Suspension System.

REFERENCES

1. Lawrence D. Miles, *Techniques of Value Analysis and Engineering* 2nd Ed (New York: Mc-Graw Hill Inc., 1972).
2. Kaneo Akiyama, *Function Analysis: Systematic Improvement of Quality and Performance* (Cambridge, UK; Cambridge Productivity Press, 1991).
3. D. Shetty and Z. Hamdani, "Advances in On-Line Surface Roughness Inspection for Supervisory Manufacturing," Paper presented at the Japan/USA Symposium on Flexible Automation, *ASME* vol. 1 (1996), 415–20.
4. D. Shetty and H. Neault. Method and apparatus for surface roughness measurement using laser diffraction pattern. U.S. Patent 5,189,490, filed on Sep 27, 1991 and issued on Feb 23, 1993.
5. H.Y. Kim, Y. F. Shen, and J. H. Ahn, "Development of a Surface Roughness Measurement System Using Reflected Laser Beam," *Journal of Materials Processing Technology* (2002), 130–31.

6. John C. Stover, *Optical Scattering: Measurement and Analysis*, 2nd ed (Bellingham, WA: SPIE Optical Engineering Press, 1995), 29–57.
7. E. Marx and T. V. Vorburger, 1990, "Direct and Inverse Problems for Light Scattering by Rough Surfaces" *Applied Optics*, 29, no. 25 (1990), 3613–626.
8. P. Beckmann and A. Spizzichino, *The Scattering of Electromagnetic Waves from Rough Surface* (Oxford, U.K.: Pergamon, 1963).
9. D. Shetty, A. Fast, and C. Campana. Ambulatory Suspension and Rehabilitation Apparatus. U.S. Patent No. 7462138 B2, filed on Jul 1, 2005 and issued on Jan 4, 2007.
10. PNEUtera Technologies. http://www.pneumex.com.
11. Biodex. http://www.biodex.com/physical-medicine/products/pbws/unweighing-system.
12. Lifetec Inc., http://www.lifetecinc.com.
13. C. L. Wilson. Lifting and Walking Jacket. U.S. Patent No. 3252704, (May 22, 1963).
14. D. Shetty, N. Poudel et al." Portable Harness Ambulatory System", Independent Research Project, University of the District of Columbia, Washington D.C., May 2014.

EXERCISES

4.1. What are the key factors in identifying customer needs?

4.2. Consider an example of a consumer product whose function diagram can identify the aspects of a recycling and environmentally appropriate product. Identify the design approach to the product you have chosen.

4.3. Assuming that you are creating a new supermarket cart, generate three concepts and a concept selection chart. Identify the concept screening data based on these criteria: (1) ease of handling, (2) portability, (3) weight of material that can be carried, and (4) durability.

4.4. Assume that you are designing a mountain bike for optimal safety and ease of maintenance. Develop an FFT structure.

4.5. Assume that you are working on a new generation of X-ray CT scanner that contributes to patients' well-being by minimizing radiation and providing improved image recognition. Determine the overall function that needs to be accomplished, and create subfunction descriptions and a concept selection chart.

4.6. Electric toothbrushes are being used in the jewelry industry for polishing. However, the continuous use of the electric toothbrush causes electric motors to burn out. Develop a step-by-step product architecture for a handheld electric polishing device for jewelry that can be operated 8 hours per day.

4.7. How do ethical issues impact product architecture? Identify three major technological innovations in the past 20 years. For each, list any ethical issues they pose for engineers and scientists. Comment on how they can be rectified by product redesign.

4.8. Generate a design concept for securing a coffee cup near the driver's seat of a vehicle. The device should prevent the cup from spilling and not interfere with the vehicle's proper operation. It should be universally adaptable to a wide variety of vehicles.

4.9. Develop the product architecture of a device that will allow the inside and outside surfaces of windows to be cleaned from the inside only.

4.10. According to a survey by the Department of Transportation's National Highway Traffic Safety Administration (NHTSA), only about 14% of drivers in the United States know how to find their tire pressure and maintain it properly. Consider creating a new product that would be a smart air pump station for automobile tires. This system should contain the car manufacturer's recommended tire pressure in an internal database without requiring more input from the customer than a basic knowledge of the vehicle's make, model, and year.

CHAPTER 5

Design Evaluation: Assessing Design Assembly

CHAPTER OUTLINE

OBJECTIVES

This chapter examines the methods used to evaluate the product design processes by highlighting their weaknesses and strengths. The main focus is on design for assembly. Several methods are discussed with examples of the calculation of product efficiency: analyze a design, investigate how the design components are processed and assembled, reduce cost, and increase product quality. The procedures outlined here evaluate initial design's efficiency and examine its validity by using comparison charts. These techniques are applied in a case study.

5.1 Introduction: Design for Manufacturing Methodology

A product's cost is essentially determined at the design stage, including the costs of fabrication, assembly, and inspection. The designer should be aware of the nature of assembly processes and should always have sound reasons for requiring separate parts—resulting in higher assembly costs—rather than combining several parts into one manufactured item. The designer should always keep in mind that each combination of two parts into one will eliminate at least one operation in manual assembly or an entire section of an automated assembly machine.

Several techniques and tools enable designers to deal with the design process and avoid pitfalls when developing a new product. Among these techniques, Hitachi, Boothroyd–Dewhurst, and Lucas methods are used in many industries for product design.

These methods are basically used to analyze a design and investigate how its components are processed and assembled with the goal of increasing product quality and reducing cost and time to market. Most of these methods evaluate the efficiency of the current design and use comparison charts to determine the design validity. This section also discusses the choice of the assembly method.

Design for Assembly and Manufacturing

Specifying parts that require secondary and other operations instead of parts that require only one operation can seriously limit a company's ability to generate profit. Product design for easy manufacture should consider each part of every product's life cycle. The central issue in the design-for-manufacture system is determining the proper use of design guidelines to minimize the number of parts. The designer's ability to apply these rules is a key factor in design for manufacturing.

Common procedures used by designers as guidelines are listed in **Table 5.1**.

Guidelines	Poor Design	Improved Design
Avoid parts that tangle or nest.	Springs with open loops can tangle.	Closed-ended springs will not tangle.
Avoid parts that interconnect.	Parts interconnect.	Design barriers to prevent interconnection.
Design parts to prevent nesting.	Avoid nesting parts that have a "locking" angle.	Adding ribs prevents nesting.

TABLE 5.1 Best Practices in Part Handling Concepts

5.1.1 General Guidelines for Optimizing the Number of Parts in Design

Part Count Reduction

The designer should examine the assembly process and each part to determine whether the part can be eliminated or combined with another part or its function can be performed in another way.

To determine the theoretical minimum number of parts, ask the following:

- Does the part move relative to all other moving parts?
- Must the part absolutely be of a different material than the other parts?
- Must the part be different to allow possible disassembly?

Simplify and reduce the number of parts because each part has the potential to be defective and cause an assembly error. The probability of a perfect product decreases exponentially as the number of parts increases, raising the costs of fabrication and assembly. Reduction in the number of parts lowers both the cost of inventory and purchasing. Use of manufacturing processes such as injection molding, extrusion, and additive manufacturing can reduce the number of parts required.

Standardization

Standard components are less expensive than customized components. Use of common parts can result in lower inventories, reduced costs, and higher quality.

- Standardize and use common parts and materials to facilitate design activities.
- Minimize the amount of inventory in the system, and standardize handling and assembly operations.

Simplification of the operator learning process and maximization of production volumes and operation standardization increase the opportunity for automation.

Ease of Fabrication

An optimal combination of material and production processes should minimize manufacturing costs.

- Avoid unnecessarily high tolerances that are beyond the natural capability of the manufacturing processes.
- Avoid tight tolerances on multiple, interconnected parts. Tolerances on multiple assembled parts can cause assembly difficulty.
- Use preformed components for molded and forged parts to minimize the machining and processing efforts.

Robustness

Design a product's individual components so that part variation does not compromise total performance. Electronic products can be designed to have self-test and/or diagnostic capabilities. For mechanical products, verifiability can be achieved with simple inspection tools or assembly procedures.

- Develop a modular design approach, and design parts for multiple applications.
- Products should be designed to avoid postassembly adjustments.
- Incorporate design verifiability into the product and its components.

Ease of Orientation

Design parts for easy orientation so that components can be handled with minimum difficulty and time. Minimize non-value-added manual effort in orienting and merging parts by adding notches, asymmetrical holes, and steps to prevent mistakes in the assembly process.

- Parts must be designed to consistently orient themselves when fed into a process.
- Design must avoid using parts that can become tangled, wedged, or disoriented.
- Incorporate symmetry, low centers of gravity, easily identifiable features, guided surfaces, and tools for easy handling in the components.

By using the general principles of part handling, the designer can minimize non-value-added efforts and uncertainty in orientation and assembly. Part-handling common practices are given below:

- Ensure that parts can be separated from bulk.
- Ensure that parts can be automatically transported.
- Provide grippers for part handling.
- Use part-handling feeding mechanisms.
- Deliver parts in convenient orientations.

Minimal Flexible Parts

- Avoid flexible parts such as belts, gaskets, cables, and wire harnesses because flexibility can lead to difficulty in material handling and assembly.
- Use plug-in boards and back panels to avoid the use of wire harnesses.

Ease of Assembly

Design for ease of assembly by utilizing simple patterns of movement and minimizing the rotation of the assembled product. Avoid complex orientation and assembly movements in various directions during assembly.

- Part features such as chamfers and tapers should be provided.
- The design should enable the assembly to begin with a base component that has a large relative mass and a low center of gravity on which other parts are added.

Assembly should proceed vertically with one part added on top of another and positioned with the aid of gravity. This minimizes the need to reorient the assembly and reduces the need for temporary fastenings and more complex fixturing. A product that is easy to assemble manually is normally easily assembled by automation.

Efficient Joining and Fastening

Threaded fasteners (screws, bolts, nuts, and washers) are time consuming to assemble and difficult to automate.

- Use standardization and minimize variety while using fasteners such as self-threading screws and washers.
- Consider the use of snap-on fit-to-replace welded joints.
- Evaluate bonding techniques such as adhesives. Match fastening techniques to material and product requirements.

Guidelines Summarized

Information in **Tables 5.1, 5.2, 5.3, 5.4,** and **5.5** pertains to examples of common practices. Common practices for part insertion (**Table 5.2**) are to

- Provide a layered product design.
- Minimize secondary operation.

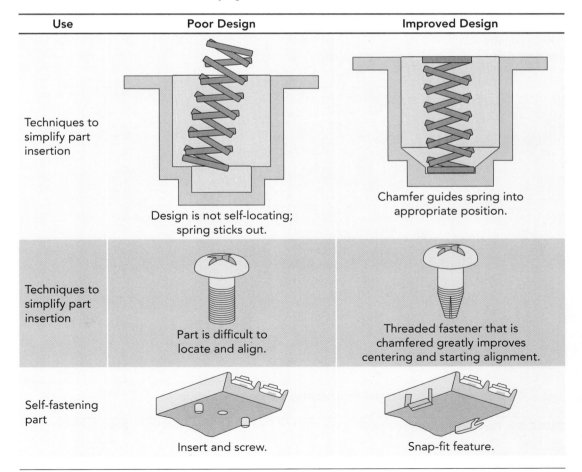

Use	Poor Design	Improved Design
Techniques to simplify part insertion	Design is not self-locating; spring sticks out.	Chamfer guides spring into appropriate position.
Techniques to simplify part insertion	Part is difficult to locate and align.	Threaded fastener that is chamfered greatly improves centering and starting alignment.
Self-fastening part	Insert and screw.	Snap-fit feature.

TABLE 5.2 Part Insertion

- Make parts self-aligning, self-locating, and self-fastening.
- Avoid unsecured parts.
- Provide access to insertion locations.
- Avoid use of flexible parts and special tools.

  Part orientation (**Table 5.3**) should be

- Symmetrical
- Exaggerated nonsymmetrical
- Asymmetrical defined by one main feature

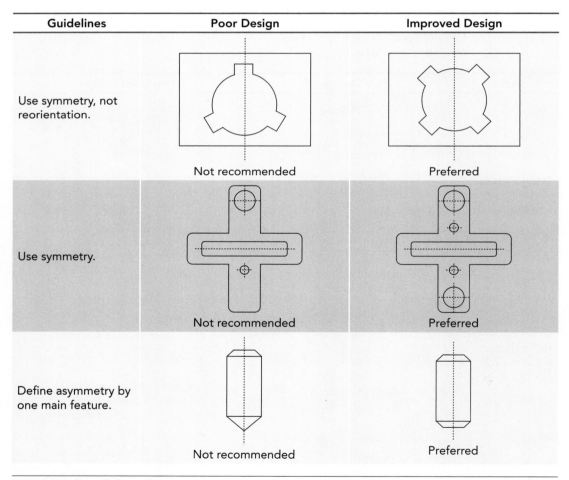

Guidelines	Poor Design	Improved Design
Use symmetry, not reorientation.	Not recommended	Preferred
Use symmetry.	Not recommended	Preferred
Define asymmetry by one main feature.	Not recommended	Preferred

TABLE 5.3 Part Orientation

Guidelines	Poor Design	Improved Design
Minimize the number of components in an assembly.		

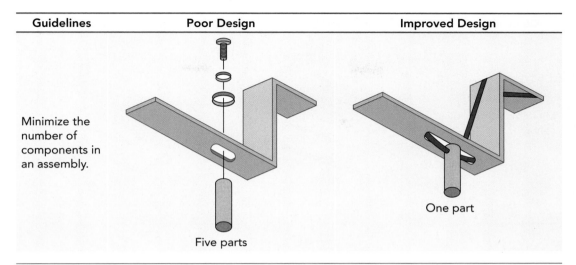

Five parts · One part

TABLE 5.4 Part Reduction

Common practices for part reduction (**Table 5.4**) recommend:

- Reduce the part count by using multifunctional parts
- Identify unnecessary parts

Examples of Design for Assembly

When design function permits, make parts with functionally superfluous features that facilitate handling during assembly (**Table 5.5**).

Guidelines	Poor Design	Improved Design
Reduce number of parts.		Snap fastening to the frame
Minimize part count.	A B	

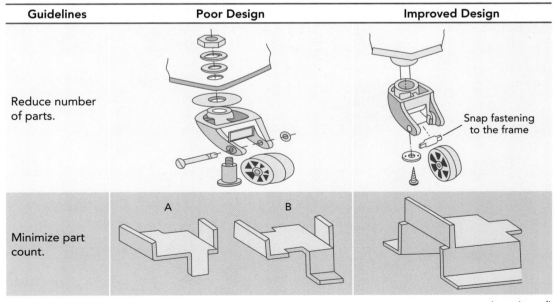

(continued)

Guidelines	Poor Design	Improved Design

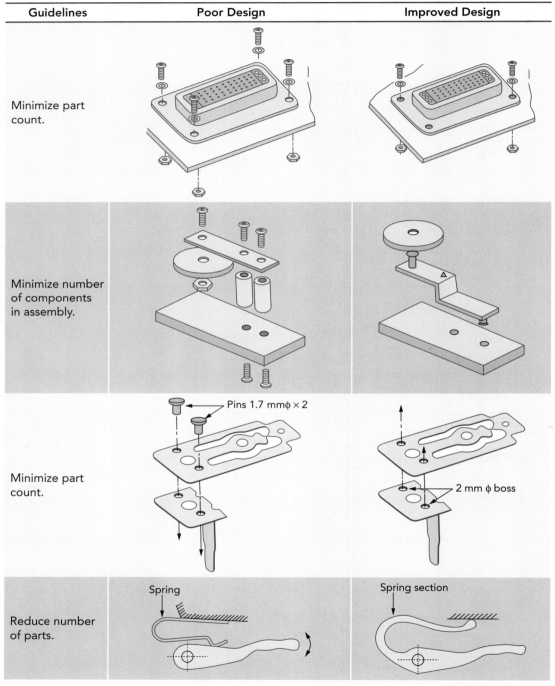

Minimize part count.

Minimize number of components in assembly.

Minimize part count.

Pins 1.7 mmφ × 2

2 mm φ boss

Reduce number of parts.

Spring

Spring section

(continued)

Guidelines	Poor Design	Improved Design

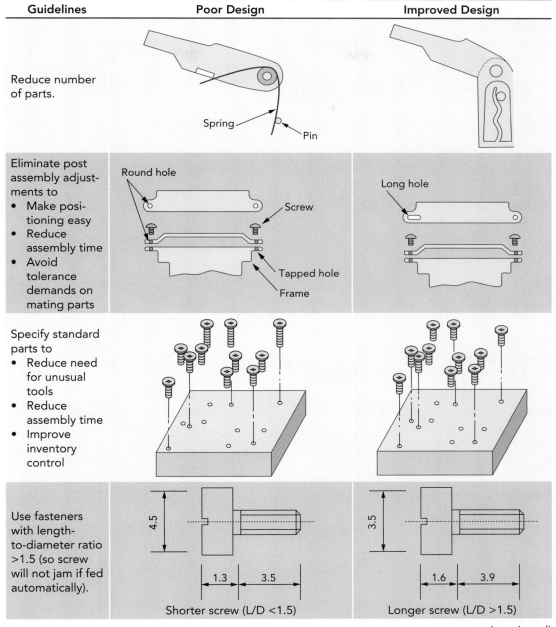

Reduce number of parts.

Spring — Pin

Eliminate post assembly adjustments to
- Make positioning easy
- Reduce assembly time
- Avoid tolerance demands on mating parts

Round hole — Screw — Tapped hole — Frame

Long hole

Specify standard parts to
- Reduce need for unusual tools
- Reduce assembly time
- Improve inventory control

Use fasteners with length-to-diameter ratio >1.5 (so screw will not jam if fed automatically).

4.5 — 1.3 — 3.5

Shorter screw (L/D <1.5)

3.5 — 1.6 — 3.9

Longer screw (L/D >1.5)

(continued)

Guidelines	Poor Design	Improved Design

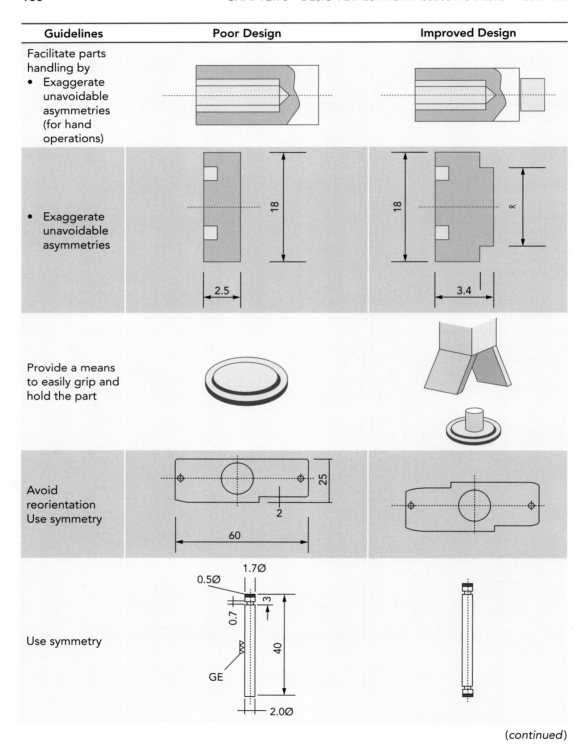

Facilitate parts handling by
- Exaggerate unavoidable asymmetries (for hand operations)

- Exaggerate unavoidable asymmetries

Provide a means to easily grip and hold the part

Avoid reorientation
Use symmetry

Use symmetry

(*continued*)

Guidelines	Poor Design	Improved Design

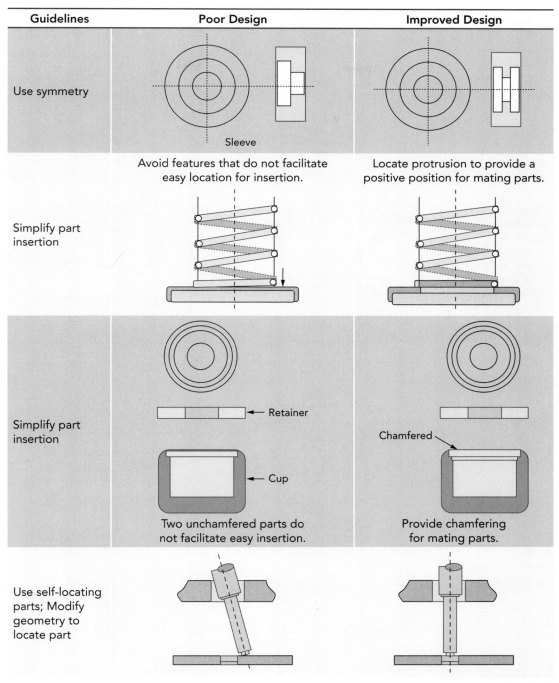

Use symmetry

Sleeve

Simplify part insertion

Avoid features that do not facilitate easy location for insertion.

Locate protrusion to provide a positive position for mating parts.

Simplify part insertion

Retainer

Cup

Two unchamfered parts do not facilitate easy insertion.

Chamfered

Provide chamfering for mating parts.

Use self-locating parts; Modify geometry to locate part

(continued)

Guidelines	Poor Design	Improved Design
Use shaft length exceeding free length of spring and thrust washer thickness to prevent parts from loosening in a vibrating atmosphere.		
Use a compensating slot for stacking tolerances. A long length X, such as $+\Delta X$, eliminates the need for tight tolerance for a quality assembly.		$x + \Delta x$
Use self-locating parts	Enlarged hole for position adjustment	Position the embossing pins into the holes and adjustment is not needed
Use self-locating parts		

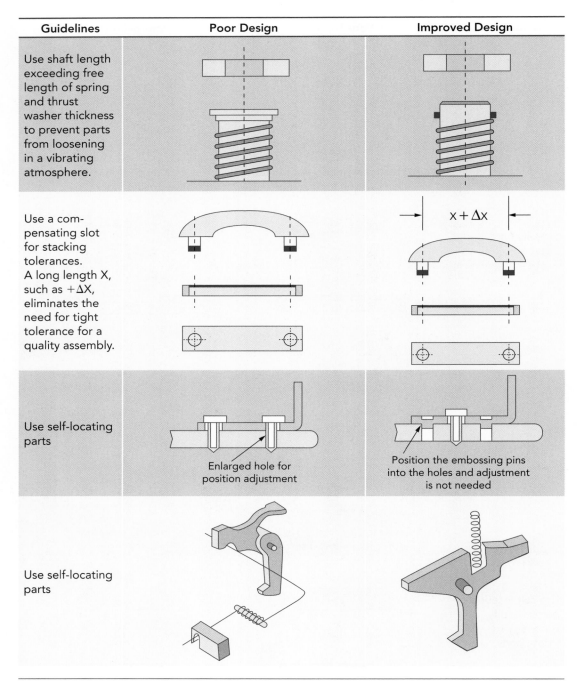

TABLE 5.5 Examples of Design for Assembly

5.1.2 **Definitions and Terminology**

Ease of alignment and position: A part is easy to align and position if its placement is established by locating its features or those on its mating parts and it has well-designed chamfers or similar features.

Resistance to insertion: The resistance experienced by a part while it is being inserted can be caused by small clearances, jamming or wedging, or acting against a large force. Examples of these parts are press fits or self-tapping screws.

Tangle: The components may entwine if reorientation is required to separate them from the bulk layout.

Severe tangle: Components are said to be severely entwined if they require manipulation to specific orientations and a force is required to separate them. Operators may need to use both hands to untangle.

Flexibility: Parts can deform substantially during assembly and manipulation. Operators might need to use both hands in handling paper, belts, felt gaskets, and cable assemblies, for example.

Handling difficulties: Components can present part-handling difficulties if they nest, tangle, or stick together because of magnetic attraction, grease coatings, slipperiness, delicacy, or heat/coldness.

Obstructed access: The space available for the assembly operation determines the assembly time. An obstruction can limit reach or access by fingers or tools, significantly increasing the time required for assembly.

Restricted vision: This causes an operator to depend on tactile sensors during assembly.

Holding down requirement: Holding down is considered another operation needed to maintain the position and orientation of a part during assembly or pre- or post-assembly. To finally secure an unstable part after placement or insertion can require gripping, realignment, or holding down during subsequent operations.

Located: A part is said to be *located* if it is partially placed and does not require holding down or realignment to accomplish the next step.

Envelope: This is the smallest cylinder or rectangular prism that can completely enclose a part. Size is the length of its longest side, and thickness is the length of its shortest side.

5.1.3 **Influence of Part Characteristics on the Assembly Process**

Part Handling

The manual handling process involves grasping, transporting, and orienting parts or subassemblies before they are inserted into or added to the work fixture or are partially built-up assembly.

A subassembly is considered a *part* if it is added during assembly. However, adhesives, fluxes, fillers, and such used for joining parts are not considered to be parts. The time spent on assembly increases if the part must be held down because it is unstable after placement or insertion or during subsequent operations. The part requires gripping, realigning, or holding down before it can be finally secured. *Holding down* also describes a situation that maintains the position and orientation of a part already in place.

Handling Difficulties Parts can present handling difficulties if they nest or tangle, stick together because of magnetic force or grease coating, are slippery, require careful handling, or are flexible. Parts can nest, tangle, or interlock when in bulk but can be separated by one simple manipulation of a single part. Severely nested or tangled parts require both hands to separate. Interlocking items include paper cups, closed-end helical springs, and circlips. Slippery parts easily come detached from ringers or standard grasping tools because of their shape or surface condition. Parts that require careful handling can be fragile or delicate, have sharp corners or edges, or present other hazards to the operator. Flexible parts substantially deform during manipulation and necessitate the use of two hands. Examples of such parts are paper and felt gaskets, and rubber bands or belts.

Part Location

A part is considered to be *located* if it does not require holding down or realignment for subsequent operations. A part is easy to align and position if its placement is established by locating the features on it or its mating part and well-designed chamfers facilitate the insertion. Insertion resistance during part insertion can be caused by small clearances, jamming or wedging, hang-ups, or placement against a large force. See the examples in **Table 5.3**. For example, a press fit is an interference fit that requires a strong force for assembly. The resistance encountered with self-tapping screws is also insertion resistance. Assembly time can vary depending on whether the parts have clear or obstructed access. Obstructed access causes a significant increase in the assembly time, and restricted vision causes the operator to rely on tactile sensors during assembly.

Part Orientation and Symmetry along Axis of Insertion

Alpha and Beta Symmetry Orientation of a part involves aligning it properly when it is inserted relative to a corresponding element. There are two types of symmetry definition for a part. α symmetry is the rotational balance of a part about an axis perpendicular to the axis of insertion. For parts with one axis of insertion, end-to-end orientation is necessary when $\alpha = 360°$; otherwise, $\alpha = 180°$. β symmetry is the rotational balance of a part about its axis of insertion or equivalently about an axis that is perpendicular to the surface on which the part is placed during assembly.

The magnitude of rotational symmetry is the smallest angle through which a part can be rotated and can repeat its orientation. For a cylinder inserted into a circular hole,

$\beta = 0$; for a square part inserted into a square hole, $\beta = 90°$, and so on. *Thickness* is the length of the shortest side of the smallest rectangular prism that encloses the part. However, parts that are cylindrical or have a regular polygonal cross-section with five or more sides have a thickness defined as the radius of the smallest cylinder that can enclose the part. *Size* is the length of the longest side of the smallest rectangular prism that can enclose the part (**Tables 5.6** and **5.7**).

Alpha (α) Symmetry	Beta (β) Symmetry
The rotational symmetry of a part about an axis perpendicular to the axis of insertion, depending on the angle through which a part must be rotated to repeat its orientation.	The rotational symmetry of a part about its axis of insertion or equivalently about an axis that is perpendicular to the surface on which the part is placed, depending on the angle through which a part is rotated about the axis of insertion.
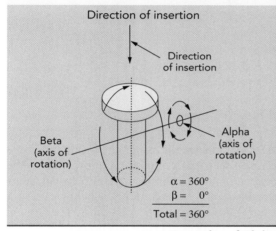	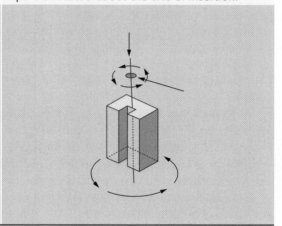

Examples of Alpha and Beta Symmetry

Beta symmetry

A beta symmetric part is one that does not require orientation about the axis of insertion.

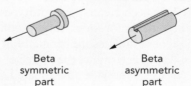

TABLE 5.6 Alpha and Beta Symmetry Definition

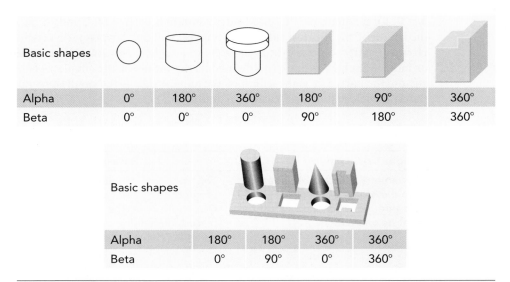

Basic shapes	○					
Alpha	0°	180°	360°	180°	90°	360°
Beta	0°	0°	0°	90°	180°	360°

Basic shapes				
Alpha	180°	180°	360°	360°
Beta	0°	90°	0°	360°

TABLE 5.7 Examples of Alpha and Beta Symmetry

5.2 **Design for Assembly Methodology**

The Boothroyd–Dewhurst method (BDM) completes the design and evaluation procedure by selecting an assembly process and then analyzing it and improving it. The BDM distinguishes between manual and automatic or robotic assembly by considering the cost and time saving in automated assembly. Design improvement focuses on reducing the number of parts and shortening the associated process times. The Hitachi method suggests three basic steps for the design process: (1) product design, (2) assembly evaluation, and (3) design comparison. The Lucas method consists of six steps: (1) product specification, (2) design for assembly, (3) functional analysis, (4) handling analysis, (5) fitting analysis, and (6) redesign. The design improvement process results in a reduction in the number of parts and an improvement in the assembly time.

The BDM is an analytical procedure used to evaluate a new product design after engineering drawings have been created or prototypes have been developed. It can be used to re-evaluate the process for an existing product or evaluate the potential for automating it. The designer evaluates the geometry of each component in the product or its subassemblies and then determines the degree of difficulty in handling and inserting each part. The analysis can provide an estimated assembly cost and a direction for redesign to improve the design.

The main goal of BDM is to minimize product cost within the constraints imposed by design features. The best way to achieve this goal is to reduce the number of components used and then to ensure that they are easy to install or assemble. In the early stage of design, the designer must evaluate the assembly cost, which requires being familiar with the nature of the assembly processes.

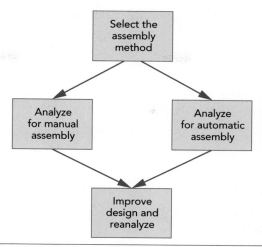

FIGURE 5.1 Stages in Assembly Analysis of Designs

BDM offers a way to judge the design efficiency in terms of assembly. The entire procedure (**Figure 5.1**) consists of three basic steps:

▸ **Step 1. Select the assembly method.**

▸ **Step 2. Analyze the assembly.**

▸ **Step 3. Improve the design.**

The differences in the ability of human operators and automated assembly result in a significant change in assembly. A product's assembly cost is related to both its product design and assembly. Minimum product cost can be achieved when the appropriate assembly method is used.

Redesign must produce a better product with increased design efficiency. The most effective way to improve design efficiency is by reducing the number of parts. With manual assembly, reducing operation time is another way to improve design efficiency. The feeding and orienting efficiency of automatic assembly needs to be evaluated.

5.2.1 **Step-by-Step Design Efficiency Calculation**

A design's features should be examined in a systematic manner and a "design efficiency" calculated. This efficiency is then used to compare different designs.

Design efficiency is determined by using the appropriate formula. The manual assembly design efficiency is obtained by using the following equation:

$$E_m = \frac{3N_m}{T_m}$$

where

E_m = manual design efficiency

N_m = theoretical minimum number of parts

T_m = total assembly time

The automatic assembly design efficiency can be determined by using the formula

$$E_F = \frac{0.09 N_m}{C_A}$$

where

E_F =   automatic design efficiency

N_m =   theoretical minimum number of parts

C_A =   total assembly cost

The BDM involves two important steps for each part in the assembly:

1. Decide whether the part can be eliminated or combined with other parts in the assembly.
2. Estimate the time required to grasp, manipulate, and insert the part.

▸ **Step 1. Disassembly**
Each part is assigned a number. Any subassembly items are first treated as parts and are subsequently analyzed.

▸ **Step 2. Reassembly**
The part with the highest identification number is assembled first to the work fixture and then the remaining parts are added as per the plan. It should not be assumed that a part can be grasped in each hand and then put together before placing all parts in the partially completed assembly. Finally, the assembly is analyzed according to the information in **Table 5.8**.

	1	2	3	4	5	6	7	8	9	10
Number	Part Description	Number of Identical Operations	Symmetry	Manual Handling Code	Manual Handling Time	Manual Insertion Code	Manual Insertion Time	Total Assembly Time $2 \times (5+7)$ $= T_M$	Theoretical Minimum Number of Parts (Nm)	Costs

TABLE 5.8 Manual Assembly Analysis Worksheet

‣ **Column 1.** Enter the part identification number and part description.

‣ **Column 2.** Enter the number of times that the identical part is used at only this level in the assembly. For example, assuming that all screws are inserted individually, if this level in the assembly has six half-inch diameter machine screws, it has six operations. Screws and washers are considered as separate parts.

‣ **Column 3.** Use **Table 5.9** to estimate the ease of handling of the part. Always look at the part in relation to the Z axis assembly direction to determine α and β. Enter the two-digit handling process code in Column 3.

‣ **Column 4.** Enter the handling code for this operation (*found in* **Table 5.9**) in the proper column (**Table 5.8**).

‣ **Column 5.** Enter the handling time (in seconds) for this operation in this column.

‣ **Column 6.** Estimate the ease of assembly from **Table 5.10**, which shows estimated times for insertion. Carefully consider the attributes, hold down, align, and position, insertion resistance, in determining insertion times. Enter the two-digit insertion process code in Column 6.

‣ **Column 7.** Enter the insertion time (in seconds) for this operation in this column.

‣ **Column 8.** The total assembly operation time in seconds is calculated by adding the handling and insertion in Columns 4 and 6, and multiplying this sum by the number of repeated operations in Column 2.

‣ **Column 9.** An estimate of the theoretical minimum number of parts for the assembly is determined for each part by answering each question below.

‣ **Column 10.** The total operation cost in cents is obtained by multiplying the operation time in Column 8 by a labor and overhead rate.

- During operation of the product, does the part move relative to all other parts already assembled? Only gross motion should be considered.
- Must the part be of a different material or be physically isolated from all other parts already assembled?
- Must the part be separate from all other parts already assembled? If no, necessary assembly or disassembly would be impossible.

If the answer to any of these questions is yes, place a 1 in Column 9 except where multiple identical operations are indicated in Column 2; in this case, place the number of parts that must be separate in Column 9.

The remaining parts are added one by one to the assembly, and the manual assembly analysis worksheet is completed for each additional part.

When all the rows are completed, add the numbers in Column 8 to find the estimated assembly time. Add the numbers in Column 9 to find the theoretical minimum number of parts and add the numbers in Column 10 to find the total assembly cost for the complete assembly.

Redesign

The initial design for assembly analysis provides the designer useful information for the product redesign in two areas:

1. Data in Column 9 of the worksheet (**Table 5.8**) indicates where it might be possible to reduce the number of parts.

2. Columns 4 and 6 indicate those parts that are difficult to handle or insert.

The redesign procedure requires three steps:

▸ **Step 1. Examine Column 9 on the worksheet.**
When the number in this column is less than that in Column 2, elimination of parts is possible. A reduction in part count is usually the most effective way to improve assembly greatly, significantly enhancing design efficiency. Creative techniques should be applied to groups of parts in the assembly that can be combined, and alternate group parts should be sketched. Examine Columns 4 and 6; the numbers in them indicate parts for which handling time or insertion time has the potential to decrease. Review the design of the parts that pose a difficulty in handling and insertion.

▸ **Step 2. Consider the way parts are handled.**
In the BDM procedure, the parts are considered to be added one at a time. However, for bench assembly and on manual assembly lines, workers often handle two parts simultaneously. Under these circumstances, the assembly time can be reduced by one-third. Thus, the design engineer can obtain a more accurate estimated time by dividing the derived number by 1.5.

▸ **Step 2. Evaluate the location of the parts.**
In preparing the time and motion studies to derive manual handling times, it is assumed that the parts are randomly oriented in bins at the assembly station. If the parts are available in the proper orientation in trays or magazines at the assembly station, this information can be incorporated in the analysis. Refer to **Tables 5.9** and **5.10** for the basic information for manual handling analysis and component insertion analysis.

As explained earlier, the design efficiency is determined by using the appropriate formula. The manual assembly design efficiency is

$$E_m = \frac{3N_m}{T_m}$$

where

E_m = manual design efficiency

N_m = theoretical minimum number of parts

T_m = total assembly time

MANUAL HANDLING — ESTIMATED TIMES (seconds)

Based on Geoffrey Boothroyd, Peter Dewhurst, Winston Anthony Knight, Product Design for Manufacture and Assembly, M. Dekker, 1994

Key: ONE HAND

parts can be grasped and manipulated by one hand without the aid of grasping tools

	parts are easy to grasp and manipulate					parts present handling difficulties (1)				
	thickness > 2 mm			thickness ≤ 2 mm		thickness > 2 mm			thickness ≤ 2 mm	
	size >15 mm	6 mm ≤ size ≤15 mm	size < 6 mm	size > 6 mm	size ≤6 mm	size >15 mm	6 mm ≤ size ≤15 mm	size < 6 mm	size > 6 mm	size ≤6 mm
	0	1	2	3	4	5	6	7	8	9
0 — $(\alpha + \beta) < 360°$	1.13	1.43	1.88	1.69	2.18	1.84	2.17	2.65	2.45	2.98
1 — $360° \leq (\alpha + \beta) < 540°$	1.5	1.8	2.25	2.06	2.55	2.25	2.57	3.06	3	3.38
2 — $540° \leq (\alpha + \beta) < 720°$	1.8	2.1	2.55	2.36	2.85	2.57	2.9	3.38	3.18	3.7
3 — $(\alpha + \beta) = 720°$	1.95	2.25	2.7	2.51	3	2.73	3.06	3.55	3.34	4

ONE HAND with GRASPING AIDS

parts can be grasped and manipulated by one hand but only with the use of grasping tools

	parts need tweezers for grasping and manipulation								parts need standard tools other than tweezers	parts need special tools for grasping and manipulation
	parts can be manipulated without optical magnification				parts require optical magnification for manipulation					
	parts are easy to grasp and manipulate		parts present handling difficulties (1)		parts are easy to grasp and manipulate		parts present handling difficulties (1)			
	thickness >0.25 mm	thickness ≤0.25 mm	thickness >0.25 mm	thickness ≤0.25 mm	thickness >0.25 mm	thickness ≤0.25 mm	thickness >0.25 mm	thickness ≤0.25 mm		
	0	1	2	3	4	5	6	7	8	9
4 — $\alpha \leq 180°$, $0 \leq \beta \leq 180°$	3.6	6.85	4.35	7.6	5.6	8.35	6.35	8.6	7	7
5 — $\alpha \leq 180°$, $\beta = 360°$	4	7.25	4.75	8	6	8.75	6.75	9	8	8
6 — $\alpha = 360°$, $0 \leq \beta \leq 180°$	4.8	8.05	5.55	8.8	6.8	9.55	7.55	9.8	8	9
7 — $\alpha = 360°$, $\beta = 360°$	5.1	8.35	5.85	9.1	7.1	9.55	7.85	10.1	9	10

TWO HANDS for MANIPULATION

parts severely nest or tangle or are flexible but can be grasped and lifted by one hand (with the use of grasping tools if necessary)(2)

	parts present no additional handling difficulties					parts present additional handling difficulties (e.g., sticky, delicate, slippery, etc.)(1)				
	$\alpha \leq 180°$			$\alpha = 360°$		$\alpha \leq 180°$			$\alpha = 360°$	
	size >15 mm	6 mm ≤ size ≤15 mm	size <6 mm	size >6 mm	size ≤6 mm	size >15 mm	6 mm ≤ size ≤15 mm	size <6 mm	size >6 mm	size ≤6 mm
	0	1	2	3	4	5	6	7	8	9
8	4.1	4.5	5.1	5.6	6.75	5	5.25	5.85	6.35	7

TWO HANDS required for LARGE SIZE

two hands, two persons or mechanical assistance required for grasping and transporting parts

	parts can be handled by one person without mechanical assistance								parts severely nest or tangle or are flexible (2)	two persons or mechanical assistance required for parts manipulation
	parts do not severely nest or tangle and are not flexible									
	part weight < 10 lb				parts are heavy (> 10 lb)					
	parts are easy to grasp and manipulate		parts present other handling difficulties (1)		parts are easy to grasp and manipulate		parts present ther handling difficulties (1)			
	$\alpha \leq 180°$	$\alpha = 360°$	$\alpha \leq 180°$	$\alpha = 360°$	$\alpha \leq 180°$	$\alpha = 360°$	$\alpha \leq 180°$	$\alpha = 360°$		
	0	1	2	3	4	5	6	7	8	9
9	2	3	2	3	3	4	4	5	7	9

TABLE 5.9 Handling Table

MANUAL INSERTION — ESTIMATED TIMES (seconds)

Based on Geoffrey Boothroyd, Peter Dewhurst, Winston Anthony Knight, Product Design for Manufacture and Assembly, M. Dekker, 1994

Key: PART ADDED but NOT SECURED

addition of any part (1) where neither the part itself nor any other part is finally secured immediately

		after assembly no holding down required to maintain orientation and location (3)				holding down required during subsequent processes to maintain orientation or location (3)			
		easy to align and position during assembly (4)		not easy to align or position during assembly		easy to align and position during assembly (4)		not easy to align or position during assembly	
		no resistance to insertion	resistance to insertion (5)	no resistance to insertion	resistance to insertion (5)	no resistance to insertion	resistance to insertion (5)	no resistance to insertion	resistance to insertion (5)
		0	1	2	3	6	7	8	9
part and associated tool (including hands) can easily reach the desired location	0	1.5	2.5	2.5	3.5	5.5	6.5	6.5	7.5
due to obstructed access or restricted vision (2)	1	4	5	5	6	8	9	9	10
due to obstructed access and restricted vision (2)	2	5.5	6.5	6.5	7.5	9.5	10.5	10.5	11.5

PART SECURED IMMEDIATELY

addition of any part (1) where the part itself and/or other parts are being finally secured immediately

		no screwing operation or plastic deformation immediately after insertion (snap/press fits, circlips, spire nuts, etc.)		plastic deformation immediately after insertion						screw tightening immediately after insertion (6)	
				plastic bending or torsion			rivetting or similar operation				
		easy to align and position with no resistance to insertion (4)	not easy to align or position during assembly and/or resistance to insertion (5)	easy to align and position during assembly (4)	not easy to align or position during assembly		easy to align and position during assembly (4)	not easy to align or position during assembly		easy to align and position with no torsional resistance (4)	not easy to align or position and/or torsional resistance (5)
					no resistance to insertion	resistance to insertion (5)		no resistance to insertion	resistance to insertion (5)		
		0	1	2	3	4	5	6	7	8	9
part and associated tool (including hands) can easily reach the desired location and the tool can be operated easily	3	2	5	4	5	6	7	8	9	6	8
due to obstructed access or restricted vision (2)	4	4.5	7.5	6.5	7.5	8.5	9.5	10.5	11.5	8.5	10.5
due to obstructed access or restricted vision (2)	5	6	9	8	9	10	11	12	13	10	12

SEPARATE OPERATION

		mechanical fastening processes (part(s) already in place but not secured immediately after insertion)				non-mechanical fastening processes (part(s) already in place but not secured immediately after insertion)				non-fastening processes	
		none or localized plastic deformation				metallurgical processes					
							additional material required				
		bending or similar processes	rivetting or similar processes	screw tightening (6) or other processes	bulk plastic deformation (large proportion of part is plastically deformed during fastening)	no additional material required (e.g. resistance, friction welding, etc.)	soldering processes	weld/braze processes	chemical processes (e.g. adhesive bonding, etc.)	manipulation of parts or sub-assembly (e.g. orienting, fitting or adjustment of part(s), etc.)	other processes (e.g. liquid insertion, etc.)
		0	1	2	3	4	5	6	7	8	9
assembly processes where all solid parts are in place	9	4	7	5	3.5	7	8	12	12	9	12

TABLE 5.10 Insertion Table

EXAMPLE 5.1

Electric Switch

The first case study examines the design of an electric switch. The design of an electric switch influences not only the cost of its fabrication but also the cost of its installation. Using the BDM procedure, an existing switch is first analyzed before redesign.

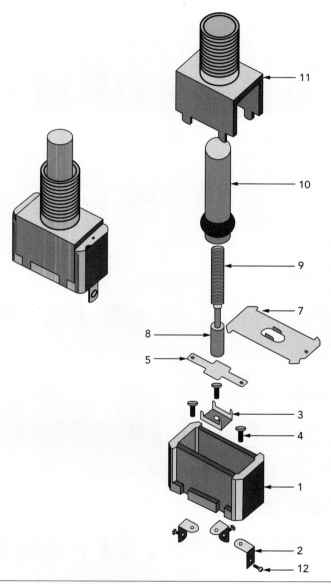

FIGURE 5.2 Single-Pole Double-Throw Switch (SPDT)

Initial Design Evaluation

During the analysis, the changes in the number of parts and design efficiency were calculated. The original design for the switch contained a total of 14 separate parts and operations. The total assembly time was found to be 183.11 seconds, and design efficiency was calculated as 15%. The analysis results indicated that this switch assembly was a good candidate for the application of DFA tools for a more efficient redesign (**Figure 5.2** and **Table 5.11**).

Approach to Redesign

The goal of redesigning the switch was to reduce the number of parts to meet the design's theoretical minimum number of parts, while maintaining the functionality of the original design. The basic guidelines for DFA analysis were applied for each part. A brief description of the new design on a part-by-part basis is as follows (**Table 5.12** and **Figure 5.3**).

Switch Base The plastic switch base was modified to incorporate a snap fit into the switch cover. This eliminated the existing bent tabs on the metal switch cover used to attach it to the base. Other modifications were the addition of snap-fit sockets to hold the two metal wire contact terminals and the center terminal rocker in place. Additional fabrication charges for this piece are not foreseen. A new plastic mold must be created.

Number	Parts Name	Dimensions (mm)	
		Thickness	Size
1	Switch base	15	29
2	Terminal	9	8
3	Center terminal contact rocker	6	8
4	Terminal rivet	7	4
5	Contact rocker	4	22
6	Addition of grease	—	—
7	Base cover	3	29
8	Switch plunger	4	16
9	Switch spring	3	20
10	Switch toggle	9	35
11	Mounting thread	12	14
12	Mounting cover	13	19
13	Mounting hardware	2	16
14	Terminal screw	7	7

TABLE 5.11 Components of the Electric Switch (Initial Design)

Number	Item Type	Parts Name	Repeat Count	Minimum Number of Parts	Tool Acquisition Time	Item Handling/Acquisition Time	Item Insertion Time	Total Operation Time	Total Operation Cost
1	Part	Switch base	1	1	0	2.73	1.5	4.23	0.0352
2	Part	Terminals	3	3	0	5.10	7.4	37.50	0.3125
3	Part	Terminal center contact	1	0	0	4.80	7.4	12.20	0.1016
4	Part	Terminal rivets	3	0	2.9	4.80	11.2	50.90	0.4241
5	Part	Contact rocker	1	1	0	4.35	7.4	11.75	0.0977
6	Oper.	Addition of grease	1	–	3	–	–	7.00	0.0583
7	Part	Base cover	1	0	0	2.73	5.2	7.93	0.0660
8	Part	Switch plunger	1	1	0	2.06	3.0	5.06	0.0421
9	Part	Switch spring	1	0	0	5.60	6.5	12.10	0.1008
10	Part	Switch toggle	1	1	0	1.50	2.6	4.10	0.0341
11	Part	Mounting cover	1	0	0	1.80	2.6	4.40	0.0366
12	Part	Mounting thread	1	0	0	1.80	5.2	7.00	0.0583
13	Part	Mounting washer	1	0	2.9	2.06	7.5	12.46	0.1038
14	Part	Terminal screws	1	0	2.9	1.8	9.2	13.90	0.1158

Design efficiency = 15%

TABLE 5.12 Calculation of Original Assembly Time and Design Efficiency

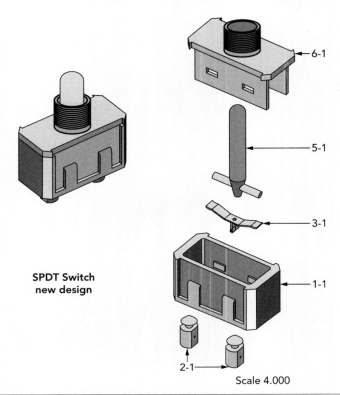

SPDT Switch
new design

Scale 4.000

FIGURE 5.3 Redesigned Electric Switch

Wire Clinch Terminal This replaces the terminals, terminal rivets, and terminal screws in the original design. It performs the dual function of holding the stranded wires and providing contact points for the terminal rocker. Its wires are held in place within the terminal by a metal-locking spring action. The two wire clinch terminals formed of rolled brass sheets snap into the plastic switch base. Additional tooling and fabrication charges would be incurred to create this specialized part.

Center Terminal Contact Rocker This part replaces the original design's center terminal contact, center terminal rivet, contact rocker, and switch spring. Like the wire clinch terminal, this piece formed of brass sheet metal snaps into the plastic switch base and provides a flexible interface at the switch toggle. This part would incur extra tooling and fabrication charges.

Addition of Grease This separate operation remains the same as in the original switch design. The addition of grease lubricates the switch toggle/rocker interface and thereby increases the life of the switch during normal operation.

Switch Toggle This was modified extensively from the original design. A molded plastic piece with snap-fit posts replaces the cast aluminum piece. The plastic design of the new toggle incorporates the original switch plunger piece into the toggle itself. No extensive charges are foreseen in the fabrication of the new part; the original casting process will be replaced with a plastic mold process.

Switch Cover This is another extensive redesign. It replaces the base cover, mounting threads, and mounting cover in the original design. This piece requires a complicated fabrication process requiring several machining operations, making the switch cover one of the most expensive parts in the new design. The overall shape is a metal casting. The new switch cover allows for a snap fit at the switch base interface, and the posts on the toggle snap into the inner diameter of the threaded portion.

Mounting Hardware This was not changed from the original design. The switch assembly was redesigned to keep the original hardware functionality and the way in which it is mounted to the electrical panel, chassis, and so on.

Design Efficiency Discussion

The design changes just described led to a faster and more efficient assembly of the SPDT switch. The number of parts and operations decreased from 14 to 7 as shown in **Tables 5.13** and **5.14**, respectively. This led to an assembly time of 64.68 seconds, approximately 2.83 times faster than for the original design. The assembly efficiency was calculated as 42%, an increase of 15%. These changes to the switch design will lead to a more efficient assembly and less assembly time, but the overall cost of production might not be reduced because of the modifications that were made to the parts to decrease the part count. New tooling and fabrication processes must be developed to create the required specialized, combination parts. A design engineer, therefore, must calculate the higher tooling and manufacturing charges to determine whether they would compromise the cost of production.

Number	Parts Name	Dimensions (mm)	
		Thickness	Size
1	Switch base	15	29
2	Wire clinch terminals	5	12
3	Center terminal/rocker	15	22
4	Addition of grease	—	—
5	Plastic switch toggle	9	45
6	Switch cover	26	29
7	Mounting hardware	2	16

TABLE 5.13 Components of the Redesigned Electric Switch

Number	Item Type	Parts Name	Repeat Count	Minimum Parts	Tool Acquisition Time	Item Handling/ Acquisition Time	Item Insertion Time	Total Operation Time	Total Operation Cost
1	Part	Switch base	1	1	0	1.95	1.5	3.45	0.0287
2	Part	Wire clinch terminal	2	2	0	1.80	5.0	13.60	0.1133
3	Part	Terminal center contact	1	1	0	1.80	5.0	6.80	0.0566
4	Oper.	Application of grease	1	—	3	—	—	7.00	0.0583
5	Part	Plastic switch toggle	1	1	0	1.80	2.6	4.40	0.0366
6	Part	Switch cover	1	1	0	1.80	1.8	3.60	0.0300
7	Part	Mounting hardware	1	0	2.9	1.69	7.5	12.09	0.1007

Design efficiency = 42%

TABLE 5.14 Calculation of Redesigned Assembly Time and Design Efficiency

EXAMPLE 5.2

Electric Motor

This example considers the mechanical components for ease of assembling an electric motor without reviewing the design of electromagnetic components. The initial design efficiency index of 0.061 increased to 0.110. The number of components decreased from 16 to 10 (**Figures 5.4** and **5.5**, **Tables 5.15** and **5.16**).

In spite of the redesign of the motor, the design efficiency index improved only marginally. The size of the motor is a major contributing factor. Its components are heavy and big in size; fitting the parts together requires precision and delicate handling.

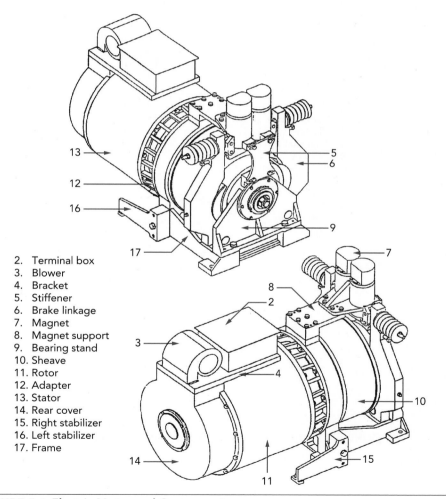

2. Terminal box
3. Blower
4. Bracket
5. Stiffener
6. Brake linkage
7. Magnet
8. Magnet support
9. Bearing stand
10. Sheave
11. Rotor
12. Adapter
13. Stator
14. Rear cover
15. Right stabilizer
16. Left stabilizer
17. Frame

FIGURE 5.4 Electric Motor and Components

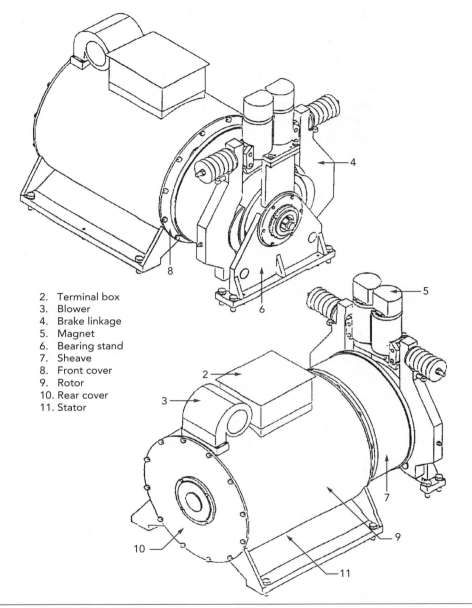

2. Terminal box
3. Blower
4. Brake linkage
5. Magnet
6. Bearing stand
7. Sheave
8. Front cover
9. Rotor
10. Rear cover
11. Stator

FIGURE 5.5 Redesigned Electric Motor and Components

Number	Part Description	Identical Operations	Manual Handling Code	Manual Handling Time	Manual Insertion Code	Manual Insertion Time	Total Assembly Time (2)[(4) + (6)] = T_m	Theoretical Minimum Number of Parts (N_m)
		1	2	3	4	5	6	7	8
2	Terminal box	1	95	4.0	38	6	20	1
3	Blower	1	91	3.0	38	6.0	18.0	1
4	Bracket	1	95	4.0	49	10.5	29.0	0
5	Stiffener	1	70	5.1	39	8.0	26.2	0
6	Brake linkage	1	99	9.0	92	5.0	28.0	1
7	Magnet	1	99	9.0	38	6.0	30.0	1
8	Magnet support	1	99	9.0	38	6.0	30.0	0
9	Bearing stand	1	99	9.0	37	8.0	34.0	1
10	Sheave	1	99	9.0	92	5.0	28.0	1
11	Rotor	1	99	9.0	92	5.0	28.0	1
12	Adapter	1	99	9.0	49	10.5	39.0	0
13	Stator	1	99	9.0	49	10.5	39.0	1
14	Rear cover	1	99	9.0	39	8.0	34.0	1
15	Right stabilizer	2	95	4.0	38	6.0	20.0	0
16	Left stabilizer	2	95	4.0	38	6.0	20.0	0
17	Frame	1	99	9.0	00	1.5	21.0	0

$T_m = 444.2$
$N_m = 9$
Design efficiency index = 0.06

TABLE 5.15 Design for Manufacturing (DFM) Worksheet for Original Motor

0	1	2	3	4	5	6	7	8
	Part Description	Identical Operations	Manual Handling Code	Manual Handling Time	Manual Insertion Code	Manual Insertion Time	Total Assembly Time $(2) \times [(5) + (7)] = T_m$	Theoretical Minimum Number of Parts (N_m)
2	Terminal box	95	4	38	6.0	20	8.0	1
3	Blower	91	3	38	6.0	18	7.2	1
4	Brake linkage	99	9	92	5.0	28	11.2	1
5	Magnet	99	9	38	6.0	30	12.0	1
6	Bearing stand	99	9	39	8.0	34	13.6	1
7	Sheave	99	9	92	5.0	28	11.2	1
8	Front cover	99	9	39	8.0	34	13.6	1
9	Rotor	99	9	92	5.0	28	11.2	1
10	Rear cover	99	9	39	8.0	34	13.6	1
11	Stator	99	9	00	1.5	21	8.4	1

$T_m = 275$
$N_m = 10$
Design efficiency index = 0.11

TABLE 5.16 DFM Worksheet for the Redesigned Motor

EXAMPLE 5.3

Hydraulic Shuttle Valve

The aerospace industry is a major user of shuttle valves for controlling the flow of various fluids. The shuttle valve in **Figure 5.6** is designed to isolate the normal fluid flow from the emergency hydraulic system during normal operation. The pressure-actuated shuttle valve is completely self-contained, three-port valve. At each end of the valve is a supply port. In typical applications, one port provides normal flow. When the pressure is lost in the normal system and the emergency pressure is applied, the poppet shuttles across to block the normal port. Fluid flows out the center (third) port of the shuttle valve through a series of windows cut at the outer diameter of the valve's center (**Figure 5.7**).

Inside the valve body is a spring-loaded poppet that normally closes the emergency port, allowing fluid to flow from the normal supply port to the center discharge port. Each port is screened to prevent contamination from passing into the valve. The shuttle valve must be constructed of aerospace-grade stainless steel except for the external nose seal, which is normally made from aerospace-grade aluminum.

During *normal operation*, the normal and emergency ports have equal pressure. The poppet is forced against the emergency valve seat, sealing it with an internal spring force. The poppet valve in this position provides a free path for fluid to flow from the normal supply port to the discharge port at the center.

In the event of a loss of normal operating system pressure and flow, the emergency port pressure overcomes the poppet spring force and normal supply pressure, forcing

Based on http://www.leesl.it/images/i_amh-start.jpg, The Lee Company

FIGURE 5.6 Typical Shuttle Valve

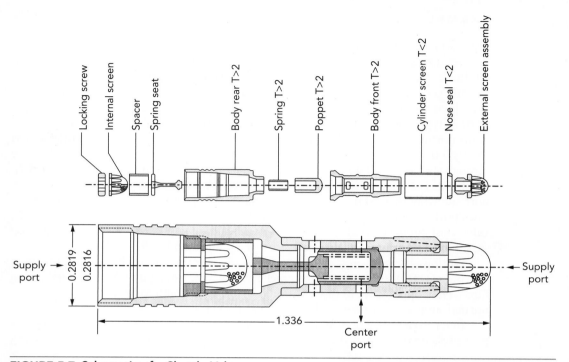

FIGURE 5.7 Schematic of a Shuttle Valve

the poppet to move against the normal port valve seat. This action moves the shuttle from the emergency port valve seat to provide a means of automatic transfer of control from normal hydraulic systems to redundant emergency ones such as in an aircraft landing system.

Initial assembly: The initial valve in **Figure 5.7** has 11 individual parts and was assembled in two parts: the front body and the rear body. See **Table 5.17** for the complete data on the nature of the components, their handling time, insertion time, and parameters needed to calculate the efficiency index.

The results of the analysis are as follows:

- Design efficiency (percent): 24
- Total assembly time (seconds): 97
- Total number of parts: 11
- Theoretical minimum number of parts: 8

The DFA analysis of the shuttle valve indicates the possibility for some assembly time elimination. Approximately 40% of the 107-second time can be reduced. The three components that took the longest to assemble were the retainer (15.6 seconds), the barrel screen (13.88 seconds), and the spring seat mounted on the rear body (14.5 seconds).

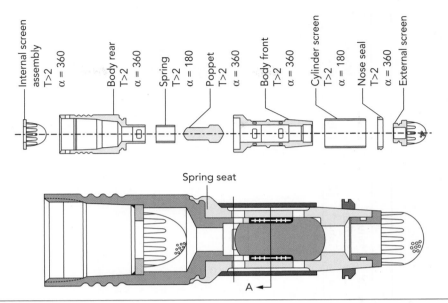

FIGURE 5.8

Redesign: This process focuses on eliminating unnecessary parts and operations and on combining parts (**Figure 5.8**). The most significant component redesign involves the poppet valve and return spring. The spring is seated against the poppet and has a separate spring seat mounted in its rear body. An additional spacer is required to support the spring seat over the bleed screen (a valve component).

The assembly is then attached to the rear body through a locking screw that retains the internal screen, spacer, seat, and spring components. By combining parts, the redesign eliminates the spacer and seat by creating a spring seat surface on the nose of the rear body, where it is swaged into the front body.

The locking screw is eliminated by using a screen that is brazed to the rear body. The poppet has been redesigned to have a stepped diameter with a smaller outer diameter (OD) that moves within the rear body, and the larger OD moves within the front body. The step surface is used as a seat by which the spring force is applied to the poppet (See **Table 5.17** and **Table 5.18**).

The redesigned product showed a significant reduction in assembly time (**Table 5.19**). As a result, design efficiency has increased. The redesigned product has two sets of windows that allow flow to discharge from the valve; one set is for the normal flow path, and the second is for the emergency flow path. Although these two ports discharge into the same downstream annulus, they are internally isolated by the poppet valve except for minimal lapped leakage. The redesign can also cause additional operations. The rear body of the redesigned valve is swaged inside the front body. The valve moves inside the rear body to close the normal discharge flow path during an emergency, requiring a discharge window cut into both the rear and front bodies.

Number	Part Description	Number of Identical Operations	Symmetry	Manual Handling Code	Manual Handling Time	Manual Insertion Code	Manual Insertion Time	Total Assembly Time (2) × [(5) + (7)] = T_m	Theoretical Minimum Parts (N_m)
0	1	2	3	4	5	6	7	8	9
1	Retainer	1		82	5.10	49	10.5	15.6	0
2	Rear screen	1		12	2.25	02	2.5	4.75	1
3	Spring seat sleeve	1		02	1.88	0	1.5	3.38	0
4	Spring seat	1		89	7.00	23	7.5	14.50	0
5	Rear body	1		11	1.80	96	12.0	13.80	1
6	Spring	1		2	1.88	1	2.5	4.38	1
7	Poppet	1		12	2.25	0	1.5	3.75	1
8	Front body	1		11	1.80	0	1.5	3.30	1
9	Front screen	1		12	2.25	49	10.5	12.75	1
10	Nose seal	1		16	4.80	1	2.5	7.30	1
11	Barrel screen	1		02	1.88	96	12.0	13.88	1
									8

$T_m = 97.39$
$N_m = 11$
Design efficiency index = 0.246

TABLE 5.17 DFM Chart for Initial Valve Data

Number	Part Description	Number of Identical Operations	$\alpha + \beta$	Manual Handling Code	Manual Handling Time	Manual Insertion Code	Manual Insertion Time	Total Assembly Time $(2) \times [(5) + (7)] = T_m$	Minimum Number of Parts (N_m)
0	1	2	3	4	5	6	7	8	9
1	Rear screen	01		12	2.25	90	4.0	6.25	01
2	Rear body	01		11	1.80		4.0	5.80	01
3	Spring	01		02	1.88	01	2.5	4.38	01
4	Poppet	01		12	2.25	00	1.5	3.75	01
5	Nose seal	01		16	4.80	01	2.5	7.30	01
6	Barrel screen	01		02	1.88	96	12.0	13.88	01
7	Front screen	01		12	2.25	90	4.0	6.25	01
8	Body front	01		11	1.80	00	1.5	3.30	01

$T_m = 50.91$
$N_m = 8$
$T_o = 8$
Design efficiency index = 0.471

TABLE 5.18 DFM Chart for Redesigned Valve

195

DFA Metric	Initial	Redesign	Improvement (percent)
Design efficiency index	0.24	0.47	97%
Total assembly time (seconds)	97	51	38
Total number of parts	11	8	27
Theoretical minimum number of parts	8	8	
Necessary parts (percent)	73	100	27

TABLE 5.19 Comparison of two designs

On assembly of two bodies, the windows must line up perfectly to avoid blocking the discharge port. This operation adds additional assembly movement because the α and β angles in the redesign would be different and the components would require manipulation. During the assembly process, an additional feature is created in the rear and front of the body for perfect alignment of the windows.

EXAMPLE 5.4

Mechanical Hand Press

Presses compress and shape components by exerting a high pressure on them. The screw press rotates a screw spindle in a fixed nut, transmitting a longitudinal force through the spindle to the work piece. On large presses, the upper end of the screw spindle has a large fly wheel that contains a large reserve of stored energy when rotating. Mechanical presses use various drive systems. Consider the example of a mechanical press to illustrate the use of this DFA methodology.

The assembly layout of the mechanical press is shown in **Figure 5.9**. The present design uses a rack and pinion combination and a long lever arm. It has 13 components and 17 parts in total such as base, column, table, and so on. The design uses six components that require screwing. The components of the press and their dimensions are given in **Table 5.20**. The press assembly involves securing a series of components, including a column, to its base and connecting a handle to the column. The data for the completed analysis of the tabulated list of operations and the related handling and insertion times and assembly times are given in **Table 5.21**. Each assembly operation is divided into handling and insertion, and the corresponding codes for each process are given.

Step-by-Step Assembly Operation

The assembly operation starts by placing the base on an assembly table and attaching the parts to the base. As an example, consider the process of handling the column (shown as part 2) and attaching it to the base of the press.

Handling code: The insertion axis for the column is vertical in the direction of beta symmetry. The symmetry for the column for alpha is 360° and is the same for beta. Thus,

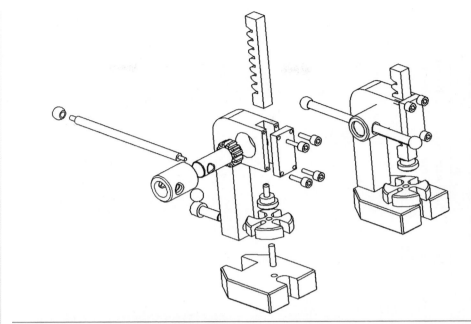

FIGURE 5.9 Assembly View of the Mechanical Press

Number	Item	Qty	Length (mms)	Width (mms)	Thickness (mms)	Diameter (mms)
1	Base	1	85	60	25	
2	Column	1	150	75	25.0	
3	Machine screw	1	25			6.00
4	Table	1			50.0	12.00
5	Table pin	1	50			6.00
6	Sleeve	1	40			27.00
7	Handle	1	120			40.00
8	Ball end	2				15.00
9	Gear	1			50.0	27.00
10	Rack	1	110	12	14.0	
11	Cover plate	1	45	25	0.4	
12	Cap screw	4	15			6.00
13	Rack pad	1			25.0	0.90

TABLE 5.20 Components of the Mechanical Press

Number	1 Part Description	2 Number of Items	3 α + β Symmetry	4 Manual Handling Code	5 Manual Handling Time	6 Manual Insertion Code	7 Manual Insertion Time	8 Total Assembly Time (2) × [(5) + (7)] = T_m	9 Theoretical Minimum Number of Parts, N_m
1	Base	1	540	91	3	00	1.5	4.50	1
2	Column	1	720	30 Easy to align	1.95	09 Holding down; Not easy to align	7.5	9.45	0
3	Machine screw	1	360	18	3	02	2.5	5.50	0
4	Fastening	Operation.	–			92	5.0	5.00	
5	Table	1	180	00	1.13	02	2.5	3.63	1
6	Table pin	1	180	00	1.13	00	1.5	2.63	0
7	Sleeve	1	540	20	1.80	31	5.0	6.80	1
8	Handle	1	180	00	1.13	00	1.5	2.63	1
9	Ball end	1	180	01	1.43	00	1.5	2.93	0
10	Gear	1	540	20	1.80	00	1.5	3.30	1
11	Rack	1	720	30	1.95	09	7.5	9.45	1
12	Cover plate	1	540	20	1.80	08	6.5	8.30	1
13	Cap screw	4	360	10	1.50	00	1.5	12.00	0
14	Fastening	Operation	—			92	5.0	5.00	
15	Rack pad	1	360	10	1.50	31	5.0	6.50	1

Total 13
$T_m = 87.62$
$N_m = 8$

TABLE 5.21 Compilation of Assembly Time for the Mechanical Press

the total angle of symmetry is 720°. **Table 5.21** provides the database for the column's handling time. The column is handled and manipulated with one hand without the aid of tools. Because the total angle of alpha and beta is 720°, the first digit of the handling code is 3. The column presents no handling difficulties and can be separated from the bulk easily. Its thickness is >2mm and size is >15mm; therefore, the second digit is 0, giving a handling code of 30.

Handling time: A handling time of 1.95 seconds corresponds to the handling code of 30 (**Table 5.21**).

Insertion code: The column is not secured on insertion. It is fastened in the next operation. Because there is no vision restriction, the first digit of the insertion code is 0. Holding down is necessary while the subsequent operations are performed; alignment of the column is not easy and must be completed with screw holes. There is no resistance to insertion. Therefore, the second digit of insertion code is 9.

Insertion time: An insertion of 7.50 seconds corresponds to the insertion code of 09 (**Table 5.21**).

Total operation time: This is the sum of the handling and insertion times multiplied by the number of items. For the column, the total operation time is 9.45 seconds. The cost of assembly depends on the hourly operator cost and any overhead cost of the organization and varies from region to region.

Minimum number of parts: Identification of the theoretical minimum number of parts is a way to determine whether the specific part might be eliminated.
The process is as follows:

1. During operation of the product, the column does not move relative to any other part already assembled.
2. The column cannot be made of a different material than or physically isolated from any other part already assembled.
3. The column need not be separate from all other parts already assembled to facilitate assembly and disassembly. Because "no" is the answer to all of these questions, a 0 is placed in Column 9.

The remaining parts are added one by one to the assembly and the manual assembly analysis (**Table 5.22**) is completed for each additional part. The total number of parts is 13, and there are two operations (items 4 and 14). The operations have no associated handling time and are considered only as a separate insertion operation. The total assembly time is 87.62 seconds. The theoretical minimum number of parts is 8; they are essential and cannot be combined or eliminated. The assembly design efficiency index for manual assembly is obtained using this equation:

$$E = 3N_{min}/T_m$$
$$E = (3)(8)/87.62$$
$$E = 0.27$$

Number	Part Description	Number of Items	Manual Handling Time	Manual Insertion Time	Total Assembly Time $(2) \times [(5) + (7)] = T_m$	Theoretical Minimum Number of Parts, N_m
1	Body	1	3	1.5	4.5	1
2	Table	1	2.5	1.5	4.1	1
3	Application of grease	1	3.0		7.0	
4	Rack	1	3.0	2.6	5.6	1
5	Rotor assembly	1	4.8	3.0	7.8	4
6	Align screw	1	2.55	7.5	12.9	1

Total
$T_m = 41.9$
$N_m = 8$

TABLE 5.22 Assembly Time for the Mechanical Press

Original	Redesign	Improvement
Total number of parts	13.00	8.0
Assembly time	87.62	41.9
Design efficiency index	0.27	0.6

	38%
	51%

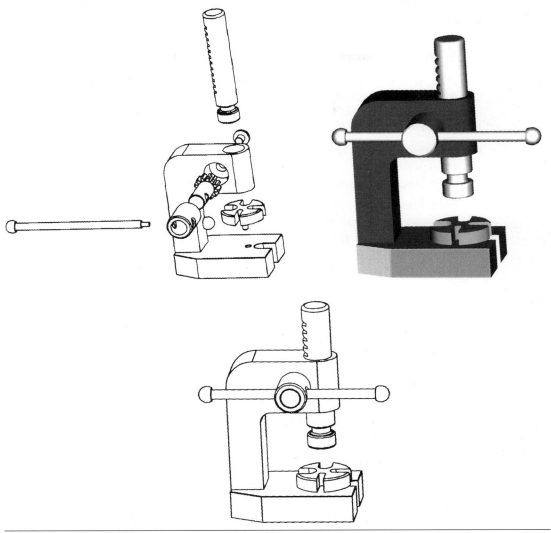

FIGURE 5.10 Redesigned Mechanical Press

Some components used in this design can be eliminated because they lack functional purpose. In the redesign, the rack and pinion combination is still being used, but there are only 8 components. The column and cover plate are combined into one single machinable part. The subassembly rotary set includes the sleeve, gear, handle, and ball end. **Figure 5.10** and **Table 5.22** provide the design details and assembly data for the redesigned press.

5.3 **Additional Assembly Evaluation Methods**

5.3.1 **Hitachi Method**

The Hitachi method, also known as the *assembly evaluation method (AEM)*, is used to assess the manufacturability of a product. Manufacturability to a large extent depends on the design and material, processing, and indirect costs. Some of the features of Hitachi methodology are to:

- Provide for comparison of designs and consideration of advantages among concept designs
- Rank concept designs and compare them to competitors' products
- Rank product in terms of assemblability
- Facilitate product design improvements
- Identify key points that need improvement
- Estimate effects of improvement
- Estimate assembly cost

The objectives of the assembly evaluation method can be reached by using the steps in **Figure 5.11**.

The assembly evaluation is generally completed using the product design drawings. However, the evaluation of the conceptual design is also a part of process improvement. Design improvement is then performed based on data obtained by reviewing the

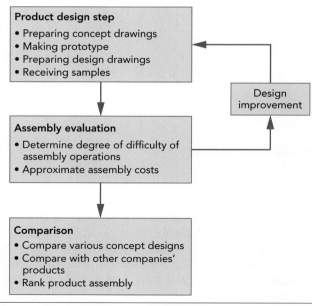

FIGURE 5.11 Steps of Assembly Evaluation Method

evaluation results. The design after improvement is again subject to the assemblability evaluation process to identify the effects of the improvements. It is important to point out that AEM does not distinguish between manual and automatic assembly; this is based on the belief that a strong correlation exists between the degree of manual assembly difficulty and that of automatic assembly. The evaluation procedure consists of the steps in **Figure 5.12**.

▸ **Step 1. Preparation**
This step begins by collecting drawings, conceptual and completed samples, and so on. If more precise drawings and data are available, the evaluation results will be more accurate.

▸ **Step 2. Attachment**
After the attachment sequence has been determined, the names and numbers of parts with corresponding symbols are entered on the evaluation forms in the same order as the assembly sequence.

▸ **Step 3. Calculation of the Evaluation Index**
Simple calculations are used to calculate the evaluation indices (parts and product assemblability evaluation scores).

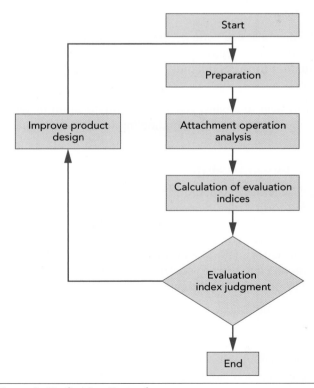

FIGURE 5.12 Steps in Evaluation Procedure

▸ **Step 4. Judgment of Evaluation Index**
The assemblability evaluation indices are compared to the target values. Lower assembly costs are obtained when the product assemblability evaluation score is higher than the target value. Higher scores indicate possible opportunities for assembly automation. If the product assemblability evaluation score is lower than the target value, product improvement is needed.

▸ **Step 5. Improvement of Product Design**
The improvement process consists of:

- Identifying weak points and the causes of low scores
- Creating alternative designs that will eliminate weak points
- Evaluating the effects of improvement by comparing the assembly evaluation scores of the improved product to those of the original product

Advantages of using AEM follow:

Reduction in assembly labor: Easy identification of weak design points enables rapid improvement, resulting in a reduction in manual assembly time.

Early decision making: The evaluation and improvement of the product can be reached at an early stage of design, resulting in simpler production and assembly operations.

Reduction in design period: The time for design is significantly shortened because the evaluation is achieved during the early stages. The designer should perform the cost evaluation for assemblability; doing this contributes greatly to reducing the design process.

Improvement in product reliability: Simplification of parts production and assembly operations and use of automation improve product reliability.

5.3.2 Lucas Design Method

The Lucas design method enables a designer to identify any nonfunctional and difficult-to-fit design elements that should be scrutinized before finalizing the design. It offers the ability to change a system very rapidly between batches of different products. The technique highlights the use of nonessential elements resulting in the reduction of parts, inventory, and assembly time. It also can lead to suggestions for simple handling and tooling and ways to control production costs.

The objectives of the Lucas methodology are to:

- Use standard parts within the product and across its range to maximize tooling and minimize variety.
- Assemble parts from the same direction and in the same sequence, eliminating the need for duplicate tooling.
- Use common handling and feeding features on large components to minimize the need for specific handling tools.

The Lucas design process has the following steps:

1. Product design specification
2. Evaluation procedure
   - Functional analysis
   - Handling analysis
   - Fitting analysis
3. Redesign

‣ **Step 1. Product Design Specification (PDS)**
The design specification is a crucial document in the analysis. It includes all requirements, including customer and business data that the product must satisfy to be successful. A well-researched PDS provides a solution for frequently conflicting requirements (customer needs and component functionality). PDS is considered a reference point for an emerging design in which every component must be present for a specific purpose and that purpose must be in the form of a specification.

A major factor in the Lucas procedure is the determination of whether the product has a unique design or whether opportunities for rationalization and standardization of parts and procedures are possible. The Lucas methodology becomes very efficient if the organization can establish a set of *product family themes* in which identical components are used across a range of products. The product family theme enables grouping products to create an overall demand high enough for automated production and assembly. Assembly system designs can avoid obsolescence by providing the ability to assemble new products designed in the original product profile.

‣ **Step 2. Evaluation Procedure**
Evaluation consists of three analytical steps (**Figure 5.13**):

1. Functional analysis
2. Handling analysis (manual or automatic)
3. Fitting analysis (manual or automatic)

Functional analysis: The first element of the design iteration procedure, functional analysis is repeated until a satisfactory level of "design efficiency" is reached. Using this technique, one can evaluate the alternatives and direct the design team to make the best choice. Functional analysis may be undertaken in the early design stage. Every component must be itemized by name and number in a logical sequence for product assembly. Functional analysis is carried out in four steps:

1. Determine the product's functional requirements.
2. Decide whether the product can be considered as a whole or as a series of functional subsections. Consideration of the product as a whole is preferred to avoid the duplication of parts that could be in adjacent subsections.
3. Divide components into two categories.
   - A: Functional components, such as drive shafts, that are vital to the product's performance
   - B: Nonfunctional components, such as fasteners or locator, that are not critical to the function

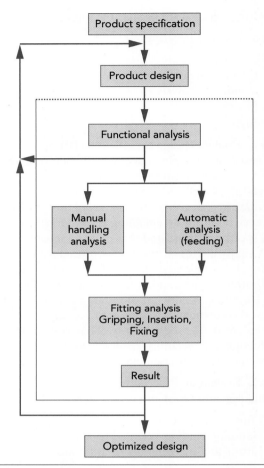

FIGURE 5.13 DFA Evaluation Procedure

4. Categorize mating components in a logical progression until every component has been considered.

Determination of **design efficiency (E)** is used to assess the product design's functionality by using the following formula:

$$E = \frac{A_N}{C_t} \times 100\%$$

where

A_N = number of A components

C_t = total number of components

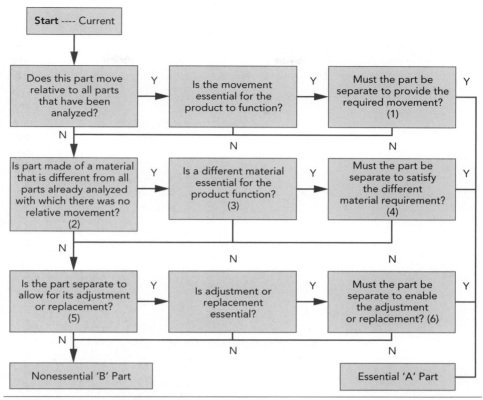

FIGURE 5.14 Function Analysis Chart

The Lucas method suggests that design efficiency must be at least 60% to make a good new product. While performing the functional analysis, it is important to assess what the design *is*, not what it *should be*. A significant factor should be classified as **B** if there is any doubt of a component's category.

The goal of this analysis is to determine which components are necessary for the product to function (classification A) and which are theoretically nonessential parts (classification B). There is little point in simplifying the assembly if the manufacture of redesigned components will not result in any savings (**Figure 5.14**).

Handling analysis: As shown in **Figure 5.15**, this analysis helps the designer to assess whether the product design with a satisfactory level of functionality is acceptable for assembly. The next important step in the analysis is to determine how to assemble components and subassemblies. There are two modes to be considered: manual handling and automated feeding. In the case of manual assembly, the manual handling index must be calculated, and in the case of automated assembly, the automated feeding index (see the following equation) must be determined. If the index does not meet the expectation, a redesign should be considered for a possible index reduction.

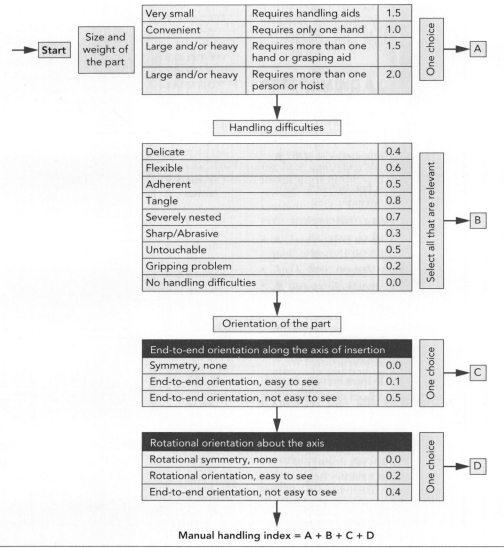

FIGURE 5.15 Handling Analysis

Manual handling: The manual handling index analysis is based on

- Size and weight of the part
- Difficulties handling the part
- Orientation of the part

A less complex process is used for the manual assembly than for the automated feeding process.

Automated feeding: For an automated feeding assembly, the calculation of the relative handling cost is the same as the calculation of the feeding ratio. A useful measure of overall effectiveness of product design for automated feeding is the feeding ratio. Automatically assembled parts are subjected to a three-step analysis for obtaining the feeding ratio:

- Determine whether the components are best transported in a retained orientation or in the form of bulk supply that is reoriented at the input point.
- Assess the general physical properties of the components that will not be transported with retained orientation.
- Examine the suitability of a detailed design of the components proposed for automatic feeding.

Feeding indices: The feeding ratio is expressed as

$$\text{Feeding ratio} = \frac{F_C}{A_N}$$

where

F_C = total relative feeding cost

A_N = number of A parts

The total relative feeding cost can be obtained by summing all individual feeding indices A, B, and C. Stage A indices provide information on parts not suitable for mechanical orientation (**Figure 5.16**). Stage B provides information on parts that can be mechanically oriented and those with end-to-end orientations. Stage C provides information on parts with rotational orientations. Large numbers of experiments have indicated that the feeding ratio for an acceptable design is generally less than 2.5.

Fitting analysis: This analysis follows the handling analysis and identifies the type of gripping, insertion, and fixing operations that are required. Each operation is rated, and the entire assembly task produces a fitting ratio. The analysis is primarily intended for automated assembly, but if manual assembly is required, the relative manual rating is used. The fitting process analysis determines the relative ease or difficulty in performing each assembly task to complete the product from its constituent parts. This determination varies, depending on whether the process is a manual or automated method, which will be reflected in the respective costs. Individual index values, experimentally determined as 1.5 or higher, indicate the presence of a fitting problem.

The fitting process analysis determines the appropriate surface that enables a component to be carried at the required gripping force. A *surface* is said to be available when it is possible for the component to be assembled satisfactorily without the gripper obstructing the insertion process. The characteristics of a part (center of mass, gripping area, and so on) include the difficulty in holding it securely during transport accelerations and decelerations.

The gripping assessment examines how secure each part is for transportation from the point of presentation within the automatic assembly system to the completed

Start

Stage 1

Parts not suitable for mechanical orientation

If more than one property is present, select the largest value

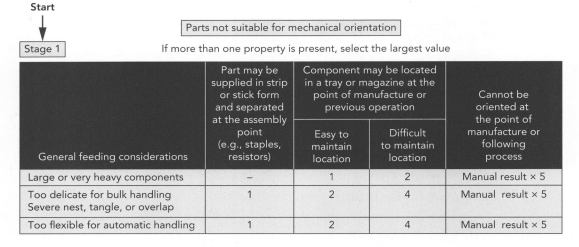

General feeding considerations	Part may be supplied in strip or stick form and separated at the assembly point (e.g., staples, resistors)	Component may be located in a tray or magazine at the point of manufacture or previous operation		Cannot be oriented at the point of manufacture or following process
		Easy to maintain location	Difficult to maintain location	
Large or very heavy components	–	1	2	Manual result × 5
Too delicate for bulk handling Severe nest, tangle, or overlap	1	2	4	Manual result × 5
Too flexible for automatic handling	1	2	4	Manual result × 5

If the component has any of these properties, exit with the result ⟶

If the component has none of these properties, go to stage 2.

Stage 2

Parts that can be mechanically oriented

A

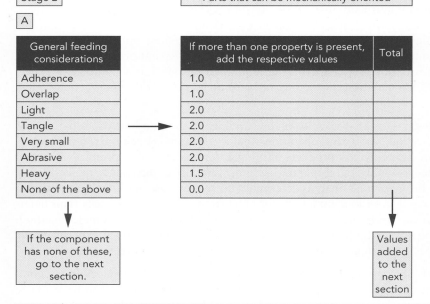

General feeding considerations	If more than one property is present, add the respective values	Total
Adherence	1.0	
Overlap	1.0	
Light	2.0	
Tangle	2.0	
Very small	2.0	
Abrasive	2.0	
Heavy	1.5	
None of the above	0.0	

If the component has none of these, go to the next section.

Values added to the next section

FIGURE 5.16 Automatic Feeding Analysis Chart (*continued*)

B		End-to end orientation				
Directional orientation of major axis	End-to-end orientation not required	Can be fed in a slot supported by its head	External features	Internal features		Nongeometric features too small for mechanical orientation
L/I >5	MT 0.5	MT 0.7	MT 1.6	MT 1.6		LT 3
L/I <5	MT 0.5	MT 0.7	MT 1.0	Manual handling × 5		LT 3

C		Rotational side-to-side orientation			
		Rotational side-to-side orientation not required	Features seen in end view/silhouette	Features not seen in end view/silhouette	Nongeometric features or features too small for mechanical orientation
Poor stability at major axis	L/I >5	MT 0.5	MT 0.8	MT 1.5	Manual Handling × 5 (A and B ignored)
	L/I <5	MT 0.5	MT 0.9	MT 0.8	Manual Handling × 5 (A and B ignored)
Rotational stability at major axis	L/I >5	MT 0.5	MT 0.8	MT 1.2	LT 3 (ignore B)
	L/I <5	MT 0.5	MT 0.9	MT 1.2	LT 3 (ignore B)

MT = mechanical tooling
LT = laser tooling
M = manual orientation (3)
L = longest dimension
I = intermediate dimension

Feeding index = A + B + C

FIGURE 5.16 Automatic Feeding Analysis Chart

insertion stage. This applies primarily to automated assembly applications. The fitting process analysis examines the overall effectiveness of the product design. In nonassembly situations, the procedure considers any individual nonassembly process and identifies a relative cost that contributes to the total assembly cost. Examples of such processes are tightening pre-placed screws, welding, or using adhesive bonding for parts. The fitting ratio (F_r) is calculated as

$$F_r = \frac{G_c + I_i + N_i}{A_n}$$

where

F_r = fitting ratio
G_c = gripping cost index
I_i = insertion and fixing cost index
N_i = nonassembly cost index
A_n = number of A parts

For an acceptable design, the fitting ratio is generally <2.5. The goal is to minimize this factor. The gripping cost index is primarily based on the availability of a suitable grip component (**Figure 5.17**).

PROCEDURE				
Component has appropriate gripping surface				Component has no suitable gripping surface
Surface is available during the insertion process		Surface is not available during the insertion process		
Component is easy to grip securely enough during transport	Component is not easy to grip securely enough during transport	Component is easy to grip securely enough during transport	Component is not easy to grip securely enough during transport	
Index 0	0.5	1	1.5	2.5

FIGURE 5.17 Gripping Cost Index Analysis

The fitting ratio is a measure of the effectiveness of the product design. For a good design, the ratio is <2.5, although the goal is to minimize this factor. The fitting ratio is based on the gripping, insertion, and nonassembly costs. After analyzing the product, certain aspects of the design, part feeding, and assembly are re-examined for more evaluation. Attention is given to the areas of efficiency in feeding components. All B category parts should be eliminated or combined with A category components. During the concept design stage, the emphasis should be on increasing the design efficiency rating by considering the suitability of the component part design for handling and feeding. The result of the redesign should be consistent with gripping provisions if automation is to be applied. In all events, the product redesign stage must consider the task of actually assembling the parts into their final position.

Functional Analysis of an Electric Motor

The section addresses the functional analysis of an electric motor according to the Lucas method, including analysis of functional handling and fitting, as outlined in the DFA evaluation procedure (**Figure 5.13**). The charts for functional, handling, and fitting analysis are drawn. A function analysis chart (**Figure 5.14**) is used to identify essential parts (A) and nonessential parts (B). The ratio of the number of A components to the total number of components provides the design efficiency. **Table 5.23** is a list of 16 components of which 9 are classified as A parts. The design efficiency in functional analysis is identified as 56%.

The design efficiency, manual handling, and fitting ratios are calculated for the initial and improved electric motor designs (**Tables 5.23, 5.24, 5.25, 5.26, 5.27,** and **5.28**).

Design efficiency = Number of A parts/Number of total parts = 9/16 (56%)

Analysis of handling an electric motor: As outlined in **Figure 5.15**, the handling analysis takes care of the part's size, weight (A), handling difficulties (B), and orientation (C and D). The handling index is the sum of A, B, C, and D. **Table 5.24** shows individual as well as cumulative handling difficulties for all 16 parts of an electric motor.

1	Terminal box	Connects to power source	A
2	Blower	Cools the motor	A
3	Bracket	Holds blower and terminal box	B
4	Stiffener	Stiffens magnet	B
5	Brake linkage	Transmits braking force	A
6	Magnet	Creates braking	A
7	Magnet support	Provides support	B
8	Bearing stand	Forces from rotor	A
9	Sheave	Provides motion and traction	A
10	Rotor	Enables torque creation	A
11	Adapter	Provides grip between stator and frame	B
12	Stator	Provides magnet circuit creation	A
13	Rear cover	Supports end bearing	A
14	Right stabilizer	Stabilizes motor	B
15	Left stabilizer	Stabilizes motor	B
16	Frame	Grips to stator and stand	B

TABLE 5.23 Function Analysis Chart for an Electric Motor

Number	Part Description	Size and Weight	Handling Difficulties	Orientation	Manual Handling Index
1	Terminal box	1.5	0.4	0.4	2.3
2	Blower	1.5	0.4	0.1	2.0
3	Bracket	1.5	0.0	0.1	1.6
4	Stiffener	1.5	0.0	0.1	1.6
5	Brake linkage	2.0	0.2	0.1	2.3
6	Magnet	2.0	0.4	0.1	2.5
7	Magnet support	2.0	0.0	0.1	2.1
8	Bearing stand	2.0	0.2	0.1	2.3
9	Sheave	2.0	0.2	0.1	2.3
10	Rotor	2.0	0.4	0.1	2.5
11	Adapter	2.0	0.0	0.5	2.5
12	Stator	2.0	0.4	0.7	3.1

(continued)

Number	Part Description	Size and Weight	Handling Difficulties	Orientation	Manual Handling Index
13	Rear cover	2.0	0.0	0.3	2.3
14	Right stabilizer	1.5	0.0	0.1	1.6
15	Left stabilizer	1.5	0.0	0.1	1.6
16	Frame	2	0.0	0.1	2.1

Total = 34.7
Manual handling index = 34.7

TABLE 5.24 Handling Analysis of an Electric Motor

Number	Part Description	A	B	C	D	E	F	Insertion and Fixing (Fitting Index)	Non-assembly Index
1	Terminal box	2	0	0	0	0.7	0	2.7	5.5
2	Blower	2	0	0	0	0	0	2.0	5.5
3	Bracket	2	0	0	1.5	0.7	0	4.2	5.5
4	Stiffener	2	0.1	0	0	0.7	0	2.8	5.5
5	Brake linkage	2	0.1	0.7	0	0	0	2.8	5.5
6	Magnet	2	0	0	0	0	0	2.0	4.0
7	Magnet support	2	0	0	0	0	0	2.0	4.0
8	Bearing stand	1	0.1	0	1.5	0.7	0	3.3	5.5
9	Sheave	2	0.1	0	1.5	0.7	0	3.3	1.5
10	Rotor	1	0.1	0	1.5	0.7	0	4.3	5.5
11	Adapter	2	0.1	0	0	0.7	0	2.8	4.0
12	Stator	2	0.1	0	1.5	0.7	0	4.3	4.0
13	Rear cover	2	0.1	0	0	0.7	0	2.8	5.5
14	Right stabilizer	2	0.1	0	0	0.7	0	2.8	5.5
15	Left stabilizer	2	0.1	0	0	0.7	0	2.8	5.5
16	Frame	1	0	0	0	0	0	1.0	0
								Total = 45.9	72.5

Insertion and fixing index = 45.9
Nonassembly index = 72.5
Gripping index = 0
FR = Gripping cost index + Insertion and fixing cost index + Nonassembly cost index
Number of A parts

TABLE 5.25 Fitting Ratio Analysis Table for an Electric Motor

Number	Part Description		Rating
1	Terminal box	Connects to power source	A
2	Blower	Cools motor	A
3	Brake linkage	Transmits braking force	A
4	Magnet	Creates braking	A
5	Bearing stand	Withstands forces from rotor	A
6	Sheave	Provides motion & traction	A
7	Rotor	Enables torque creation	A
8	Front cover	Takes rotor force	A
9	Rear cover	Supports end bearing	A
10	Stator	Creates magnet circuit	A

Design efficiency = Number of A parts/Number of total parts
= 10/10
= 100%

TABLE 5.26 Functional Analysis of Redesigned Electric Motor

Number	Part Description	Size and Weight	Handling Difficulties	Orientation	Manual Handling Index
1	Terminal box	1.5	0.4	0.4	2.3
2	Blower	1.5	0.4	0.1	2.0
3	Brake linkage	2.0	0.2	0.1	2.3
4	Magnet	2.0	0.4	0.1	2.5
5	Bearing stand	2.0	0.2	0.1	2.3
6	Sheave	2.0	0.2	0.1	2.3
7	Rotor	2.0	0.4	0.1	2.5
8	Front cover	2.0	0	0.5	2.5
9	Rear cover	2.0	0	0.3	2.3
10	Stator	2.0	0.4	0.7	3.1

Total = 23.9
Manual handling index = 23.9

TABLE 5.27 Handling Analysis of Redesigned Electric Motor

Number	Part Description	A	B	C	D	E	F	Fitting Index	Nonassembly Index
1	Terminal box	2	0	0	0	0.7	0	2.7	5.5
2	Blower	2	0	0	0	0	0	2.0	5.5
3	Brake linkage	2	0.1	0.7	0	0	0	2.8	5.5
4	Magnet	2	0	0	0	0	0	2.0	4.0
5	Bearing stand	1	0.1	0	1.5	0.7	0	3.3	5.5
6	Sheave	2	0.1	0	1.5	0.7	0	3.3	1.5
7	Rotor	1	0.1	0	1.5	0.7	0	4.3	5.5
8	Front cover	2	0.1	0	0	0.7	0	2.8	5.5
9	Rear cover	2	0.1	0	0	0.7	0	2.8	5.5
10	Stator	2	0.1	0	1.5	0.7	0	4.3	4.0
							Total =	30.3	48.0

Insertion and fixing index = 30.3
Nonassembly index = 48.0
Gripping index = 0
Fitting ration = 7.83

TABLE 5.28 Fitting Ratio Analysis of Redesigned Electric Motor

Fitting ratio analysis of an electric motor: As in **Figure 5.18** for the electric motor, the fitting analysis takes care of assembly directions, part placing, part fastening, aligning, and inserting possibilities. The nature of the fitting process (A), process direction (B), process volume (C), access to the process (D), alignment difficulties (E), and insertion difficulties (F) are identified. Refer to **Table 5.25** for the individual and cumulative values of insertion and fixing index for all 16 parts of an electric motor.

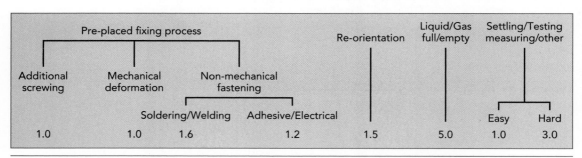

FIGURE 5.18 Nonassembly Processes Cost Index

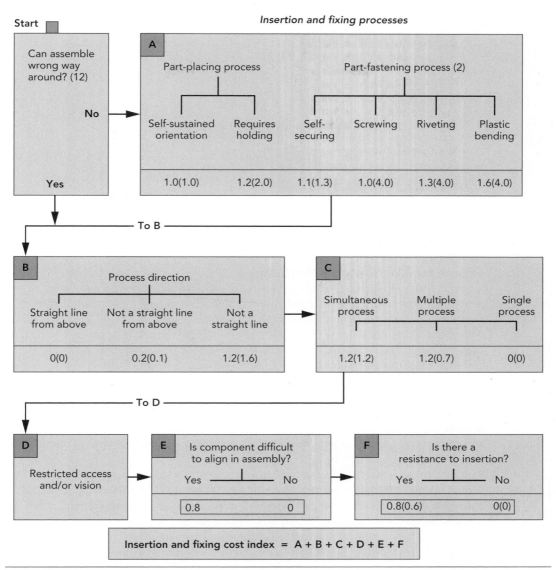

FIGURE 5.19 Insertion and Fixing

The fitting ratio calculates the sum of the cost index for gripping, insertion, fixing cost index, and nonassembly. In this case, the gripping cost index is zero because the component has an appropriate gripping surface that is available during insertion process (**Figure 5.17**).

When the gripping index is 0, the insertion and fixing index is 45.9 and the nonassembly index is 72.5. A components in this product are 9, and FR is calculated to be 13.15.

This example shows that the electric motor before redesign has the following data:

Number of subassemblies = 16
Design efficiency = 56%
Manual handling ratio = 34.7
Fitting ratio = 13.15

After the motor redesign, the following were observed.

Number of subassemblies = 10
Design efficiency = 100%
Manual handling ratio = 23.9
Fitting ratio = 7.83

Although the design efficiency has improved substantially, the manual handling ratio and fitting ratios have not improved proportionally.

Design for Manufacturing Methods

The preceding techniques provide a systematic and disciplined way to signify the importance of manufacturing and assembly in a designer's mind. They were developed early in the design stage to create a product that is easy to manufacture and assemble before expending much effort and cost in pursuing another design that might not be discovered until much later to be unnecessarily expensive to manufacture and assemble. In the product development process, the techniques provide the basis from which to develop an integrative prospective of design, manufacture, and assembly.

The methods discussed here except the axiomatic method apply mainly to mechanical assemblies of a size that can be conveniently joined on a desktop. Typical products include tape- and video recorders, car alternators, and water pumps. The methods are not appropriate for large products such as complete vehicles because of the lack of data for large products.

The Hitachi, BDM, and Lucas methods are supported by computer software systems that provide step-by-step instructions for their use. The software's advantage is the aid it gives the user during the evaluation procedure by providing help screens in context and by conveniently documenting the analysis. The user can quickly assess the effect of a proposed design change by editing a current analysis.

BDM provides the most complete calculation of design efficiency by considering part reductions and improving handling and insertion. The Hitachi method calculates the design efficiency based on the insertion process only, whereas the Lucas method focuses on the reduction of part numbers.

On the other hand, the Hitachi technique gives a process overview of the assembly sequence and insertion operations. There is no explicit criterion for a minimum part count. The Hitachi method does not offer direct analysis for parts feeding and orientation. For this reason, the design for automated assembly is not an option because its assessment is sensitive to part configuration and is difficult to handle precisely at the early design stages. These aspects should be dealt with at later design stages.

The Lucas method is based on a symbolic logic programming paradigm. Its advantage is that it is easier to encode and derive the DFA rules embodied in the method and provide the user generalized suggestions for possible design changes as the evaluation proceeds. The Lucas method adopts aspects of both Hitachi and BDM by dealing with handling and insertion with some consideration of automation and some emphasis on the fitting (insertion) processes. It provides a good overview of the assembly process. The design efficiency of the Lucas method is based solely on reducing the number of parts in the product design and is not as comprehensive as the BDM.

CASE 5.1
Portable Coffeemaker: Design and Evaluation

Conceptual design: A conceptual design of the portable coffeemaker is made keeping in mind the subfunctions listed in the morphological chart according to the voice of the customer.

The main elements of the design are a water compartment, threaded lid, heating unit, ground reservoir with a permanent filter, and mug as shown in **Figure 5.20**. All of these components are joined except for the water reservoir that is attached to the heating unit with a snap-fit mechanism. An O-ring is placed in between the two components

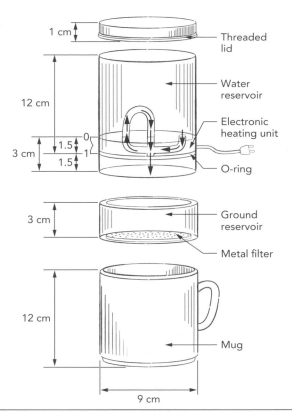

FIGURE 5.20 Conceptual Sketch of Portable Coffeemaker

to maximize the sealing function. Joining all of the components by a threaded fastener prevents leaks and minimizes the risk of hazard (importance ranking, 5). The mug is designed to be detachable to meet the need for interchangeability and size. The ability to disassemble the parts facilitates maximum convenience for transportation (importance ranking, 4). As for parts modularity, the removable filter allows for easy maintenance. The drip-brew mechanism offers good coffee quality. The number, shape, and size of components influence manufacturing costs to make the price appealing to the consumer (importance ranking, 4).

Principle of operation: The portable coffeemaker requires water, ground coffee, and power source. Its step-by-step procedure is given below:

1. Disassemble the ground reservoir from the top part of the machine.
2. Place the filter in place.
3. Fill the filter with coffee grounds.
4. Fill the water reservoir.
5. Connect the top section to the ground reservoir.
6. Replace the lid.

When the coffeemaker is powered, the heating unit is activated, and cold water is conveyed by a tube at the bottom as shown in **Figure 5.21**. As the water comes in touch with the heating element, it boils and is pushed up through a one-way valve into a tube that goes up to the water reservoir before being released into the ground reservoir. This mechanism also allows for preheating the water in the reservoir. The switch turns off when the coffee is ready in the container.

Design for manufacturability: The portable coffeemaker is designed to be cylindrical in shape (about 9 cm in diameter). The water reservoir is designed to hold a cup of 8 cm in diameter. The six components identified before (water reservoir lid, water reservoir,

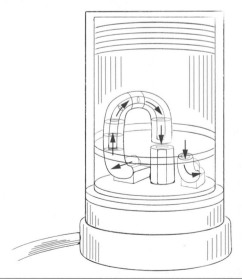

FIGURE 5.21 Detailed View of the Coffeemaker Heating Element

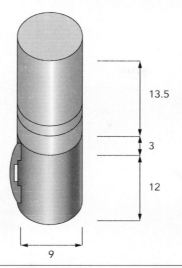

FIGURE 5.22 External View of Portable Coffeemaker Mug

electronic heating unit, ground reservoir, filter, and cup) are threaded to one another. **Figure 5.22** shows an exploded view. The overall length when the device is brewing coffee is 28.5 cm.

Specifications: Some components of the prototype are fabricated and some come off the shelf. The majority of the parts are built from FDA-approved plastic.

The engineering parameters are:

Overall volume for assembled coffeemaker = 2300 cm³

Water reservoir = 500 cm³

Ground reservoir volume = 40 cm³

Overall weight = 0.8 kg

Temperature of water when brewing = 93°C

Design for assembly: Figure 5.23 shows an exploded view of the coffeemaker subassemblies. See **Table 5.29** for dimension information and characteristics of each part and **Table 5.30** for the design for assembly analysis. The total assembly time for the current design, calculated as the sum of manual handling time and assembling time for each component, is 48.07 seconds.

Based on the analysis, the design efficiency could be calculated as follows:

$$\text{Design efficiency} = \frac{3 \times 8}{48.07} = 50\%$$

Redesign: To improve the design efficiency, consider some alternatives to the original design.

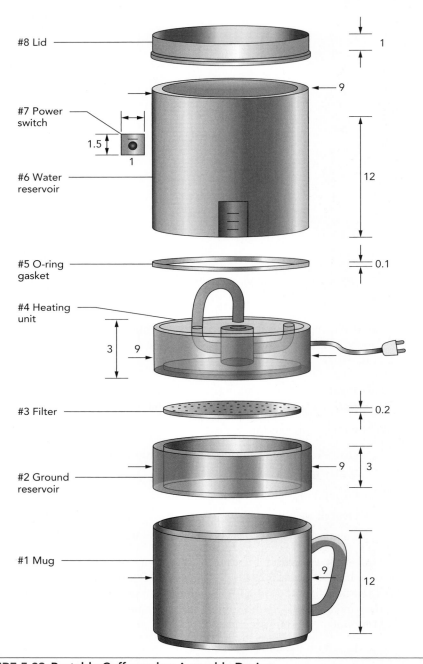

FIGURE 5.23 Portable Coffeemaker Assembly Design

Component Number	Component Name	Component Type	Dimensions	Quantity	Characteristics
1	Mug	Part	12 cm height, 9 cm diameter	1	✓ Added but not secured
2	Ground reservoir	Part	3 cm height, 9 cm diameter	1	✓ Secured
3	Metal filter	Part	8.5 cm height, 0.2 cm thickness	1	✓ Not secured
4	Electronic heating unit	Part	3 cm height, 9 cm diameter	1	✓ Secured ✓ Power cord tangles
5	O-ring gasket	Part	0.1 cm height, 9 cm diameter	1	✓ Not secured
6	Water reservoir	Part	12 cm height, 9 cm diameter	1	✓ Secured
7	Power switch	Part	1.5 cm height, 1 cm thickness	1	✓ Secured ✓ Not easy to align
8	Lid	Part	1 cm height, 9 cm diameter	1	✓ Secured

TABLE 5.29 Coffeemaker Subassembly Components and Their Characteristics

Use self-fastening parts. Snap mechanisms can be used instead of screw features to secure the ground reservoir to the mug, the electronic heating unit to the ground reservoir, and the lid to the water reservoir. This would save 3.5 seconds for each operation (2.5 seconds to snap the part versus 6 seconds to screw it). However, to prevent water/coffee spilling, O-rings should be assembled. This would increase the total number of parts and the total assembly time for each O-ring to approximately 4.5 seconds. A delta comparison shows that replacing screwing operations with a snap mechanism and an O-ring is not convenient because it calls for one more second of assembly time for each part for a total of three additional seconds.

Part count reduction: The filter and the ground reservoir could be unified in one part. This would eliminate the need to manage production for two separate parts during manufacturing and to handle them during assembly. Also, having just one part might be a more desirable feature so that the consumer must manage only one component instead of two. This will also prevent the consumer from worrying about losing the filter during coffeemaker transportation/use outside the home. On the other hand, a removable filter allows for easier manufacturing, which translates into cost and time savings. From the consumer point of view, a detachable filter might be easier to maintain and thus be more acceptable. Additional market studies might be conducted to assess the desirability of each option, and a cost/benefit analysis can provide the feasibility of options for the manufacturer.

Number	1 Part Description	2 Identical Operations	3 $\alpha + \beta$	4 Man. Handling Code	5 Man. Handling Time	6 Man. Insertion Code	7 Man. Insertion Time	8 Total Assembly Time $(2) \times [(5) + (7)] = T_m$	9 Theoretical Min Number of Parts
1	Mug	1	360	1-0	1.5	0-0	1.5	3.00	1
2	Ground reservoir	1	360	1-0	1.50	3-8	6.0	9.50	1
3	Metal filter	1	360	1-3	2.06	0-0	1.5	3.56	1
4	Electronic heating unit	1	360	1-0	1.50	3-8	6.0	7.50	1
5	O-ring gasket	1	540	2-3	2.36	0-2	2.5	4.86	1
6	Water reservoir	1	720	3-0	1.95	0-1	2.5	4.45	1
7	Power switch	1	720	3-2	2.70	3-1	5.0	7.70	1
8	Lid	1	360	1-0	1.5	3-8	6.0	7.50	1
		8						48.07	

TABLE 5.30 Coffeemaker Design for Assembly Analysis

Summary

The design steps of the portable coffeemaker case study has been discussed in Chapters 2, 3, and 5. The initial step identified the market needs. Studies have shown that the U.S. coffee market is growing, and there is a demand for low-cost, convenient devices for brewing coffee outside the homes. Chapter 3 addressed the statement-restatement techniques to define the problem. The Duncker diagram procedure was used to define a path to proceed to the desired state. The design proceeded through the steps of function diagram and morphological analysis. The best combinations of subassemblies were identified. Customer requirements for each subfunction were analyzed and were translated into engineering specifications. Based on these findings, a preliminary sketch and a conceptual drawing were made. As the concept was generated, the manufacturing and assembly studies were performed, identifying the steps involved in computing assembly efficiency. Additional techniques for product enhancement include the techniques of failure mode effect analysis to test the robustness of the design against potential hazards and failure modes.

REFERENCES

1. Barry Hyman, *Fundamentals of Engineering Design* (Upper Saddle River, NJ: Prentice Hall, 2002).
2. Stuart Pugh, *Total Design-Integrated Methods for Successful Product Engineering* (Boston, Addison-Wesley, 1991).
3. N. F. M. Roozenburg and J. Eekels, *Product Design: Fundamentals and Methods* (Chichester, U.K.: John Wiley and Sons, 1995).
4. Sherif El Wakil, *Processes and Design for Manufacturing*, 2nd ed. (Boston: PWS, 1998).
5. George E. Dieter and Linda C. Schmidt, *Engineering Design*, 5th ed. (New York: McGraw-Hill, 2012).
6. J. Corbert, M. Dooner, J. Meleka, and C. Pym, *Design for Manufacture, Strategies, Principles and Techniques* (Workingham, England: Addison-Wesley, 1991).
7. S. G. Shina, *Successful Implementation of Concurrent Engineering Products and Processes* (New York: Van Nostrand Reinhold, 1994).
8. N. P. Suh, *The Principles of Design* (New York: Oxford University Press, 1990).
9. S. Miyakawa and T. Ohashi, "The Hitachi Assemblability Evaluation Method (AEM)," *Proceedings, First International Conference on Product Design for Assembly* (April 1986).
10. G. Boothroyd and L. Alting, "Design for Assembly and Disassembly," *Annals of the CIRP* 41, no. 2 (1992), 625–36.
11. G. Boothroyd and P. Dewhurst, *Product Design for Assembly Handbook* (Wakefield, RI: Boothroyd Dewhurst, 1987).
12. J. Corbert, M. Dooner, J. Melaka and C. Pym, "Design for Assembly" (Hull, UK: Lucas Engineering & Systems Ltd., University of Hull, December 1990.
13. Edward Magrab, *Integrated Product and Process Design and Development* (Boca Raton, FL: CRC Press, 2010).
14. J. Usher, U. Roy, and H. Parsaei, *Integrated Product and Process Development* (Hoboken, NJ: John Wiley and Sons, 1998).

EXERCISES

5.1. List three reasons why reducing the number of parts in a product might reduce production costs. Also list reasons that costs might increase.

5.2. Consider the following 10 "design rules" for electromechanical products. Do these seem to be reasonable guidelines? Under what circumstances could one rule conflict with the other one? How should such a trade-off be settled?
   a. Minimize part count
   b. Use modular assembly
   c. Stack assemblies
   d. Eliminate adjustments

e. Eliminate cables
f. Use self-fastening parts
g. Use self-locating parts
h. Eliminate reorientation
i. Facilitate parts handling
j. Specify standard parts

5.3. Is it practical to design a product with 100% efficiency (DFA index = 1.0)? What conditions must be met to achieve this efficiency? Can you think of any products with very high (>75%) assembly efficiency?

5.4. What are the major characteristics of each of the following three assembly analysis methods: Boothroyd–Dewhurst (BDM), Lucas, and Hitachi?

5.5. The following five rules represent ideas for efficient design for manufacturing. Explain these ideas with an example of your own.
a. Facilitate easy handling and orientation
b. Standardize parts
c. Reduce part count
d. Eliminate assembly adjustments
e. Use self-locating screws

5.6. With the following information for a cartwheel assembly, perform a design for assembly analysis.

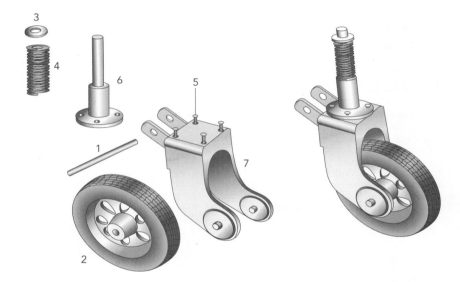

Component Number	Component Name	Component Type	Dimensions	Quantity
1	Axle	Part	170 length, 15 diameter	1
2	Tire	Part	250 diameter, 60 thickness	1
3	Front spring plate	Part	60 diameter, 5 thickness	1
4	Front spring	Part	135 length, 50 diameter	1
5	Hex bolt	Part	25 length, 18 diameter	4
6	Front pillar	Part	290 length, 120 diameter	1
7	Wheel bracket	Part	310 × 220 × 140	1

Recommended alternate designs (dimensions in mm)

5.7. Select a product with less than 15 parts; estimate the total product cost. What types of products are more suitable for direct marketing immediately after the design phase rather than after the conventional customer test with prototypes of products?

5.8. Briefly describe interrelationships, if any, among product architecture and its material selection, manufacturing-process selection, and life cycle maintenance and reliability. Give examples of products for which only one or two of the factors are dominant.

5.9. List five items that must be considered when designing a machined part to ensure efficient manufacturing and easy assembly. What are the considerations when designing a section for extrusion (a metal-forming operation)? What three questions form the criteria for eliminating a part from the assembly or for combining it?

5.10. Consider the assembly of a product.
 a. Why is standardization important?
 b. What suggestion should be offered if the designer cannot make the part exactly symmetrical?
 c. Can screws be considered an essential part of a product?
 d. What design attributes of LEGO blocks make them popular?
 e. Select a product from the following list and offer a proposal for its design by outlining a step-by-step design procedure.
  - Food processor
  - Wheelbarrow
  - Quadcopter
  - Steerable knee walker as an alternative to crutches

CHAPTER 6

Product Evaluation: Assessing Design for Disassembly and Maintenance

CHAPTER OUTLINE

OBJECTIVES

This chapter discusses factors concerning a product's disassembly and maintenance. The ease of disassembly influences its service and maintenance as well as the recycling of its parts. The principles of design for disassembly (DFD) identify the ease with which products can be assembled, maintained, serviced, and recycled. Applying DFD principles can reduce production, maintenance, and recycling costs. This chapter concludes by examining an automobile example, where the procedure for DFD, DFMA, and DFA are applied using a common platform.

6.1   Introduction: Product Design for Disassembly

Few products are designed for easy dismantling or disassembly. As manufacturers recognize their responsibility for products at the end of their operational lives, the dismantling of products has emerged as a serious consideration in manufacturing.

DFD is the proactive step incorporating disassembly characteristics into a product during its initial design. A product's design for ease of disassembly influences its service, maintenance, and recycling for parts and material. The intent is to save resources by prolonging a product's useful life. Disassembly is generally achieved by taking apart individual components or subassemblies of a product. The strategy is to use DFD guidelines at the early stage of product design.

Basically, there are two disassembly methods:

1. Reversible Disassembly This involves the removal of fasteners during disassembly and all screwed-in items.

2. Irreversible Disassembly Otherwise called *brute force disassembly*, this method separates parts depending on the complexity of the assembly process. This is necessary primarily because of the presence of soldered joints and welded parts. These parts could require special tools to separate them that are unlikely to be available at disassembly.

Two questions concerning joining must be answered, regardless of how it is done:

- Which method of disassembly will be used?
- Are the fastening points accessible?

The standardization of threaded fasteners can minimize the number of attachments required to disassemble a product. Metal inserts, if used in the design, create additional parts for disassembly. In this case, not only must the fastener be removed from the product, but also the metal insert must be removed. Certain critical assemblies may require the use of adhesives. Adhesives typically create more problems than they solve. Unless adhesives are water or solvent soluble, brute force is the only method of disassembly. Snap-on fixtures are the ideal fasteners. Snap-on connections require no additional parts for separation and can be removed quickly.

6.1.1 **Best Practices in Design for Disassembly**

While planning for DFD, designers should consider the requirements of shape, size, geometry, tooling, and nature of handling and manipulation. Some considerations are given below:

- Size and shape of parts
- Special tooling
- Handling and manipulation needs
- Expertise for disassembly
- Cleaning operations and requirements
- Product fragility
- Component reusability
- Disposal of liquids

It is critical that designers weigh many factors and options before going forward with a specific design. Specifically, designers must understand each part's basic purpose, where it is to be installed, and whether it can be reused. These factors include a focus on materials, fasteners, snap-on connections, and product architecture. **Tables 6.1** and **6.2** list the general guidelines for selecting materials and fasteners. **Table 6.3** shows guidelines for choosing product architecture.

Guideline	Rationale for Guideline
Minimize the number of different types of material.	Simplifies the recycling process.
Make subassemblies and connected parts from the same material.	Reduces the need for disassembly to improve assembly time.
Use materials that can be recycled.	Minimizes waste and increases the product's end-of-life value.
Use materials that have been recycled.	Stimulates the market for recycling.
Ensure standardization and easy identification of materials.	Maintains maximum value of recovered materials.

Based on Tracy Dowie & Matthew Simon, "Guidelines for designing for disassembly and recycling," WKD –ICED 1993, The Hague, The Netherlands, August 1993, pages 3-6, URL: http://sun1.mpce.stu.mmu.ac.uk/pages/projects/dfe/dfe.htm

TABLE 6.1 General DFD Guidelines and Rationale for Materials

Guideline	Rationale for Guideline
Minimize the number of fasteners and tools.	Impacts disassembly time.
Use fasteners that are easy to remove.	Reduces disassembly time.
Use fastening points for access.	Avoids slow awkward movements during disassembly.
Correctly locate and dismantle snap fits.	Availability of special purpose tools minimizes disassembly.
Use fasteners of material compatible with that of parts.	Assess ease of assembly and disassembly.
Eliminate adhesives unless they are compatible with parts joined.	Avoids contamination of materials
Minimize the number and length of inter-connecting wires or cables.	Reduces time to remove flexible elements.

Based on Tracy Dowie & Matthew Simon, "Guidelines for designing for disassembly and recycling," WKD –ICED 1993, The Hague, The Netherlands, August 1993, pages 3-6, URL: http://sun1.mpce.stu.mmu.ac.uk/pages/projects/dfe/dfe.htm

TABLE 6.2 General Guidelines and Rationale for Fasteners

Guideline	Rationale for Guideline
Minimize the number of parts.	Reduces cost.
Make designs as modular as possible.	Allows service and upgrade options.
Locate parts with the critical functionality in easily accessible places.	Enables partial disassembly for optimum return.
Design for easy separation and handling.	Speeds up the process.
Avoid molded metal inserts or reinforced elements in plastic parts.	Avoids the need for separation.
Make logical access and break points in a structure.	Speeds disassembly and training.
Locate nonrecyclable parts that can be quickly discarded in one area.	Speeds disassembly.

Based on Tracy Dowie & Matthew Simon, "Guidelines for designing for disassembly and recycling," WKD –ICED 1993, The Hague, The Netherlands, August 1993, pages 3-6, URL: http://sun1.mpce.stu.mmu.ac.uk/pages/projects/dfe/dfe.htm

TABLE 6.3 DFD Guidelines and Rationale for Selecting Product Architecture

6.1.2 Recyclability in Product Design

One aim of design for disassembly is to provide the maximum amount of disassembly and to save more from recyclable parts than the cost involved in manufacturing the accessible parts. Often costs exceed revenue. Design for recycling (DFR) is a proactive step that incorporates recyclability characteristics in a product. It can reduce recycling costs at the end of a product's life cycle.

The automobile industry has used DFD principles for some time. Automobile vehicles are among the most highly recycled products. About 75% by weight of materials, which include most of the metal components, is recovered and recycled. Automobile junkyards and the availability of a wide range of rebuilt automotive components have facilitated recyclability. The automobile industry uses certain rating criteria for recyclability and the ability to separate materials. Recyclability of 1 indicates the part is remanufacturable (for example, starters and alternators). If the material (for example, metal) is recyclable, the rating is 2. Materials that cannot be recycled have a rating of 3, and those that can be recycled with additional techniques have a rating of 4. Organic materials (for example, wooden components) have a recyclability rating of 5.

6.2 Evaluation of Design for Disassembly Aspects in Products

6.2.1 Activity-Based Costing (ABC) Method

A management tool known as activity-based costing provides the organization with value-based information to make effective decisions. The data that it provides include:

- Product
- Process
- Uncertainty

These data consider the ability of a product's material to be disassembled, reused, and recycled in the disassembly process. ABC methodology does not address the issue of potential damage to parts. These data must be continuously updated because the recycling information depends on the volume of material used and its cost.

Product information: ABC product information is collected on a spreadsheet-like *action chart* that presents quantifiable data on a product's disassembly, reusability, and material recyclability. The disassembly information includes evaluating a component's accessibility, the force needed, and disassembly time. Time information reflects the time needed to disassemble a product's components. The reusability includes an assessment of a given component's amount of fatigue and the resultant damage involved in removing it. Recycling information for each component includes the material type, volume of material, and recyclability, as well as the percentage recovered.

Assembly/Disassembly Process information: Process information that is embedded in the general disassembly process model that is provided by ABC includes demanufacturing activities such as product's storage, dismantling, and sorting as well as transporting reusable and recyclable materials.

Uncertainty information: Uncertainty is inherent for any model, especially when dealing with ecological issues for which there is a lack of hard data. Uncertainty is represented in the model as a standard deviation of the mean disassembly time.

ABC data output: Once the entire product, process, and uncertainty information is defined, design teams can use ABC data to calculate assembly and disassembly costs. The *assumptions* about the product and process are translated into forecasts about the demanufacturing cost based on sampling a value from each assumption.

6.2.2 **Fishbone and Reverse Fishbone Diagrams**

Drawing a fishbone diagram forces designers to identify the costs of assembly tasks and paths that could lead to defects. This diagram is an essential part of the evaluation of assembly difficulties.

Fishbone Diagram

The *fishbone diagram* is a popular design tool used to plan for disassembly. It is an effective way to enhance product design for ease of both assembly and disassembly.

Figure 6.1 is a fishbone diagram for an automobile. Designers critically analyze a product including its disassembly and related costs. They also must understand the post-disassembly process of a component. Before starting a fishbone diagram, designers make a preliminary analysis of the product's service information. The construction of the diagram begins with the main part to which other parts are attached. In **Figure 6.1**, the design of the automobile is represented by the central spine, and

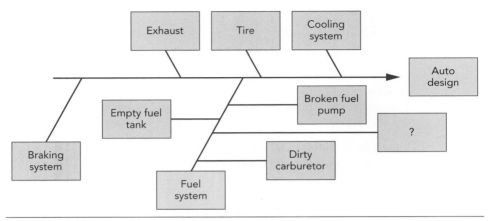

FIGURE 6.1 Fishbone Diagram of an Automobile

each rib in the design represents a specific subsystem or component, such as fuel and exhaust systems. Any identified potential problems and flaws are listed for each secondary rib.

Reverse Fishbone Diagram Method

When creating reverse fishbone diagrams, the identification of disassembly steps is important. The reverse fishbone diagram graphically characterizes the difference between independent disassembly and sequence-dependent disassembly. The analysis could reveal that additional work is required to make the disassembly process more sequence independent and help designers to identify the strategic components that could be retired. The features of the reverse fishbone diagram allow designers to generate specific information about a design's performance under different product retirement scenarios on the basis of results.

See **Figure 6.2** for a reverse fishbone diagram of a coffeemaker. The diagram schematically shows the product's disassembly steps and specifies retirement content for each component. The fishbone diagram promotes DFD by forcing designers to identify assembly difficulties and decide how to remedy them. Each assembly step indicates fixturing needs, reorientation, and insertion directions. An assembly rating is computed based on these values (see the discussion under the heading "Disassembly Evaluation"). The diagram can include symbols to indicate time factors such as necessary inspection and testing. By knowing the intended placement of each part, designers can construct the reverse fishbone diagram and determine the sequence of the disassembly process and identify any parts that need not be disassembled.

Constructing the reverse fishbone diagram normally starts at the top end of the product design and goes down step-by-step, physically disassembling it. This characterizes each disassembly step according to its fixturing needs (the symbol F), reorientation requirements (circular arrow), and removal directions (straight and rotational

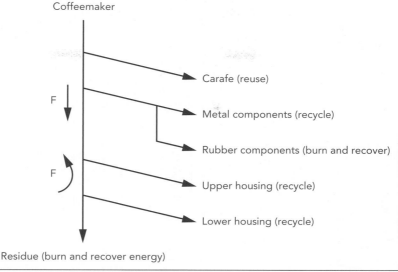

FIGURE 6.2 Reverse Fishbone Diagram of a Coffeemaker

arrows). The diagram can include symbols indicating component fate category, time factors, and connection separation method (break, pop, or unscrew). The symbols for tool requirement and removal difficulty should be included to facilitate the rapid visual evaluation of any disassembly difficulty. The initial steps of the reverse fishbone include a short series of disassembly steps as the first set of fasteners, and then removal of the product housing.

Reverse fishbone analysis can provide designers early guidance in the following areas:

- Identification of retirement components
- Matching of a retirement component with market demand for reused components
- Identification of potential improvements in disassembly steps.
- Identification of intercomponent connections that pose disassembly difficulties
- Retirement cost/revenue stream projections
- Identification of special disassembly tooling requirements

6.2.3 **Chart Method for Evaluating DFD Efficiency (Kroll)**

Charts can be used to evaluate the efficiency of DFD. Ehud Kroll proposed a method based on the degree of difficulty of and the types of tools used in disassembly. Kroll's chart (**Figure 6.3**) uses a methodology using spreadsheet software for recording the disassembly data in redesign. The chart has two components: First, an informative

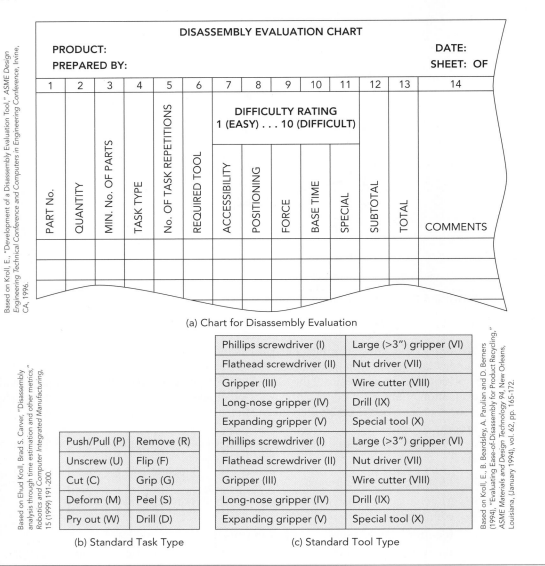

Based on Kroll, E., "Development of a Disassembly Evaluation Tool," ASME Design Engineering Technical Conference and Computers in Engineering Conference, Irvine, CA, 1996.

(a) Chart for Disassembly Evaluation

Based on Ehud Kroll, Brad S. Carver, "Disassembly analysis through time estimation and other metrics," Robotics and Computer Integrated Manufacturing, 15 (1999) 191-200.

(b) Standard Task Type

(c) Standard Tool Type

Based on Kroll, E., B. Beardsley, A. Parulian and D. Berners (1994), "Evaluating Ease-of-Disassembly for Product Recycling," ASME Materials and Design Technology 94, New Orleans, Louisiana, (January 1994), vol. 62, pp. 165-172.

FIGURE 6.3 (a–c) Disassembly Evaluation Chart

section describes the operations, directions, and other factual information, and second, it provides data to use in subjectively evaluating the degree of difficulty associated with each part's removal. The procedure calls for disassembling the product while entering information in the chart.

Disassembly evaluation: As partially shown in **Figure 6.3 (a)**, each row on the chart corresponds to a separate disassembly task. Tasks are sequentially recorded and assessed

Part Number	Minimum Number of Parts	Number of Repetitive Tasks	Task Type	Task Direction	Tool Required	Tool Accessibility	Position	Force	Time	Special	Sub Total	Total
1	0	4	U	Z	I	1	2	2	1	1	7	28, (Screws)
2	0	1	F	Z	II	1	1	1	1	2	6	6
3	1	1	P	Z	III	1	1	3	1	1	7	7
4	1	1	C	Z	IV	1	1	3	1	1	7	7
5	1	1	R	Y	III	1	1	1	1	1	5	5

Tasks:

P = push/pull   U = unscrew   I = Phillips-head screwdriver

C = cut   W = pry out   II = large gripper

R = remove   F = flip   III = gripper

G = grip   IV = wire cutter

V = flat-head screwdriver

Tools:

TABLE 6.4 Kroll's Evaluation Ratings

during the actual or simulated disassembly process. Several rows may correspond to the disassembly of a single part if multiple operations are required to remove it. Each column contains data pertaining to different aspects of the disassembly. The task difficulty scores are based on the fundamental assumption that performance time is a valid indicator of disassembly effort. The fact that time is related to cost is a further incentive to measure difficulty in this way.

The informative part of the disassembly evaluation chart (see **Figure 6.3 (a)**) consists of a number assigned to each part in the order of disassembly (Column 1). The number 1 or 0 in Column 2 denotes whether the part is theoretically required to exist as a separate component. The number of times each disassembly task is performed is added in Column 3; the code for each task type from **Figure 6.3 (b)** indicating the axis and which part is removed from the assembly is in Column 5; and the code from **Figure 6.3 (c)** indicates required tooling in Column 6. Columns 7 to 11 reflect the subjective judgments of the difficulty of performing each disassembly task in five categories. Each is scored on a scale of 1 to 10 in the increasing order of difficulty.

Table 6.4 shows the evaluation rating for parts installation and disassembly. The first column identifies part number, and the second column identifies the minimum number of parts required. Each part is evaluated to determine whether it is theoretically required to exist as a separate component.

6.2.4 **Rating Chart Method (Shetty)**

The DFD Rating Chart Method proposed by Shetty records a step-by-step procedure using a spreadsheet and focuses on evaluating a design according to both assembly and disassembly difficulties. The main idea of a spreadsheet is to record data for DFA and DFD procedures that reflect a product's complexity in a single software program. This allows designers to quickly evaluate those data for the initial design before moving forward.

This method is generally known as AR^3T^3, which stands for access (A); removal, reuse, and recycle (R^3); and tool, task, and time (T^3). It uses a common point system from 0 (bad) to 9 (good). The final results obtained are assembly and disassembly rating scores. The procedure evaluates not only the product's overall design but also other factors such as service, repair, and end-of-life status. Designers can study the design from many viewpoints to see how the design is rated in these different aspects.

1. Task/Damage Rating Ratings are based on the nature of the task to be accomplished and the damage to the part considered for removal. The rating for removing a part without any damage is from 0 (difficult) to 9 (easy).

Task/Damage Description

Task easily accomplished with minimum concern for part damage	9
Task accomplished with very little determinable damage to the part	8
Task accomplished with destruction only method that could be reused	7

(continued)

Task accomplished through destructive assembly; part not suitable for reuse	6
Task easily accomplished; part needs additional attention.	5
Task easily accomplished with extra care	4
Task easily accomplished; very fragile part and needs utmost care	3
Task accomplished with considerable damage to part	2
Part completely damaged during the task	1
Part damaged during the removal; needs care to minimize additional damage	0

2. Reuse Rating Ratings are based on the *reuse* level of the part being removed. Reconditioning is necessary if the part suffers damage in the removal process.

Reuse Description

Reuse with no conditioning	9
Reuse with minor conditioning; part in good condition	8
Possible reuse of degraded part after some reconditioning	7
Reused only after reconditioning	6
Reused with simple tools for separation followed by reconditioning	5
Reused with technical expertise on tool's separation followed by reconditioning	4
Reused with highly rated technical separation techniques from the assembly	3
Reused with skilled separation techniques and time-consuming reconditioning	2
Reused only after high-cost reconditioning	1
No reuse; reconditioning techniques are very expensive and time consuming	0

3. Removal Rating *Removal* is defined as the process of removing the part from the assembly without damaging it or its surrounding parts.

Removal Description

Removed with no damage to the adjacent parts and itself	9
Removed safely with minor damage to the adjacent part(s)	8
Removed with minor damage to other parts; requires reconditioning	7
Removed with potential damage to adjacent part(s) requiring some reconditioning	6
Removed with damage to the surrounding part(s) for which reconditioning is required	5

(continued)

Removed with required major reconditioning of adjacent part(s); potential reuse possible	4
Removed with difficulty; reconditioning of adjacent part(s), potential reuse less likely	3
Removed with no reconditioning or reuse of adjacent part(s)	2
Removed with major damage to adjacent part(s); no reconditioning	1
Removed with complete damage; no reconditioning for part	0

4. Recyclability Rating *Recyclability* can be defined as the ease with which a material can be reused safely and economically. Ratings are provided based on the lead content and recyclability of the materials used in the assembly.

Recyclability Description

Made of one nonhazardous material and lead free, easily recyclable	9
Made of one nonhazardous material; special recycling equipment needed	8
Made of dissimilar lead-free nonhazardous materials; no separation needed	7
Made of dissimilar lead-free nonhazardous materials; separable and recyclable	6
Made of dissimilar lead-free nonhazardous, recyclable materials; special tooling needed	5
Made of dissimilar lead-free nonhazardous, recyclable materials difficult to separate	4
Made of lead-free dissimilar nonhazardous, inseparable materials	3
Made of one or more hazardous materials; not recyclable even if separated	2
Made of one or more hazardous and nonhazardous materials	1
Made of one or more hazardous materials	0

5. Disassembly Time Rating The time required to disassemble a component either for repair, reuse, or recycling is the *disassembly time rating*. It includes human motions. A rating is based on the number of fasteners and the availability access to the assembly and the required use of special tools and skills.

Disassembly Time Description

Disassembly fast and easy by hand(s)	9
Disassembly time less; requires hand(s) and simple hand tools	8
Disassembly time moderate; one or more universal tool(s) at one stage	7

(*continued*)

Disassembly time moderate; uses dissimilar fasteners, more accessible subassemblies	6
Disassembly time moderate; fastener material incompatible with attached part	5
Disassembly time moderate; has more fasteners, limited access to subassemblies	4
Disassembly time high; has critical joints in subassemblies requiring special tools	3
Disassembly time high; has critical joints, readily unavailable tools; requires skill	2
Disassembly time high; has corroded fasteners, limited access to subassemblies	1
Disassembly time very high; has permanent fastenings, limited access to subassemblies	0

Disassembly Time Rating Includes Access and Tool Ratings.
Access rating is another metric that is often overlooked. It is the availability of access area during disassembly. This variable, if critical, influences other variables as well. Access is defined as the ease with which a part can be approached and removed. The rating is given in the range of 0 to 9 with 0 being hard to access and 9 easy.

Description [Access Area Rating Chart]   **Rating**

Area easy to work in, tools easily accessible	9
Restricted access; part can be removed without damage	6
Restricted access and restricted vision; special care is needed	3
Difficult to access area and tooling; extreme care is needed to prevent damage	0

Tool rating is the ability to acquire proper tools. The size of the tool and the distance to be traveled to acquire it are important factors. The best rating occurs when no tool is required, and the worst rating is when the assembler must travel to another location to obtain the tool.

Description [Tool Rating Chart]   **Rating**

Tool not required; task accomplished by hand	9
Common hand tool required	6
Special tooling/equipment required; no delay to obtain the tool	3
Special tooling required; there is delay to acquire the tool	0

6. Rating Score The general rating score can be calculated by adding the ratings for task, reuse, removal, and recyclability. Disassembly time should be divided by the sum of the maximum rating of each column (i.e., 9 + 9 + 9 + 9 + 9 + 9 = 54) and multiplied by the weighting of each part (i.e., the number of parts of each component in the assembly divided by the total number of components in the assembly).

7. Time The time taken to disassemble a part from the assembly is measured in seconds using a timer. It includes time for the operator's hand movement to perform the operation and return to the original position after finishing the task.

8. Total Time The total time taken for every removal of a part used more than once is measured in seconds. For example, if the removal of a wheel requires unscrewing four nuts, the total time is the time taken to remove one nut multiplied by 4.

9. Time Rating Parts that take near the average normal removal time are not penalized, whereas parts that are slightly (10%) higher than the average are. The calculations are performed as follows: (a) If the time measured (TM) is more than twice the average, use the ratio of TM/Average. (b) If the TM is less than 1.1 times (within 10%), use 1.0. This desensitizes the rating when TM is within 10% of the average. (c) If 1.1 <TM 2 × Average, use 2.0. This sensitizes the rating when TM is just above the average.

11. Estimated Time Rating It may not be possible to measure the time required to perform a step. This requires making a quick estimate using the following criteria to determine an overall value for the time: (a) Use 1 when the time required to accomplish the step is "average" in comparison to the other steps performed. (b) Use 3 if the time required to complete the step is above average;

Estimated Weighted Time Rating Multiply the estimated time by the parts' weighted average.

EXAMPLE 6.1

DISASSEMBLY FOR THE HOUSEHOLD JUICE EXTRACTOR

See **Figure 6.4 (a)** for the design of the juice extractor. **Figure 6.4 (b)** demonstrates the use of Rating Chart method for the household juice extractor. The overall rating of this product is very good. The parts are designed for easy assembly and access; reassembly without damaging any parts is easy. Most parts are reusable and recyclable, with the possible exception of the cutter/strainer. No tools are needed to assemble or disassemble the juice extractor.

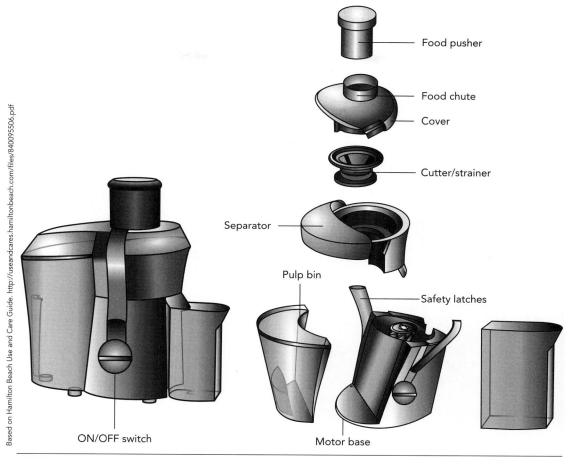

Food pusher

Food chute

Cover

Cutter/strainer

Separator

Pulp bin

Safety latches

ON/OFF switch

Motor base

FIGURE 6.4 (a) Household Juice Extractor

Component Number	Component Name	Component Type	Dimensions (mm)	Quantity
1	Food pusher	Part	105 length, 72 diameter	1
2	Cover	Part	215 × 170 × 140	1
3	Cutter/strainer	Part	60 length, 130 diameter	1
4	Separator	Part	100 length, 150 diameter	1
5	Pulp bin	Part	110 × 80 × 230	1
6	Juice container	Part	60 × 80 × 150	1
7	Safety latches	Part	30 × 4 × 130	2

Parts List

Step	Item #	Name	Number of Parts Required	Rate Damage (0 = bad, 9 = Good)	Rate Tooling (0 = Bad, 9 = Good)	Rate Re-Use (0 = Bad, 9 = Good)	Rate Access (0 = Bad, 9 = Good)	Rate Removal (0 = Bad, 9 = Good)	Recycle Rating (0 = Bad, 9 = Good)	Rating Score ((E+F+G+H+I+J)/54) × Wgt	Time Measured (sec)	Total Time	Time Rating	Estimate Time < than Average = 1, > than Average = 3	Estimated Weight Time Rating
1	6	Juice Container	1	9	9	9	9	9	9	12.5%	5	5	1.00	1	0.13
2	1	Food Pusher	1	9	9	9	9	9	9	12.5%	5	5	1.00	1	0.13
3	7	Safety Latches	2	9	6	9	9	9	9	23.6%	10	20	4.00	3	0.75
4	2	Cover	1	9	9	9	9	9	9	12.5%	5	5	1.00	1	0.13
5	3	Cutter/Strainer	1	9	9	9	9	3	9	11.1%	7	7	1.00	1	0.13
6	4	Separator	1	9	9	9	9	9	9	12.5%	10	10	2.00	3	0.38
7	5	Pulp Bin	1	9	9	9	9	9	9	12.5%	5	5	1.00	1	0.13
		Totals	8	63	60	63	63	57	63	97.2%	47	57	11.00	11.00	57.14%
		Average	1.14	9.00	8.57	9	9.00	8.14	9.00		6.71	8.14	1.57	1.57	

Total Number of Steps 8

	Actual Time	Est Time
Rating Score	97.22%	97.22%
Time Score	72.73%	57.14%
Parts Score	100.00%	87.50%
Overall Design Score	**90.0%**	**80.6%**

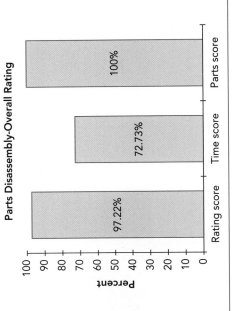

Parts Disassembly-Overall Rating

Rating score: 97.22% Time score: 72.73% Parts score: 100%

FIGURE 6.4 (b) Rating Chart Method for the Household Juice Extractor

Based on Devdas Shetty, Vishwesh Coimbatore and Claudio Campana, "Design Methodology for Assembly and Disassembly Based on Rating Factors," ASME 2005 International Mechanical Engineering Congress and Exposition Manufacturing Engineering and Materials Handling, Parts A and B Orlando, Florida, USA, November 5–11, 2005 Conference Sponsors: Manufacturing Engineering Division and Materials Handling Division.

EXAMPLE 6.2

CELL PHONE DISASSEMBLY EVALUATION USING THE RATING CHART METHOD

This example considers the use of the rating chart method to study the assembly and disassembly of a cell phone. The phone consists of eleven parts; the entire assembly was performed from the bottom to the top. The procedure requires the availability of basic tools and a step-by-step disassembly sequence for examining and recording the time it takes to acquire the tool, inspect the part, and conduct the task. **Figure 6.5** shows some stages of the disassembly sequence of cell phone A.

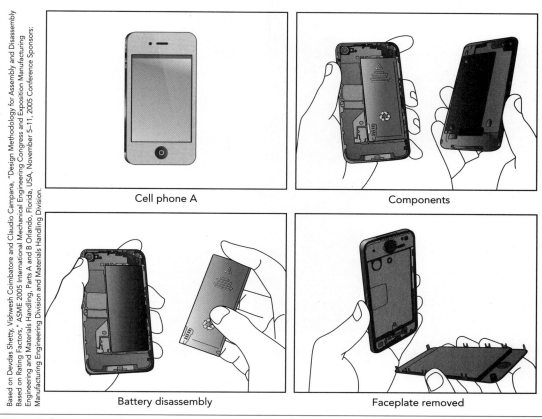

Cell phone A

Components

Battery disassembly

Faceplate removed

FIGURE 6.5 Stages of Disassembly Sequence of Cell Phone

See **Figure 6.6** for a list of the cell phone parts.

The spreadsheet shown in **Figure 6.7** provides symbols to immediately alert the designer to look more closely at the design and thus indicates the need for a redesign strategy:

- E/C indicates the part is a likely candidate for elimination or combination.
- # indicates that the part has a very low or zero score in one area of evaluation.

Figure 6.7 is a chart showing disassembly data for a cell phone.

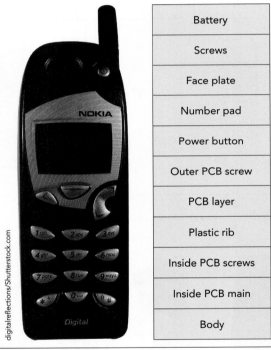

| Battery |
| Screws |
| Face plate |
| Number pad |
| Power button |
| Outer PCB screw |
| PCB layer |
| Plastic rib |
| Inside PCB screws |
| Inside PCB main |
| Body |

digitalreflections/Shutterstock.com

FIGURE 6.6 Components of Cell Phone

Step	Item #	Name	Number of Parts Required 9 = Good	Rate Damage (0 = bad) 9 = Good	Rate Tooling 0 = Bad 9 = Good	Rate Re-Use 0 = Bad 9 = Good	Rate Access 0 = Bad 9 = Good	Rate Removal 0 = Bad 9 = Good	Recycle Rating 0 = Bad 9 = Good	Rating Score ((E+F+G+H+I+J)/54) × Wgt	Time Measured (sec)	Total Time	Time Rating	Estimate Time < than Average = 1 > than Average = 3	Estimated Weight Time Rating
1	1	Battery	1	9	9	9	9	9	0	5.6%	2	2	1.00	1	0.07
2	2	Screws	4	9	6	9	9	9	9	25.2%	5	20	8.00	3	0.80
3	3	Face Plate	1	3	9	9	9	9	9	5.9%	3	3	1.00	1	0.07
4	4	Number Pad	1	9	9	9	9	9	9	6.7%	1	1	1.00	1	0.07
5	5	Power Button	1	0	6	0	9	9	9	4.1%	2	2	1.00	1	0.07
6	6	Outer pcb screw	2	9	6	9	9	9	9	12.6%	5	10	4.00	3	0.40
7	7	pcb layer	1	3	9	9	9	9	6	5.6%	4	4	2.00	3	0.20
8	8	Plastic rib	1	9	9	9	9	9	9	6.7%	3	3	1.00	1	0.07
9	9	Inside pcb screws	1	9	6	9	9	9	9	6.3%	6	6	2.00	3	0.20
10	10	inside pcb main	1	3	9	9	9	9	6	5.6%	2	2	1.00	1	0.07
11	11	Body	1	9	9	9	9	9	9	6.7%	2	2	1.00	1	0.07
		Totals	15	72	87	90	99	99	84	90.7%	35	55	23.00	19.00	48.39%
		Average	1.36	6.55	7.91	8.18	9.00	9.00	7.64	3.18	5.00		2.09	1.73	

Total Number of Steps 11

	Actual Time	Est Time
Rating Score	90.74%	90.74%
Time Score	65.22%	48.39%
Parts Score	73.33%	73.33%
Overall Design Score	**76.4%**	**70.8%**

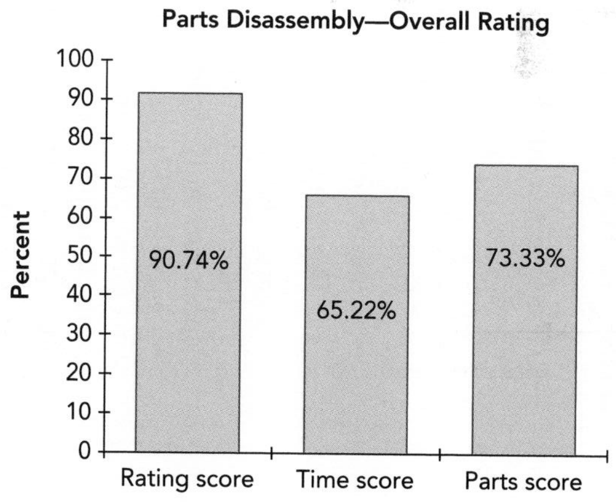

Parts Disassembly—Overall Rating

FIGURE 6.7 Overall Disassembly Rating

247

6.3 Design for Product Maintenance

The objectives of design for maintenance are to:

- Identify and prioritize maintenance requirements
- Detect and isolate faults
- Increase product availability and decrease maintenance time
- Decrease life cycle costs
- Increase customer satisfaction

Maintainability: Maintainability is the ability to perform satisfactorily throughout its useful life span with minimum expenditure of effort and money. Maintenance can be either preventive (regular or routine service required for preventing failures) or breakdown (repair service after some failure). Designing for good serviceability means providing for ease of both maintenance types. When designing for maintainability, designers must consider the trade-offs involved. Design for easy maintainability/serviceability of a product with a long life cycle is more important than the initial production cost.

Maintainability index: A product's design influences its maintainability and determines the performance of various maintenance activities. A maintainability index is useful for comparing and predicting the maintenance requirements of new products.

6.3.1 Design for Maintenance Processes

Some of the maintenance evaluation tools are given below:

- Maintainability by checklists
- Maintainability by physical mock-ups
- Maintainability by virtual approaches
- Maintability by numerical approaches

Maintainability checklists: Data that summarize design review points for the maintainability assessment are recorded on checklists as (1) general maintenance reduction, (2) standardization features, (3) design for physical accessibility, and (4) design for mechanical safety.

Maintainability evaluation using physical mock-ups: This evaluation uses physical mock-ups or models of a product on which to perform maintenance tasks using real tools and maintenance technicians. Doing this can identify areas whose maintainability could be improved.

Maintenance evaluation using virtual approaches: Virtual models of a product can be used to perform maintenance tasks in a virtual environment using virtual tools. These tasks can simulate the use of interactive virtual environments, using simulated anthropometric articulated representations of human beings conducting assembly, disassembly, and maintenance. The product's model is usually an approximation of the exact virtual model created during the design process. This process identified by using this technique is easily integrated into the design process and reduces the time delay needed to evaluate a product's maintenance requirements.

Maintenance evaluation using numerical approaches: Several criteria are used to evaluate maintainability. The process assigns each maintainability criterion a numerical value between 0 and 1.

Value	Accessibility
1	All parts directly accessible and placed in the same area
0.8	All parts directly accessible and placed in different areas
0.6	Some parts not directly accessible but are maintenance free
0.4	Some parts accessible after disassembling
0	Majority of parts accessible by disassembling one or more entities

The value for the maintenance index is obtained computing the weighted mean with the numerical values of the criteria.

The following are the different types of maintenance operations:

- Reactive/corrective
- Preventive
- Predictive
- Reliable
- Lean

Reactive/maintenance: Reactive maintenance corrects a problem when something has failed. It is detrimental to the product's operation and causes it not to be available during the time needed to perform the maintenance. This type of maintenance is basically using the product till it breaks without taking any steps to maintain it.

Preventive maintenance: *Preventive maintenance* is generally defined as actions performed on a time-or machine-run-based schedule that detect, preclude, or mitigate degradation of a component or system with the aim of sustaining or extending its useful life through controlling degradation to an acceptable level. Preventive maintenance assumes that a product has components that wear and must be replaced on a continual basis. The frequency of the maintenance is generally constant and is usually based on the expected life of the components being maintained, but no monitoring necessarily occurs during the process. Minimizing failures translates into maintenance and capital cost savings.

Advantages of preventive maintenance are:

- To be cost effective in many capital-intensive processes and to provide 15 to 20% cost savings during the maintenance program.
- To provide flexibility that allows adjustment of frequency of maintenance
- To increase component life cycle
- To save energy

Predictive maintenance: This is generally defined as measurements that detect the onset of a degradation mechanism, thereby allowing casual stressors to be eliminated or controlled prior to any significant deterioration in the component physical state. It involves performing a series of steps prior to actual maintenance to indicate a mechanism's current and future functional capability. Data collected are analyzed to create an appropriate maintenance schedule, and maintenance is performed according to the schedule. Basically, predictive maintenance differs from preventive maintenance by basing maintenance need on a machine's actual condition rather than some pre-set schedule. Preventive maintenance is time based whereas predictive maintenance provides quantified information on a device's material and equipment conditions. A well-orchestrated predictive maintenance program can eliminate catastrophic equipment failures.

Advantages of predictive maintenance are to:

- Allow for preemptive corrective actions
- Decrease equipment or process downtime
- Decrease costs for parts and labor
- Provide better product quality
- Improve worker and environmental safety

Reliability-centered maintenance (RCM): RCM is a systematic approach to evaluate a facility's equipment and resources to best match the two, resulting in a high degree of facility reliability and cost-effectiveness. This type of maintenance can be defined as an approach to maintenance that combines reactive, preventive, predictive, and proactive maintenance practices and strategies to maximize the life that a piece of equipment functions in the required manner. The RCM method focuses on creating an optimal mix of an intuitive approach and a rigorous statistical approach to decide how to maintain equipment. It recognizes that the importance of all equipment in a facility is not of equal nature in terms of performance and safety.

RCM recognizes that equipment design and operation differ and that some types of equipment have a higher failure probability as a result of different degradation mechanisms than others. It also depends on predictive maintenance but also recognizes that maintenance activities on types of equipment that are inexpensive and unimportant to facility reliability may best be left to routine maintenance.

Advantages of reliability-centered maintenance are:

- To provide an efficient maintenance program
- To lower costs by eliminating unnecessary maintenance
- To minimize frequency of overhauls
- To reduce the probability of sudden equipment failures

Lean maintenance: This is the application of lean philosophy, tools, and techniques to the maintenance function. Lean maintenance provides a focus on: eliminating wasted time, effort, and material while improving throughput and quality. Many lean systems apply to lean maintenance. These philosophies or systems include the 5-S, kaizen, jidoka (quality at the source), and just-in-time (JIT) processes. Implementing lean maintenance with a focus on eliminating waste from such processes can pay dividends in improving reliability while reducing costs.

6.3.2 **Product-Oriented Design for Maintenance**

Design for maintenance involves the designer in determining how to best design a product or subassembly for ease of maintenance specifically related to the time and resources needed to retain equipment or restore it to a specified operational condition. The processes directly affect the company resources, because they can impact operations, downtime, maintenance costs, and safety and reduce total life cost. As shown in **Figure 6.8**, design flexibility is greatest in the conceptual stage and offers low design change costs. As the product nears production, design flexibility decreases and change costs rise.

Addressing maintainability during design often involves a greater commitment to customers than is usual by reducing the end user's maintenance costs over a product's life. This commitment implies that the company accepts the responsibility for providing customers a worry-free product.

The following maintenance guidelines should be considered during system design:

- Provide easy access to serviceable items.
- Use self-test indicators to quickly isolate faults.
- Eliminate the need for adjustment.
- Identify any modules subject to a high probability of replacement and make them easily accessible.
- Reduce the number of components in the final assembly.
- Use components and subassemblies that are repairable.
- Reduce the possibility of damage to the equipment during maintenance.
- Provide a possible upgrading system with exchange of modular parts.
- Avoid having sharp edges, corners, or protrusions that could injure personnel.

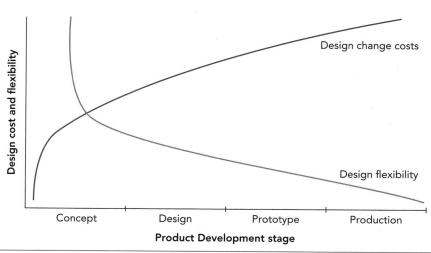

FIGURE 6.8 Product Phase versus Product Cost/Flexibility

Just like Design for Assembly and Design for Disassembly, there are certain features that influence maintenance.

1. Accessibility *Accessibility* is the relative ease with which a part or piece of equipment can be reached for service, replacement, or repair. Lack of accessibility is an important maintenance problem and a frequent cause of ineffective care leading to increased downtime and reduced revenue.

2. Standardization *Standardization* represents the choice, design, or manufacture of parts, assemblies, equipment, associated tools, service materials, and/or procedures identical to or replaceable by other items or procedures. This important design feature restricts to a minimum the variety of parts and components that a product or system could need. Standardization should be a central goal of design because the use of nonstandard parts can lead to reduced reliability and increased maintenance.

3. Interchangeability As an intentional aspect of design, *interchangeability* refers to any component, part, or unit that can be replaced in a given product or piece of equipment by any similar component, part, or unit. It is the extent to which one item can be readily replaced with an identical item without a need for recalibration. Such flexibility in design reduces maintenance work and in turn maintenance costs.

4. Replaceability Components and subassemblies should be designed so that they can be replaced in the shortest amount of time with the least amount of skill and effort. Design for Maintenance (DFMA) principles such as part reduction, modular design, and elimination of fasteners are often effective in addressing and improving replaceability. Modular component philosophy with built-in test capabilities can aid in problem diagnosis and repair. Modular design and the replacement of defective or worn modules encourage maintenance contracts.

5. Repairability This aspect of maintenance is concerned with fixing components rather than replacing them. Use of interchangeable parts can enhance system repairability. In addition, a repairability analysis should consider how to make repairs and identify the types of tools and components required. A key element that must be addressed with regard to repairability and replaceability is deciding which components to repair and which to replace.

6. Upgradeability *Upgradeability* is the ability to modify components to meet changing technology, prevent product obsolescence, and lengthen the time before becoming unusable. With the ever-increasing rate of technological advances today, this is more important than ever in a product's life. The personal computer market illustrates the attractiveness of upgradeability. As dependence and use of microelectronic technologies increase, their rapid progress requires that designers be aware of existing upgradable systems. In this regard, modular designs can directly impact system enhancement and upgrades.

7. Maintenance Frequency The regularity with which each maintenance action must be conducted is central to a product's preventive, scheduled, or corrective maintenance requirements. Frequent care can decrease downtime because it prevents repairs that cannot be done while the product is in operation.

A design for maintenance part is evaluated based on the following four criteria.

1. Cost and time: The cost/downtime involved
2. Recyclability: The recyclability of a part in a safe environment
3. Serviceability: Ease of access to the location during maintenance operation
4. Complexity: Location (on-site or off-site) in which to accomplish the task

Each evaluation criterion is rated from 0 (difficult) to 9 (easy).

1. Maintenance Cost/Time Rating This rating indicates the total cost and/or time for maintenance with little outage time. It is based on the instant availability of spare parts, tooling, skilled personnel, and short outage time.

Maintenance Cost/Time Description

Lowest maintenance cost requiring long-term planning, uninterrupted production	9
Minimal expected maintenance cost involving planned/scheduled shutdown, high service level, short outage time, no personnel training	8
Minimal expected maintenance cost requiring planned/scheduled shutdown, additional tooling, high service level, short outage time, some personnel training	7
Minimal expected maintenance cost requiring planned shutdown, medium service level, increased outage time, additional personnel training	6
Average maintenance cost requiring planned shutdown, jobs added to scheduled work, medium service level, increased outage time, personnel training	5
Average maintenance cost requiring planned shutdown, increased outage time, on-site spare-parts and tooling, no personnel training	4
Average maintenance cost requiring unplanned shutdown, additional spare parts/tooling, increased outage time, skilled personnel	3
High expected maintenance cost requiring unplanned shutdown, limited availability of spare parts and tooling, high outage time, skilled personnel	2
High expected maintenance cost requiring unplanned shutdown, no availability of spare parts/tooling, long outage time of productive equipment, personnel expertise	1
High expected maintenance cost requiring unplanned shutdown, parts meant for replacement, very long outage time	0

2. Recyclability Rating Ratings are based on the material recyclability of the parts designated for maintenance while considering the comfortable to hazardous environment in which the maintenance operations are performed.

Recyclability Description

Part(s) made of lead-free nonhazardous liquid/solid, easily separable, recyclable	9
Part(s) made of one or more nonhazardous lead-free liquids/solids and are separable, purified, or drained at room temperature	8
Part(s) made of lead-free nonhazardous, solid materials that can be recycled at monitored, safe temperatures	7
Part(s) made of lead-free nonhazardous liquid, recyclable only under elevated temperatures and pressure, safe conditions	6
Part(s) made of lead-free volatile nonhazardous liquid recyclable under high pressure/temperature; safety measures essential	5
Part(s) made of lead-free volatile nonhazardous liquid materials recyclable under enormous pressure/temperature, essential safety, personnel training	4
Part(s) made of hazardous liquid/solid materials, recyclable under room temperature/safety, mandatory training, >5ppm lead	3
Part(s) made of hazardous liquid/solid materials, recyclable under high temperature/pressure, extremely high safety measures, >5ppm lead	2
Part(s) consist of toxic liquid/solid materials, separable with time delay, unsafe work environment, >5ppm lead	1
Part(s) made of inseparable toxic materials, >5ppm lead	0

3. Service Rating Ratings are based on each part's accessibility and ease of maintenance by service personnel.

Service Description

Serviceable part, no resistance	9
Serviceable part, little resistance	8
Serviceable moving parts, some resistance	7
Serviceable hot part, careful removal	6
Serviceable part, sharp edges, care observed	5
Serviceable part, adjacent component removal required	4
Serviceable part, insufficient space for tools	3
Serviceable part, insufficient space for tools and service person	2
Heavy part, service person's vision obstructed	1
Part not within service personnel reach	0

4. Complexity Rating This represents the degree of maintenance difficulty. Ratings are based on the task's location and the ease of dismantling for maintenance without damaging the part.

Complexity Description

Task accomplished on-site; dismantling not required, part not damaged	9
Task accomplished on-site; easy manual dismantling	8
Task accomplished on-site; easy dismantling without damage	7
Task accomplished on-site; visual care required to dismantle without damage	6
Task accomplished on-site; dismantling near high voltage points without damage required	5
Task accomplished on-site; special dismantling equipment required to remove part	4
Task not accomplished on-site; part moved to a maintenance area for dismantling without damage	3
Task not accomplished on-site; heavy parts moved to access maintenance area to dismantle without damage	2
Task not accomplished on-site; permanent installations increase outage time	1
Task accomplished on-site/off-site; part damaged, no readily available spare parts	0

6.4 Integrated Design Method for Assembly, Disassembly, and Maintenance

The integrated method to evaluate the design for assembly/disassembly and maintenance uses spreadsheet-based procedure and extends the previously described AR^3T^3
rating chart method. The integrated method includes the features of design for maintenance in the analysis with DFA and DFD. The method allows for calculating design
efficiency scores for assembly, disassembly, and maintenance.

6.4.1 DFA Section (General Procedure)

The general procedure for DFA part evaluation is based on the following five criteria:

1. Handling rating: Is the difficulty level of handling a part high?
2. Insertion: How is the part inserted?
3. Operation: How is the part secured into the assembly?
4. Access: Is the part secured without any damage?
5. Tool: Is the task accomplished by hand or tools?

Rating for each evaluation criterion is from 0 (difficult) to 9 (easy). The elimination/
combination is automatically based on the three questions to be answered regarding
the fundamental reasons for the part to exist separately.

1. During the product operation, does the part move relative to all other parts
   already assembled? Only gross motion should be considered; small motions that

can be accommodated by elastic hinges, for example, are not sufficient for a positive answer.

2. Must the part be made of a different material or be isolated from all other parts already assembled? Only fundamental reasons concerned with material properties are acceptable.

3. Must the part be separate from all parts already assembled because otherwise its assembly or disassembly would be impossible?

If the answer to any of these is no, then the part automatically becomes a candidate for elimination/combination. The user is alerted with a symbol e/c noted beside the part.

1. Handling Rating The Boothroyd–Dewhurst chart provides *handling rates* for difficulties from easy to grasp (9) to requires handling equipment (0). The details of the handling ratings are given in the form of a table.

Handling Description

(Legend: OH = one hand, BH = both hands)		
Is easy to grasp (OH)	9	*Easy*
Is easy to grasp (BH)	8	
Needs orientation to be changed (OH)	7	
Needs orientation to be changed (BH)	6	
Is slippery	5	
Is flexible or small	4	
Is subject to severe nesting or tangling	3	
Requires tool to grasp	2	
Requires two people to grasp	1	
Requires handling equipment	0	*Difficult*

2. Insertion Rating Insertion rating is rated on the basis of how a part is grasped, oriented, and moved for insertion as well as how it is inserted and/or fastened. The details of insertion ratings are given below in the table.

Insertion Description

(Legend: OH = one hand, BH = both hands)		
Presents no difficulty (beta symmetry) (OH)	9	*Easy*
Is not easy to align or position	8	
Requires two hands	7	
Needs regular tool to insert	6	
Presents severe resistance	5	
Requires special tooling	4	
Requires holding down	3	

(continued)

Has obstructed access	2	
Has blind assembly	1	
Has obstructed access and blind assembly	0	*Difficult*

3. Operation Rating *Operation* refers to the method used to secure a part. The following table lists operations involved in the assembly of a part and its rating.

Operation Description

Place part	9
Use snap fit	9
Use push fit	8
Use press fit	7
Screw by hand	6
Screw with manual tool	5
Screw with power tool	4
Use riveting	3
Use crimping	2
Use soldering	1
Use welding	0

4. Access Rating *Access* is defined as the ease with which a part can be reached and removed. This rating is from 0 (difficult) to 9 (easy).

Area Access Description

Is easy to work in; has ample space for hands/tools	9	*Easy*
Has restricted access, but part can be removed without damage	6	
Is vision restricted; requires special care to remove without damage	3	
Is very difficult to access; requires special care/tooling/ techniques to remove part without damaging it	0	*Difficult*

5. Tool Rating These ratings are based on whether a task is accomplished by hand or using common hand tools or with special ones. See the following table for the ratings.

Tool Description

Is not required; task accomplished by hand	9	*Easy*
Requires common hand tool	6	
Requires special tooling/equipment without time delay	3	
Requires special tooling/equipment with time delay	0	*Difficult*

6.4.2 Design for Disassembly (DFD) Section (General Procedure)

A Design for Disassembly (DFD) part is evaluated based on the following five criteria:

1. Task/Damage Is the task easy to accomplish without damage?

2. Reuse Is the part intended for reuse?

3. Removal Does the part removal damage adjacent part(s)?

4. Recyclability Is the part recyclable?

5. Disassembly time What affects the disassembly time?

Rating for each criterion is from 0 (difficult) to 9 (easy).

Task/Damage rating: A rating is based on the nature of the task and the damage to the part considered for removal. Rating for part removal without any damage is from 0 (difficult) to 9 (easy).

Reuse rating: Ratings are based on the reuse level of the part being removed. Re-conditioning is necessary if the part is damaged in the removal process.

Removal rating: *Removal* is defined as the process of removing a part from an assembly without damaging it or its adjacent parts.

Recyclability rating: *Recyclability* is the ease with which a material can be recycled safely and economically. Ratings are based on lead content and recyclability of the materials in the assembly.

Disassembly time rating: The time required to disassemble component(s) for either repair, reuse, or recycling includes the human motions required. Ratings are based on the number of fasteners in the component(s), availability of required special tools, personnel skill, and access to the assembly.

Rating score: The *rating score* is calculated by adding the ratings for task, reuse, removal, recyclability, and disassembly time involved. This total is divided by the sum of the maximum rating of each column (i.e., $9 + 9 + 9 + 9 + 9 = 45$), and that result is multiplied by the weighting of each part (i.e., number of parts of each component in the assembly divided by the total number of components in the assembly).

6.4.3 Design for Maintenance (DFMA) Spreadsheet Section (General Procedure)

The design for maintenance DFMA part of the spreadsheet integrates data for both DFA and DFD as integral parts of the integrated design method. This part is evaluated based on the following four criteria.

1. Maintenance time/cost How much cost or downtime is involved?

2. Recyclability Is the part recyclable in a safe environment?

3. Serviceability Is the location easy to access during maintenance?

4. Complexity Is the task accomplished on-site or off-site?

Rating procedures similar to those for DFA and DFD are used. The rating for each evaluation criterion is from 0 (difficult) to 9 (easy).

Spreadsheet: This integrated tool consists of worksheets concerning DFA, DFD, and DFMA. DFD and DFMA data are placed first in the spreadsheet, and DFA information follows. The ratings description is provided in the last DFA and DFD worksheet, thereby making it easy for the user to update the spreadsheet data when a change in the ratings occurs.

1. Disassembly Rating Chart The disassembly rating chart is generated for the disassembly rating score associated with the integrated design method; refer to **Figure 6.9**. The disassembly ratings for each part are shown in the form of a chart in **Figure 6.10**.

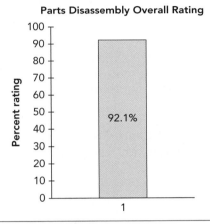

FIGURE 6.9 Overall Parts Disassembly Rating Chart

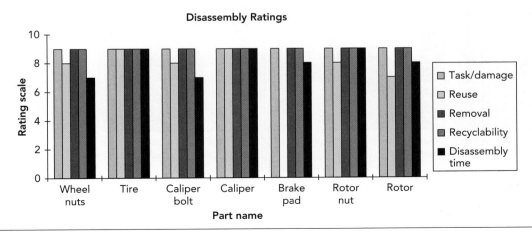

FIGURE 6.10 Disassembly Rating of Each Part

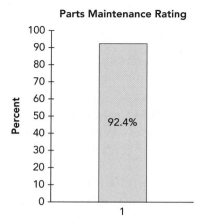

FIGURE 6.11 Parts Maintenance Rating Chart

2. Maintenance Rating Chart The maintenance rating chart is generated from the maintenance rating score associated with the new integrated design method ratings. **Figure 6.11** shows the overall parts maintenance ratings.

The maintenance ratings for each part are shown in **Figure 6.12**.

3. Assembly Rating Chart The assembly rating chart and the chart for ratings of each part are generated. See **Figure 6.13**.

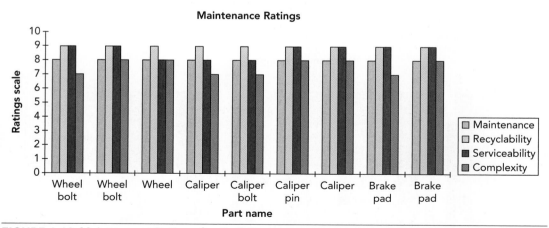

FIGURE 6.12 Maintenance Ratings for Each Part

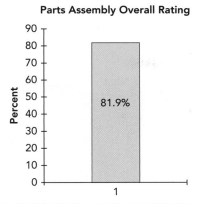

FIGURE 6.13 Overall Parts Assembly Rating Chart

The assembly ratings for each part are shown in **Figure 6.14**.

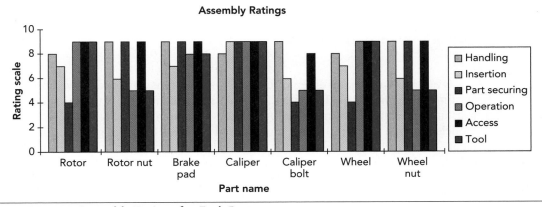

FIGURE 6.14 Assembly Ratings for Each Part

CASE STUDY 6.1
Product Design for Assembly and Disassembly of Industrial Stapler Gun

An industrial staple gun is a handheld device that pushes heavy-duty metal staples into many different materials, such as wood and concrete. These tools were developed as a variation on the traditional tabletop stapler that holds thin materials such as paper or plastic together. An industrial staple gun is used mainly in the construction industry to attach paper-backed insulation to the studs of a building, protective wrap to a building's exterior, or carpet to the floor. The industrial staple gun was developed to have high force when inserting staples. It must be robust and simple in design and application. A traditional stapler is used in offices and schools to hold multiple sheets of paper together.

Merydolla/Shutterstock.com

FIGURE 6.15 Industrial Staple Gun

Over the years, many different industrial staple guns have been developed. The existing design has many benefits. Its stainless steel components that are corrosion resistant and reliable are manufactured using a stamping process. Whereas existing models serve an industrial staple gun's basic functions, their design needs improvement. The model considered in this case study has many parts, making assembly difficult. In addition, the cartridge that holds the staples is removable and easily misplaced. Current manufacturing methods make disassembly difficult without destroying the stapler gun.

DFA of Current Design

See **Table 6.5 and Table 6.6** for the details of components during assembly and disassembly processes.

A product's design efficiency includes its disassembly processes. The current staple gun's disassembly process is time consuming and difficult, resulting in damage to many parts.

Redesign: Part Count Reduction by DFA/DFD

Some parts can be replaced or eliminated to increase ease of assembly and disassembly. Pins and C-clips can be replaced with screws to eliminate the need for special tools and extra parts, and rivets can be replaced with screws to eliminate additional steps and to make assembling and disassembling easy. The backstop can also be eliminated and incorporated into either the left or right frame. To accommodate these changes, both

Part Number	Name	Length	Width	Quantity	Alpha + Beta	Manual Handling Code	Manual Handling Time	Manual Insertion Code	Manual Insertion Time	Total Assembly Time
1	Left frame	152.4	76.2	1	720	0,3	1.95	0,0	1.5	3.45
2	Right frame	152.4	76.2	1	720	0,3	1.95	0,0	1.5	3.45
3	Top frame	73.0	22.2	1	720	0,3	1.95	0,0	1.5	3.45
4	Back stop	38.1	4.8	1	720	0,3	1.95	1,0	2.5	4.45
5	Trigger assembly	120.7	41.3	1	720	0,3	1.95	2,0	2.5	4.45
6	Staple cartridge	152.4	6.4	1	540	0,2	1.80	2,0	2.5	4.3
7	Cartridge tray	146.1	12.7	1	720	0,3	1.95	0,0	1.5	3.45
8	Compression spring	57.2	8.3	1	180	0,0	1.13	1,0	2.5	3.63
9	Large torsional spring	95.3	3.2	1	720	0,3	1.95	3,0	3.5	5.45
10	Small torsional spring	6.4	6.4	1	720	0,3	1.95	3,0	3.5	5.45
11	C-clip	0.8	6.4	4	180	8,4	7.00	3,0	3.5	42
12	Pin	25.4	3.2	4	360	0,1	1.50	0,0	1.5	12
13	Rivet	19.1	3.2	1	360	0,1	1.50	6,0	5.5	7
14	Rubber stop	19.1	3.2	1	540	0,2	1.80	3,0	3.5	5.3
15	Metal stop	19.1	3.2	1	540	0,2	1.80	3,0	3.5	5.3
16	Hanger	28.6	1.6	1	540	0,2	1.80	1,0	2.5	4.3
17	Pusher assembly	59.1	12.7	1	720	0,3	1.95	6,0	5.5	7.45
	Total									124.88

TABLE 6.5 DFA Analysis of Current Stapler Gun

0 = Difficult, 9 = Easy

Step	Name	Quantity	Weight (gm)	Damage	Reusability	Removability	Recyclability	Disassembly Time	Rating Score
1	Staple cartridge	1	4.3	9	9	9	9	9	4.3
2	C-clip	4	17.4	9	9	5	9	5	14.3
3	Pin	4	17.4	9	9	9	9	8	17.0
4	Rivet	1	4.3	3	0	3	9	5	1.9
5	Hanger	1	4.3	7	9	7	9	8	3.9
6	Top frame	1	4.3	9	9	8	9	7	4.1
7	Cartridge tray	1	4.3	9	9	9	9	6	4.1
8	Left frame	1	4.3	9	9	6	9	7	3.9
9	Back stop	1	4.3	9	9	9	9	6	4.1
10	Large torsional spring	1	4.3	9	9	5	9	4	3.5
11	Small torsional spring	1	4.3	9	9	4	9	3	3.3
12	Trigger assembly	1	4.3	9	9	7	9	5	3.8
13	Compression spring	1	4.3	9	9	6	9	6	3.8
14	Rubber stop	1	4.3	9	9	4	9	4	3.4
15	Metal stop	1	4.3	9	9	8	9	6	4.0
16	Pusher assembly	1	4.3	9	9	6	7	6	3.6
17	Right frame	1	4.3	9	9	6	9	7	3.9
	Total parts	23						Total score	86.6

TABLE 6.6 DFD Analysis of Current Stapler Gun

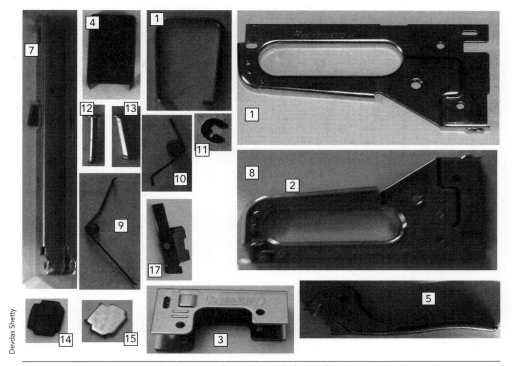

FIGURE 6.16 Disassembled Parts of an Industrial Stapler Gun

the frames must be redesigned; one side must have a secondary operation of threading. This reduction in parts causes a significant reduction in assembly time and results in an increase in disassembly score.

The information obtained by an assembly analysis from **Tables 6.5** and **6.7** indicates that assembly time has been reduced by 50.45 seconds and five parts have been eliminated. To prevent accidentally opening the staple gun, the screws can be safety screws, which are star screws with a raised cylinder in the center. These minor changes can result in major gains in the large number of quantity produced and economic advantage. These changes also affect the design for disassembly score.

The reduction in the staple gun's parts improved its disassembly scores, increasing the ease of disassembly. Refer to **Tables 6.6** and **6.8**. This allows parts to be repaired or reworked, reducing scrap.

Cartridge Redesign

The reduction in parts also improved the disassembly scores, making the staple gun easier to disassemble and allows parts to be repaired or reworked. This allows the company to reduce scrap and add a department for consumer repairs. This doesn't solve all areas for improvement; even with this design the staple cartridge is removable and

Part Number	Name	Length	Width	Quantity	Alpha + Beta	Manual Handling Code	Manual Handling Time	Manual Insertion Code	Manual Insertion Time	Total Assembly Time
1	Left frame	152.4	76.2	1	720	0,3	1.95	0,0	1.5	3.45
2	Right frame	152.4	76.2	1	720	0,3	1.95	0,0	1.5	3.45
3	Top frame	73.0	22.2	1	720	0,3	1.95	0,0	1.5	3.45
4	Trigger assembly	120.7	41.3	1	720	0,3	1.95	2,0	2.5	4.45
5	Staple cartridge	152.4	6.4	1	540	0,2	1.8	2,0	2.5	4.3
6	Cartridge tray	146.1	12.7	1	720	0,3	1.95	0,0	1.5	3.45
7	Compression spring	57.2	8.3	1	180	0,0	1.13	1,0	2.5	3.63
8	Large torsional spring	95.3	3.2	1	720	0,3	1.95	3,0	3.5	5.45
9	Small torsional spring	6.4	6.4	1	720	0,3	1.95	3,0	3.5	5.45
10	Screw	25.4	3.2	4	360	0,1	1.5	0,0	1.5	12.00
11	Screw pin	19.1	3.2	1	360	0,1	1.5	0,0	1.5	3.00
12	Rubber stop	19.1	3.2	1	540	0,2	1.8	3,0	3.5	5.30
13	Metal stop	19.1	3.2	1	540	0,2	1.8	3,0	3.5	5.3
14	Hanger	28.6	1.6	1	540	0,2	1.8	1,0	2.5	4.30
15	Pusher assembly	59.1	12.7	1	720	0,3	1.95	6,0	5.5	7.45
	Total									74.43

TABLE 6.7 DFA Analysis of Revised Design

0 = Difficult, 9 = Easy

Step	Name	Quantity	Weight (gm)	Damage	Reusability	Removability	Recyclability	Disassembly Time	Rating Score
1	Staple cartridge	1	5.6	9	9	9	9	9	22.2
2	Screw	4	22.2	9	9	9	9	9	22.2
3	Screw pin	1	5.6	9	9	9	9	8	5.4
4	Top frame	1	5.6	9	9	6	9	7	4.9
5	Hanger	1	5.6	7	9	7	9	8	4.9
6	Left frame	1	5.6	9	9	6	9	7	4.9
7	Cartridge tray	1	5.6	9	9	9	9	6	5.2
8	Large torsional spring	1	5.6	9	9	5	9	4	4.4
9	Small torsional spring	1	5.6	9	9	4	9	3	4.2
10	Trigger assembly	1	5.6	9	9	7	9	5	4.8
11	Compression spring	1	5.6	9	9	6	9	6	4.8
12	Rubber stop	1	5.6	9	9	4	9	4	4.3
13	Metal stop	1	5.6	9	9	8	9	6	5.1
14	Pusher assembly	1	5.6	9	9	6	7	6	4.6
15	Right frame	1	5.6	9	9	6	9	7	4.9
	Total parts	18						Total score	90.2

TABLE 6.8 DFD Analysis of Revised Design

267

George Adleman

FIGURE 6.17 Star screws used in staple gun

easily lost. Therefore, the staple cartridge must be redesigned. To start with the redesign, a statement-restatement technique was utilized.

The problem is redefined using statement-restatement techniques.

1. Develop a *removable cartridge* that holds and advances the staples.
2. Develop a loadable cartridge *that orients* and advances the staples.
3. Develop a *nonremovable* cartridge that is easily loadable and advances staples.
4. Develop an *easily removable,* loadable cartridge for staples.
5. Develop a *nonremovable,* easily loadable cartridge to *orient* and advance staples.

The best design statement selected is number 5. The next step in the redesign is to brainstorm different ways to perform each function in the design statement. Several ideas are introduced in a morphological chart and are then combined into any feasible combinations. The morphological chart in **Table 6.9** shows a combination of possibilities.

Many solutions can be formed by these combinations, but the most viable, cost effective and robust solution is attached by using a slider mechanism, orienting the staples on the inner wall, and advancing the staples via a compression spring.

New Design Features of Industrial Stapler Gun

The new staple gun cartridge that is on a slide allows the staples to be loaded easily and cannot be removed without full disassembly of the stapler. This prevents the cartridge from being lost and changes the assembly time and the disassembly score. On the bottom of the cartridge is a spring-loaded mechanical clip that holds the staple

Function	Solutions		
Attached to body	Cable	Pinned	Slider
Orient staples	Inner cartridge wall	Metal pusher	
Lock	Slider pin	Hook	Magnetic
Advance staples	Leaf spring	Compression spring	Linear motor

TABLE 6.9 Morphological Chart of Staple Cartridge

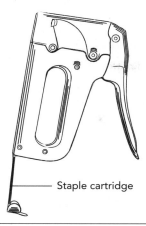

Staple cartridge

FIGURE 6.18 New Design for Stapler Gun

cartridge and advances the staples as shown in **Figure 6.18**. An assembly procedure to facilitate easy assembly without the risk of losing the cartridge must be established. The design changes to the sliding cartridge increases assembly time but by only 15 seconds per assembly. Even though this is a small increase in time, the costs associated with assembly will increase. The manufacturing costs of the new cartridge can also include new dies and increased capital cost. Although the costs increase, the organization will benefit from reduced customer complaints from the current product. These changes will make disassembly more difficult but are needed to prevent losing the staple cartridge. This change will also affect the product's disassembly score.

In the final analysis, the assembly time has been reduced by 50.3 seconds from the original design, although the disassembly score increased by 2.9. These improvements will generate savings for the organization in the long run. The product can be enhanced further as a part of continuous improvement process.

CASE STUDY 6.2
Automotive Example: Integrated Method of DFA/DFD and DFMA

Brake Pad Maintenance

As discussed earlier, the integrated method (DFA/DFD and DFMA) evaluates design for assembly, disassembly, and maintenance based on rating factors. The example considered here is that of addressing brake pad needs by preventive maintenance and extending the life of a 2003 automobile.

This case evaluates the assembly and disassembly of the rotor to access brake pads to determine its maintenance design score.

General Steps

1. Gather the required tools.
2. Perform the disassembly/maintenance/assembly.
3. Photograph each step.
4. Time each operation using a stopwatch.
5. Evaluate the scores.

This case study examines the steps for design for maintenance in three stages (DFD, DFMA, and DFA).

Each worksheet (DFD, DFMA and DFA) has a rating for each individual part in the operation and overall data for the design score. The disassembly process is described first and then disassembly for maintenance and assembly are presented step by step in detail (**Figure 6.19**).

Piston pushed

FIGURE 6.19 Disassembly for Maintenance (Replacing Brake Pads) (*continued*)

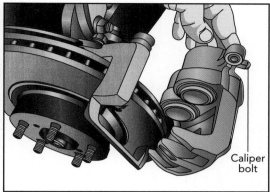

Caliper
bolt

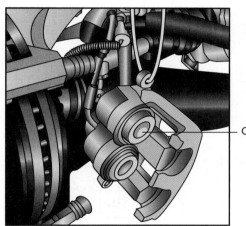

Caliper moved

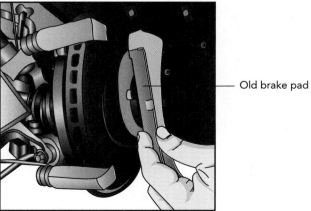

Old brake pad

FIGURE 6.19 Disassembly for Maintenance (Replacing Brake Pads)

Integrated Tool to Evaluate DFA/DFD/MAINTENANCE Based on Rating Factors

Step	Task	Part	Number of Parts	Maintenance Time/Cost Rating 0 = Difficult 9 = Easy	Recyclability Rating 0 = Difficult 9 = Easy	Service Rating 0 = Difficult 9 = Easy	Complexity Rating 0 = Difficult 9 = Easy	Maintenance Rating Score	Time Measured (sec)	Total Time	Time Rating	Estimate Time < than Average = 1 > than Average = 3	Estimated Weighted Time Rating
1	Remove wheel bolts	Wheel Bolt	4	8	9	9	7	20.4%	15	60	4.00	1	0.22
2	Greasing wheel bolt	Wheel Bolt	4	8	9	9	8	21.0%	15	60	4.00	1	0.22
3	Remove wheel	Wheel	1	8	9	8	8	5.1%	40	40	2.06	3	1.50
4	Caliper piston pressed	Caliper	1	8	9	8	7	4.9%	5	5	1.00	1	0.06
5	Remove caliper bolt	Caliper bolt	2	8	9	8	7	9.9%	60	120	12.34	3	0.33
6	Greasing caliper Bolt	Caliper pin	1	8	9	9	8	5.2%	15	15	1.00	1	0.50
7	Remove caliper	Caliper	1	8	9	9	8	5.2%	5	5	1.00	1	0.06
8	Loosen brake pads	Brake Pad	2	8	9	9	7	10.2%	5	10	2.00	1	0.11
9	Remove brake pad	Brake Pad	2	8	9	9	8	10.5%	15	30	2	1	0.11
	Total		18						175	345	29.40	13	32.14%
	Average		2					92.4%	19.44	38.33	3.27	1	
	Total Steps		9										

MAINTENANCE RATING SCORE   92.4%

PARTS MAINTENANCE RATING

	Actual Time	Est Time
Rating Score	92.4%	92.4%
Time Score	61.22%	32.14%
Parts Score	50.00%	50.00%
Overall Design Score	67.9%	58.19%

FIGURE 6.20 Integrated Method to evaluate DFMA

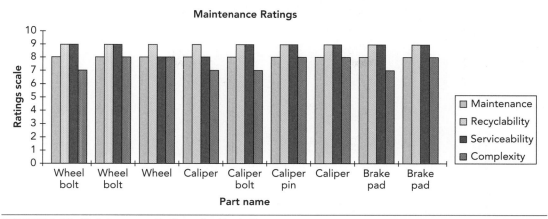

FIGURE 6.21 Maintenance Ratings for Each Part

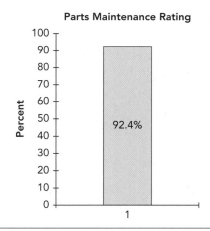

FIGURE 6.22 Parts Maintenance Rating

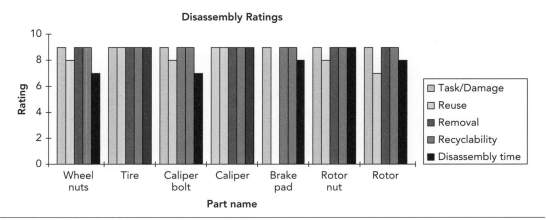

FIGURE 6.23 Disassembly Ratings for Each Part

Integrated Tool to Evaluate DFA/DFD/MAINTENANCE Based on Rating Factors

Step	Item No	Name	Number of Parts	Task/Damage Rating 0 = Difficult 9 = Easy	Re-Use Rating 0 = Difficult 9 = Easy	Removal Rating 0 = Difficult 9 = Easy	Recyclability Rating 0 = Difficult 9 = Easy	Disassembly Time Rating 0 = Difficult 9 = Easy	Disassembly Rating Score $((E+G+I+K+M)/45) \times$ Wgt	Time Measured (sec)	Total Time	Time Rating	Estimate Time < than Average = 1 > than Average = 3	Estimated Weighted Time Rating
1		Wheel nuts	4	9	8	9	9	7	24.9%	15	60	4.00	1	0.27
2		Tyre	1	9	9	9	9	9	6.7%	40	40	2.00	3	0.20
3		Caliper bolt	2	9	8	9	9	7	12.4%	60	120	10.50	3	0.40
4		Caliper	1	9	9	9	9	9	6.7%	15	15	1.00	1	0.07
5		Brake Pad	2	9	0	9	9	8	10.4%	15	30	2.00	1	0.13
6		Rotor Nut	4	9	8	9	9	7	24.9%	10	40	4.00	1	0.27
7		Rotor	1	9	7	9	9	8	6.2%	5	5	1.00	1	0.07
		Total	**15**	**63**	**49**	**63**	**63**	**55**	**92.1%**	**160**	**310**	**24.50**	**11**	**71.43%**
		Average	**2.14**	**9**	**7**	**9**	**9**	**7.86**		**22.86**	**44.29**	**3.50**	**2**	
		Total Steps	**7**											

PARTS DISASSEMBLY RATING

	Actual Time	Est Time
Rating Score	92.1%	92.1%
Time Score	61.22%	71.43%
Parts Score	46.67%	46.67%
Overall Design Score	**66.7%**	**70.08%**

DISASSEMBLY RATING SCORE   92.1%

FIGURE 6.24 Integrated Method to Evaluate DFD

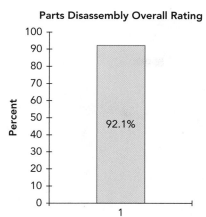

FIGURE 6.25 Overall Parts Disassembly Rating

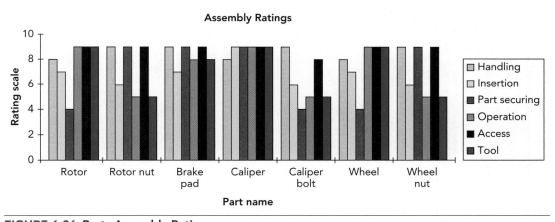

FIGURE 6.26 Parts Assembly Ratings

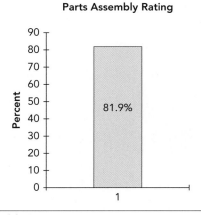

FIGURE 6.27 Parts Assembly Rating

Step	Item No	Name	eliminate/combine	Number of Parts	Theoretical minimum parts required	Relative movement; Separate material; Separate part; Y or N?	Handling Rating 0=Difficult 9=Easy	Insertion Rating 0=Difficult 9=Easy	Part secured on insertion =9 Added not secured=4 Secured separately=0	Operation Rating 0=Difficult 9=Easy	Access Rating 0=Difficult 9=Easy	Tool Rating 0=Difficult 9=Easy	Assembly Rating Score ((J+L+N+O+P+R)/54) × Wgt	Time Measured (sec)	Total Time	Time Rating	Estimated Time < than Average=1 > than Average=3	Estimated Weighted Time Rating
1	1	Rotor		1	1	y	8	7	4	9	9	9	5.7%	5	5	1.00	1	0.07
2	2	Rotor nut	e/c	4	0	n	9	6	9	5	9	5	21.2%	15	60	4.00	1	0.27
3	3	Brake pad		2	2	y	9	7	9	8	9	8	12.3%	7.5	15	2.00	1	0.13
4	4	Caliper		1	1	y	8	9	9	9	9	9	6.5%	15	15	1.00	1	0.07
5	5	Caliper bolt	e/c	2	0	n	9	6	4	5	8	5	9.1%	60	120	8.73	3	0.40
6	6	Wheel		1	1	y	8	7	4	9	9	9	5.7%	60	60	2.18	3	0.20
7	7	Wheel Nut		4	4	y	9	6	9	5	9	5	21.2%	30	120	4.00	1	0.27
		Totals		15	9								81.9%	192.5	395	22.91	11	
		Average		2.14	1.29									27.5	56.43	3.27	2	71.43%
		Total No of steps		7														

ASSEMBLY RATING SCORE   81.9%

PARTS ASSEMBLY RATING

	Actual Time	Est Time
Rating Score	81.9%	81.9%
Time Score	65.48%	71.43%
Parts Score	46.67%	46.67%
Overall Design Score	64.66%	66.65%

FIGURE 6.28 Integrated Method to Evaluate DFA

	DFA	DFD	DFMA	Existing Design
Rating score	81.00%	92.10%	92.40%	84.44%
Time (secs)	395	310	345	
Number of parts	15	15	18	

TABLE 6.10 Comparison of Results of DFA, DFD, and DMA for the Existing Design

Results

The consideration of design for maintenance (DFMA) with DFA and DFD helps designers to evaluate the design of a product or subassembly regarding its end of life and thereby identify any areas that should be redesigned to extend its life by performing necessary preventive, predictive, or reactive maintenance operations.

The integrated design method evaluates design for assembly, disassembly, maintenance, and safety based on various rating factors and offers DFA/DFD/DFMA in the same platform (refer to **Figures 6.20**, **6.24**, and **6.28**) and aids the designers to consider the maintenance issues during the initial product development stages. This method helps designers to consider how to reduce an existing product design's maintenance costs and downtime and improve the safety. The integrated method help a designer to evaluate a design for disassembly, maintenance and assembly based on rating factors which are used to score an existing product. See **Table 6.10**.

REFERENCES

1. B. R. Allenby, "Design for Environment: A Tool Whose Time Has Come," *SSA Journal* (September 1991), 5–10.
2. T. Amezquita, R. Hammond, M, Salazar, and B. Bras, "Characterizing the Re-manufacturability of Engineering Systems," *Proceedings of ASME Design Technical Conference*, 82 (Boston, MA: 1995), 271–278.
3. K. Ishii, C. Eubanks, and P. Di Marco, "Design for Product Retirement and Material Life-Cycle," *Materials and Design* 15, no. 4 (1994), 225–233.
4. Richard Crowson, *Handbook of Manufacturing Engineering*, 2nd ed. (Boca Raton, FL: CRC Press, 2006).
5. Nanua Singh, *Systems Approach to Computer-Integrated Design and Manufacturing* (New York: John Wiley, 1996).
6. M. P. Groover, *Fundamentals of Modern Manufacturing: Materials, Processes and Systems*, 5th ed. (John Wiley, 2013).
7. K. T. Ulrich and D. Eppinger, *Product Design and Development*, 5th ed. (New York: McGraw-Hill, 2011).
8. G. Boothroyd, *Assembly Automation and Product Design*, 2nd ed. (Boca Raton, FL: Taylor & Francis, 2005).
9. M. Hrinyak, Bret Bras, and W. Hoffman, *Enhancing Design for Disassembly: A Benchmark of DFD Software Tools, ASME Design Engineering Technical Conference and Computers in Engineering Conference, Irvine, CA, 1996.*
10. E. Kroll, *Development of a Disassembly Evaluation Tool, ASME Design Engineering Technical Conference and Computers in Engineering Conference, Irvine, CA, 1996.*
11. K. Ishii and B. Lee, *Reverse Fishbone Diagram: A Tool in Aid of Design for Product Retirement, ASME Design Engineering Technical Conference and Computers in Engineering Conference, Irvine, CA, 1996.*
12. Devdas Shetty, Vishwesh Coimbatore, and Claudio Campana, "*Design Methodology for Assembly and Disassembly Based on Rating Factors*," Proceedings of IMECE, Orlando, Florida, USA, 2005.

13. C. Poli, "*Design for Manufacturing: A Structured Approach*," (Woburn, MA: Butterworth-Heinemann, 2001).

14. Geoffrey Boothroyd and Peter Dewhurst, "*Design for Assembly: A Designer's Handbook*," (Amherst, MA: Boothroyd Dewhurst Inc., 1983).

15. Devdas Shetty and Ken Rawolle, "*A New Methodology for Ease of Disassembly in Product Design*," (New York: ASME Press, 2000).

16. G.P. Sullivan, R. Pugh, A.P. Melendez and W.D. Hunt, "*Operations and Maintenance Best Practices; A Guide to Achieving Operational Efficiency*," (Pacific Northwest National Laboratory and U.S department of Energy, 2004).

EXERCISES

6.1. Explain the importance of incorporating DFD guidelines during an original design. List a few benefits that are achieved by doing that.

6.2. Write briefly about any similarities or differences that you see between DFA and DFD guidelines. Explain the interaction between the guidelines. (Do they supplement or contradict each other?)

6.3. Practice making a design for disassembly evaluation chart using the spreadsheets. (The rating factors method and Kroll's disassembly evaluation chart can be developed and compared).

6.4. How can the use of self-aligning and self-locating features facilitate automatic assembly?

6.5. In what way does the method of fastening and joining a component to a product affect the feasibility of recycling?

6.6. What is the role of (a) alpha, (b) beta, and (c) preproduction prototypes in evaluating new products?

6.7. With a flow chart, explain what the term *design for manufacturing methodology* means.

CHAPTER 7
Product Architecture: The Impact on Manufacturing

CHAPTER OUTLINE

OBJECTIVES

The objective of this chapter is to examine the areas that critically impact cost-effective product development, both for new products as well as for products already in process. A close interaction between designers and those involved in manufacturing is essential for making the correct manufacturing decisions. Designers should be aware of the manufacturing consequences of their decisions because minor design changes during the early stages often prevent major problems later. The choice of a workplace structure depends on the design of the parts and lot sizes to be manufactured as well as market factors, such as the responsiveness to changes. The chapter discusses ways to structure manufacturing processes. Value Stream Mapping shows how a systematic process of the flow of material can be introduced either in existing production systems or while introducing a new product. However, without a clear understanding of various philosophies of manufacturing, such as the Toyota production system, pull systems, kanban, just in time (JIT), standard work, and one-piece flow, an organization cannot obtain full benefits from value stream mapping. The chapter's case study examines the implementation of lean philosophies to develop continuous improvement and reduce non-value-added processes.

7.1 Introduction to Processes for Efficient Manufacturing

7.1.1 Preparation for Making Manufacturing Process Decisions

Efficient manufacturing begins with the design process by identifying a market need and continues through the stages of conception, refinement, and detailing. The results of these stages lead to a set of specifications or constraints that dictate how a product should be made. In the conceptual design stage, all possible manufacturing processes should be considered. As the design progresses, more product information becomes available and can be used to identify processes capable of making the item(s). As a design reaches its final stages and becomes detailed enough to allow cost evaluation, a single manufacturing process can be selected. Joseph Datsko (1997) lists 11 design rules that should be considered for ease of production:

1. Select a material on the basis of ease of fabrication as well as function and original cost.
2. Use the simplest configuration and specify standard sizes whenever possible.
3. Use the configuration requiring the fewest separate operations.
4. Use configurations attainable with efficient manufacturing.
5. Design the fabrication processes to achieve the desired strength distribution in the finished part.
6. Provide clamping, locating, and measuring surfaces.
7. Specify tolerances and surface finish with consideration of functional requirements and processes **Figure 7.1**.

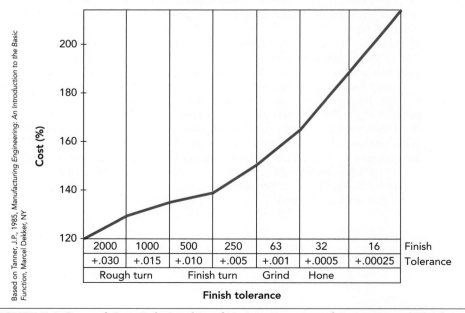

Based on Tanner, J.P., 1985, Manufacturing Engineering: An Introduction to the Basic Function, Marcel Dekker, NY

FIGURE 7.1 General Cost Relationship of Various Degrees of Accuracy and Finish

8. Determine each part's specific function.
9. Determine each part's specific feature that enables it to perform its identified function.
10. Prepare a process document for systematically listing those processes needed for each component.
11. Evaluate the preliminary design by considering any changes to it that can simplify the fabrication process.

Making correct manufacturing decisions early in the design process is extremely important; they can influence the selection of the appropriate process and the cost to make the product. A systematic procedure for the process selection should consider all processes and eliminate those that cannot satisfy the design requirements. Charts that show database information are useful in ranking contemplated processes based on their cost and other areas of interest.

During the early stages of product design, computer-aided tools can provide product modeling geometry and selection criteria for manufacturing and assembly. Close interaction among designers and those involved in manufacture is essential because minor design changes in the early stages often prevent costly major manufacturing problems at a later stage. For example, a product might be required to perform a completely new function or satisfy a need it has not previously filled. Identification of the most economical fabrication process for a new product is a major prerequisite for making a competitive product. Similarly, a product that is being modified might require the use of a different manufacturing process to satisfy new requirements.

Processes can be categorized in classes such as deformation, machining, and molding and so on (**Figure 7.2**). These processes can be subdivided into various categories, such as different casting processes as shown in the figure. Each individual process has attributes that can be compared to the design need.

Process selection generally occurs at the component level to ensure that the manufacturing process can economically produce each component. Typically, a product consists of components and their assembly. For example, an air conditioner is composed of components such as compressor, fan, cooling unit, and frame to mount, each of which is considered individually.

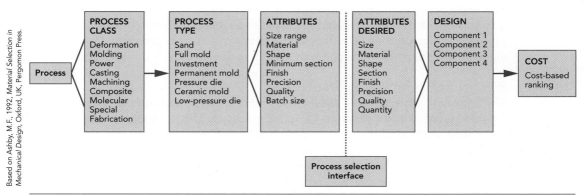

Based on Ashby, M.F., 1992, *Material Selection in Mechanical Design*, Oxford, UK, Pergomon Press.

FIGURE 7.2 Matching of Process Attributes with Design Needs

Manufacturing Process Selection

The selection of a particular manufacturing process depends on many factors. The designer must be familiar with conventional as well as specialized processes. Information about various processes can be obtained from databases and handbooks that qualitatively describe various operations including the principles, required equipment, process parameters, capabilities, and applications. However, these data are not available to facilitate comparison of the actual processes but only of their capabilities. To develop a systematic method for selecting a process, information about each should be structured to make comparison easy.

A number of product attributes determine the choice of a manufacturing process; these include size, shape, finish, strength, volume, cost, material availability, fabrication ability, useful life, physical loading, chemical environment, disposal, and recycling. The size of the component being manufactured influences the fabrication procedure, its mounting on a machine tool, and the process of production. This selection is based on functional considerations of suitable properties, strength, and ease of fabrication to the required shape. See **Table 7.1** for parameters that should be considered during the selection process.

Mechanical Parameters	Physical Parameters	Suitability for Processing
Density	Thermal conductivity	Machining
Strength	Magnetic properties	Casting
Elasticity	Melting point	Forging
Toughness	Specific heat	Drawing
Hardness		Joining
Wear resistance		
Fatigue		

Based on Hundal, Mahendra, *Systematic Mechanical Designing – A cost and management perspective*, ASME Press, NY, 1997.

TABLE 7.1 Parameters for Manufacturing Processes

Information regarding the match of manufacturing process attributes to the materials being considered can be organized in charts (**Table 7.2**). Each individual process is characterized by a set of features referred to as *process attributes* that influence the process selection decision. The design attributes that should be considered when choosing a process are listed in the columns under the title Attributes in **Table 7.3**.

Process Selection: Using Charts

The successive application of charts based on size, shape, and complexity can narrow the material choice to a short list of viable processes. The presence of additional features, such as holes, threads, undercuts, bosses, and re-entrant shapes—which cause manufacturing difficulty or require additional operations to produce—is referred to as the *complexity of the component*. Charts have obvious limitations, but they can provide an initial, graphical comparison of the capabilities of various manufacturing processes that designers can use as quick references.

TABLE 7.2 Data to Use in Matching Materials and Manufacturing Process

Material	Casting (die)	Casting (Centrifugal)	Injection Molding	Sand Casting	Investment Casting	Milling	Grinding	Electrical Discharge Machining (EDM)	Forging	Rolling	Extrusion	Powder Metallurgy	Sheet Metal working	Blow Molding
Low carbon steel	–	E	–	E	E	G	E	E	G	G	G	E	G	–
High carbon steel	–	E	–	E	E	G	E	E	G	G	G	E	G	–
Low alloy steel	–	E	–	E	E	G	E	E	G	G	G	E	G	–
Stainless steel	–	G	–	E	E	–	–	E	G	G	G	E	G	–
Malleable iron	–	E	–	E	E	G	E	E	S	S	S	E	G	–
Alloy cast iron	–	E	–	E	E	G	E	E	S	S	S	E	G	–
Zinc alloys	E	–	–	G	S	–	S	E	S	S	G	E	E	–
Aluminum alloys	E	E	–	E	E	E	G	E	E	E	E	E	E	–
Titanium alloys	–	–	–	–	S	–	S	E	G	S	S	E	–	–
Copper alloys	G	E	–	E	G	E	G	E	E	E	E	E	E	–
Nickel alloys	–	E	–	E	G	–	S	E	S	G	G	E	G	–
Tungsten alloys	–	–	–	–	G	–	S	E	S	–	–	E	–	–
Acrylonitrile butadiene styrene (ABS)	–	–	–	–	–	G	G	–	–	–	E	–	–	G
Nylons	–	–	E	–	–	G	G	–	–	–	G	–	–	G
Polystyrene	–	–	E	–	–	G	G	–	–	–	E	–	–	G
Polyvinyl chloride	–	–	–	–	–	G	G	–	–	–	E	–	–	G
Polyurethane	–	–	–	–	–	G	G	–	–	–	G	–	–	G
Polyethylene	–	–	E	–	–	G	G	–	–	–	E	–	–	E
Acrylics	–	–	–	–	–	G	G	–	–	–	S	–	–	–
Epoxies	–	–	E	–	–	G	G	–	–	–	S	–	–	–
Silicones	–	–	–	–	–	–	–	–	–	–	S	–	–	–
Polyester	–	–	–	–	–	G	G	–	–	–	S	–	–	–
Rubbers	–	–	E	–	–	–	–	–	–	–	S	–	–	–

E = Excellent; material is most suitable for the process.

G = Good; material is a good candidate for the process.

S = Seldom used in the process

– = Unsuitable; material is neither suitable nor used for the process.

Based on Ashby, M.F., 1992, *Material Selection in Mechanical Design*, Oxford, UK, Pergomon Press.

Process		Attributes					
	Surface	Dimensional Accuracy	Complexity	Production Rate	Production Run	Relative Cost	Size (Projected area)
Pressure die casting	L	H	H	H/M	H	H	M/L
Centrifugal casting	M	M	M	L	M/L	H/M	H/M/L
Compression molding	L	H	M	H/M	H/M	H/M	H/M/L
Injection molding	L	H	H	H/M	H/M	H/M/L	M/L
Sand casting	M	M	M	L	H/M/L	H/M/L	H/M/L
Shell mold casting	L	H	H	H/M	H/M	H/M	M/L
Investment casting	L	H	H	L	H/M/L	H/M	M/L
Machining	L	H	H	H/M/L	H/M/L	H/M/L	H/M/L
Grinding	L	H	M	L	M/L	H/M	M/L
Electrical Discharge Machining	L	H	H	L	L	H	M/L
Sheet metal working	L	H	H	H/M	H/M	H/M/L	L
Forging	M	M	M	H/M	H/M	H/M	H/M/L
Rolling	L	M	H	H	H	H/M	H/M
Extrusion	L	H	H	H/M	H/M	H/M	M/L
Powder metallurgy	L	H	H	H/M	H	H/M	L
Units	millimeters	millimeters		Parts/hour	Parts		m²
High (H)	>0.064	<0.13	High	>100	>5,000	High	>0.5
Medium (M)	>0.016	>0.13	Medium	>10	>100	Medium	>0.02
Low (L)	<0.0064	<1.3	Low	<100	<5000	Low	<0.5
	<0.0016	>1.3		<10	<100		<0.02

Based on Ashby, M.F., 1992, *Material Selection in Mechanical Design*, Oxford, UK, Pergomon Press.

TABLE 7.3 Manufacturing Process Attributes

Surface Finish	Root Mean Square (RMS), μm	Relative Cost
Very rough; machined	50.0	1
Rough	26.0	3
Semirough	13.0	6
Medium	6.5	9
Semifine	3.2	13
Fine	1.6	18
Coarse, ground	0.8	20
Medium	0.4	30
Fine	0.2	35
Superfine, lapped	0.1	40

TABLE 7.4 Surface Finish and Cost Comparison for Various Operations

The cost of manufacturing is a major criterion. Certain factors that influence the final cost—materials, batch size, production rate—can be built into computer-generated model information and display regarding the recommended process. Factoring in other costs such as those related to experience, idle plants, and inventory is more difficult. At the screening stage, the quality of the output of several manufacturing processes are compared as shown in **Table 7.4**. The data output of all possible processes satisfying design requirements are considered. The table gives approximate relative costs required to achieve different grades of surface finishes using various operations. The standard unit cost of rough machining operation is compared to that of other processes.

The cost increase of attaining a higher degree of accuracy and finer surface finish is illustrated by the curve rising to the right in **Figure 7.1**. However, the cost of assembly and fabrication of a manufactured component that consists of required parts is usually less if precise interchangeable parts are used that facilitate easy assembly. This is reflected in **Figure 7.3** by the curve falling to the right. The combination of two factors

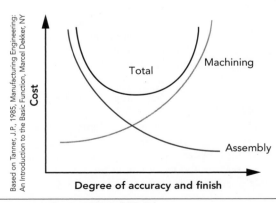

FIGURE 7.3 Cost versus Accuracy and Finish

leads to the optimum choice, the lowest overall cost of manufacture and assembly. The shape of this curve is similar to the relationship that is observed in manufacturing economics when optimum cost versus machining speed is calculated. A similar optimum relationship occurs with reliability calculations. Improved reliability also can decrease costs incurred after delivery, such as those related to warranties.

7.1.2 Best Practices for Efficient Manufacturing

A product developer who selects a manufacturing process should design not only to serve the function but also to facilitate the fabrication process. The material selected largely dictates the process of manufacturing operation. Certain manufacturing processes such as casting processes for metals, stamping and molding processes for plastics allow for the creation of complex shapes. A part's design should take full advantage of the specific manufacturing process. See **Table 7.5** for general guidelines for efficient manufacturing. **Table 7.6** provides guidelines for manufacturing processes such as machining, sheet metal working, forging, rolling, extrusion, and powder metallurgy.

Additive manufacturing processes are being increasingly used for product prototyping as well as for the final products in many situations. This topic is further discussed in Chapter 9 (Section 9.6.3 and 9.6.4).

Element	Guideline
Standardization	1. Use standard components as much as possible. 2. Preshape the workpiece, if appropriate, by casting, forging, or welding. 3. Use standard preshaped workpieces if possible. 4. Employ standard machined features if possible.
Raw material(s)	1. Choose raw material(s) that will result in minimum component cost. 2. Use raw material(s) in standard forms supplied.
Machining	1. If possible, design a component that can be produced on one machine. 2. If possible, design a component whose nonfunctional parts need no machining. 3. Design the component so that when gripped on the work's holding device, the workpiece is sufficiently rigid to withstand the machining forces. 4. Make sure that when features are machined, the part, tool, and tool holder do not interfere with each other. 5. Ensure that auxiliary holes are parallel or normal to the workpiece axis or reference surface and are related by a drilling pattern. 6. Make sure that the end blind holes are conical and that the thread for tapped blind holes does not continue to the bottom of the hole. 7. Avoid bent holes.
Cylindrical components	1. Ensure that the diameters of cylindrical surfaces increase from an exposed face of the workpiece. 2. Ensure that the diameters of internal features decrease from the exposed face of the workpiece. 3. Avoid internal features for long components. 4. Avoid components with large or very small length/diameter ratios. 5. Specify radii for internal corners on the components to be equal to the tool radius.

Element	Guideline
Nonrotational parts	1. Provide a base for holding work and reference. 2. Ensure that the exposed surfaces of the components consist of a series of mutually perpendicular plane surfaces parallel to and normal to the base. 3. Avoid cylindrical bores in long components. 4. Avoid extremely long and extremely thin components. 5. If possible, restrict plane surface machining (slots, grooves, etc.) to one surface of the component.
Assembly	1. During assembly, ensure that internal corners do not interfere with a corresponding external corner on the mating component. 2. Specify the widest tolerances and roughest surface that will give acceptable performance for operating surfaces.
Kinematics	1. Base the initial design on kinematics principles, and modify as necessary to meet the requirements of load and wear.

TABLE 7.5 General Guidelines for Efficient Manufacturing

(a) Representative Design Guidelines for Machining (Milling)

Guideline	Poor	Improved
Provide for flat surfaces.		
Arrange surfaces on one level and parallel to the clamping surface.		
Avoid undercuts.		
Make changes to reduce machining time.		

(continued)

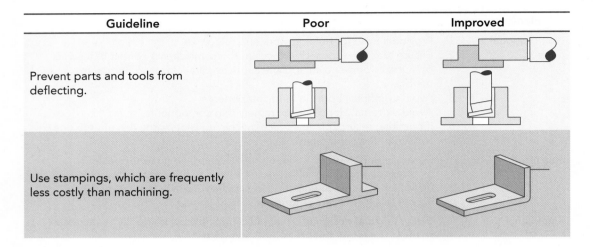

Guideline	Poor	Improved
Prevent parts and tools from deflecting.		
Use stampings, which are frequently less costly than machining.		

(b) Representative Design Guidelines for Machining (Grinding)

Guideline	Poor	Improved
Minimize surface that is to be ground and reduce part weight.		
Aim for unimpeded grinding.		
Avoid edge limitations for grinding wheels.		
Provide runouts for grinding wheels.		

(c) Representative Design Guidelines for Sheet Metal Work

Guideline	Poor	Improved
Redesign parts to improve material utilization.		
Combine parts to improve material utilization.		
Use width of stock to improve material utilization.		

(d) Representative Design Guidelines for Forging

Guideline	Poor	Improved
Avoid nonplanar parting lines.		
Provide tapers.		
Locate parting line so that metal will flow parallel to it.		

(continued)

Guideline	Poor	Improved

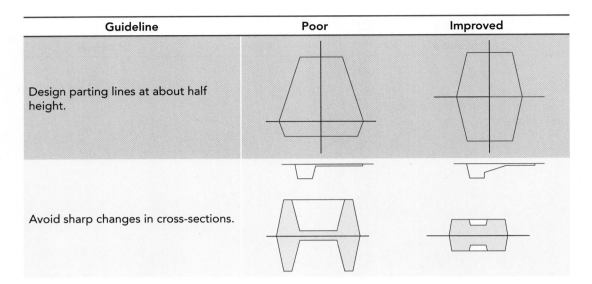

Guideline	Poor	Improved
Design parting lines at about half height.		
Avoid sharp changes in cross-sections.		

(e) Representative Design Guidelines for Rolling

Guideline	Poor	Improved

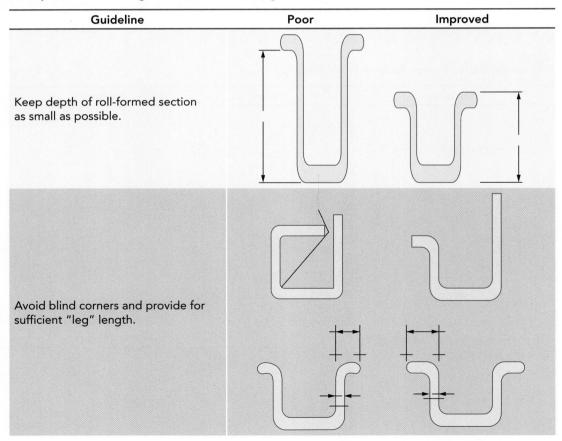

Guideline	Poor	Improved
Keep depth of roll-formed section as small as possible.		
Avoid blind corners and provide for sufficient "leg" length.		

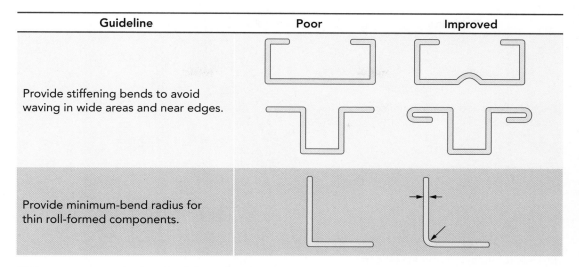

Guideline	Poor	Improved
Provide stiffening bends to avoid waving in wide areas and near edges.		
Provide minimum-bend radius for thin roll-formed components.		

(f) Representative Design Guidelines for Extrusion

Guideline	Poor	Improved
Ensure that impact-extruded parts are symmetrical.		
Avoid sharp changes in cross-section, sharp edges, and fillets.		
Avoid tapers and almost equal diameters.		

(continued)

Guideline	Poor	Improved

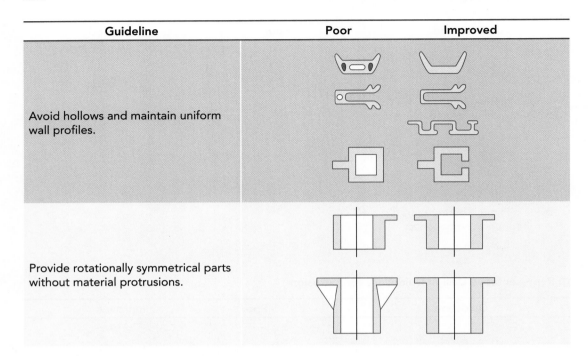

Guideline	Poor	Improved
Avoid hollows and maintain uniform wall profiles.		
Provide rotationally symmetrical parts without material protrusions.		

(g) Representative Design Guidelines for Power Metallurgy

Guideline	Poor	Improved

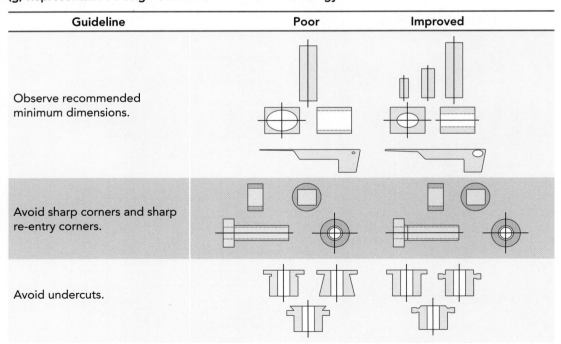

Guideline	Poor	Improved
Observe recommended minimum dimensions.		
Avoid sharp corners and sharp re-entry corners.		
Avoid undercuts.		

Guideline	Poor	Improved
Avoid blind holes with the blind end opposite a flange.	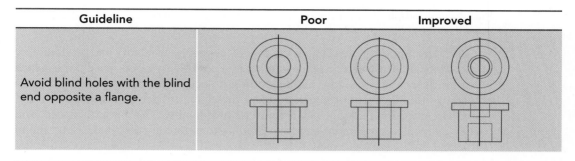	

TABLE 7.6 Guidelines for Manufacturing Process (Machining, Sheet Metal Working, Forging, Rolling, Extrusion, and Powder Metallurgy)

7.2 Designs for Life Cycle Manufacture

The major considerations in the design for life cycle manufacture (DFLC) are the following:

1. Physical concept of the product
2. Part decomposition
3. Total product quality

Breaking down the design for life cycle manufacture into these components enables designers and manufacturers to examine total product quality in terms of concept, design, and ownership (Stoll 1997). The process begins by conceiving a physical concept for the product based on customer needs and creating product specification and then a preliminary layout of the design that embodies the physical concept. The physical concept and part decomposition together determine a product's functionality and manufacturability (**Figure 7.4**).

Physical concept: For success in the marketplace, the physical concept must embody the way in which the product performs or provides its intended function. Success often is determined by consumers' identification and selection of the best physical concept for a product. This fact is a primary motivation for creativity and innovation in product design.

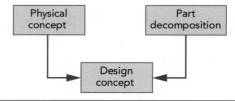

FIGURE 7.4 Design Concepts (Physical Concept + Part Decomposition)

Part decomposition: For products with large number of parts, selection of the concept as well as the sequential decomposition of components should occur concurrently.

Part decomposition determines the ease of assembly, testability and serviceability of a product. It also determines the number and complexity of designed parts, which in turn influence material and manufacturing process selection, tooling cost, and a myriad of other factors. For many products, therefore, it is decomposition into parts, more than any other factor, which determines profitability. The physical concept and part decomposition together determine product functionality and manufacturability (Ettlie and Stoll 1990).

Step-by-Step Approach for Life Cycle Manufacture

Design for life cycle manufacture (DFLC) is essentially a step-by-step method for performing the engineering design process. It focuses on systematically identifying, developing, and revising the design concept to ensure high total product quality and low total cost and time.

▸ **Step 1.** Examine all preliminary activities and create a physical concept and preliminary layout.

▸ **Step 2.** Create the part decomposition process from a functional, assembly, and quality point of view.

▸ **Step 3.** Optimize the detailed design.

The general flow of these steps is shown in **Figure 7.5**.

Design for Total Product Quality

Achieving optimal total product quality requires a conscious and systematic focus as a design objective. To facilitate this, total product quality is considered to be composed of components that help guide design decisions at each step of the DFLC approach.

Achieving total product quality requires consideration of the three primary design-related components: concept, design, and ownership and identity (**Figure 7.6**).

Decisions made during the engineering design of a product determine each of these qualities. Understanding how each element can be maximized enables the systematic design for total product quality.

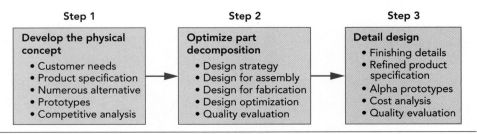

Step 1	Step 2	Step 3
Develop the physical concept • Customer needs • Product specification • Numerous alternative • Prototypes • Competitive analysis	**Optimize part decomposition** • Design strategy • Design for assembly • Design for fabrication • Design optimization • Quality evaluation	**Detail design** • Finishing details • Refined product specification • Alpha prototypes • Cost analysis • Quality evaluation

FIGURE 7.5 Three-Step Design for Life Cycle Manufacture

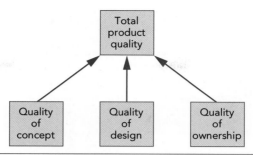

FIGURE 7.6 Three Design-Related Components of Total Product Quality

The elements considered when defining the relationship between engineering design and total product quality consist of the following:

- Determining what factors contribute to each quality component.
- Defining an acceptable quality level for each factor.
- Determining how each factor can be adjusted or controlled by design to maximize quality.

Quality of concept: In essence, *quality of concept* reflects how well the product satisfies customer requirements. For most products, characteristics of a successful product that reflects the quality of concept include performance, features, aesthetics, ergonomics, and serviceability (see **Figure 7.7**).

Quality of Design: The characteristics of a successful product involves ease of manufacture and assembly, acceptable capability from product to product, insensitivity to hard-to-control disturbances, and minimal scrap rates, rework and warranty claims. Such characteristics imply high quality of design, see **Figure 7.8**.

Quality of Ownership and Identification: This characteristic depends on customer satisfaction, reliability over its useful life. Other factors include ease of use, serviceability, maintainability, and customer service. High satisfaction is crucial because it causes a customer to become a repeat buyer and an advocate for the product. High satisfaction

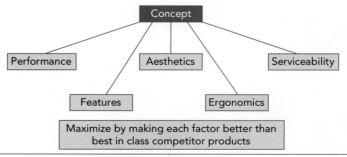

FIGURE 7.7 Quality of Concept Reflects Degree to Which a Product Satisfies Customer Requirements

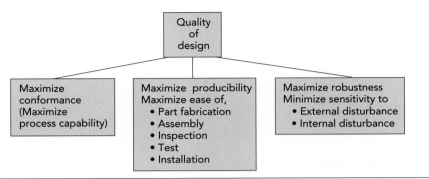

FIGURE 7.8 Maximization of Elements of Quality of Design

also contributes to and sustains a firm's reputation over time for manufacturing and selling high-quality products. These considerations can be grouped into three primary factors: (1) customer satisfaction, (2) availability, and (3) operating cost (**Figure 7.9**).

In summary, developing a successful product design depends on:

- Identifying the best physical concept for satisfying customer requirements
- Identifying the most appropriate part decomposition for ease of manufacture, assembly, and life cycle support
- Integrating these ideas into a design concept to optimize product performance, functionality, robustness, producibility, life cycle cost, reliability, durability, and maintainability

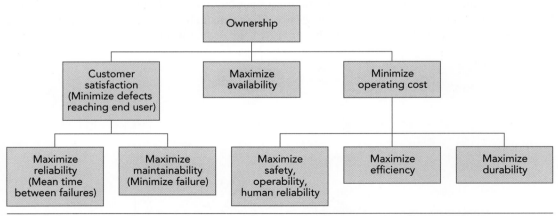

FIGURE 7.9 Quality of Ownership Factors According to Customer Satisfaction, Product Availability, and Cost

7.3 **Streamlining Product Creation**

A production line is normally associated with mass production. It consists of a series of manufacturing workstations arranged so that a product moves from one station to the next, and at each location, a portion of the total work is performed. If the number of products is very high and the work can be divided into separate tasks that can be assigned to individual workstations, a production line is the most appropriate setup. The speed of production is determined by its slowest manufacturing station, which can become a bottleneck. Transfer of the product along the line is usually accomplished by a mechanical transfer device or conveyor system, although some lines simply pass the product by hand.

7.3.1 **Workplace Structure Design**

Workplace structure affects production. The choice of workplace structure depends on the design and lot sizes of the parts to be manufactured and the ability to respond to market changes. Conventional practice includes five options: job shop, project shop, cellular system, flow line, and continuous system.

Job Shop Structure

Job shops are suitable for the low-volume production of dissimilar, multiple parts. In a job shop structure (**Figure 7.10**), machines with the same or similar material-processing capabilities are grouped. The machines are usually general-purpose types

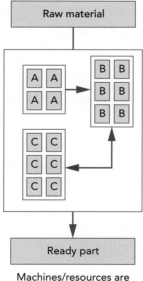

FIGURE 7.10 Schematic of a Job Shop Structure

that can accommodate a large variety of parts. In this layout, the parts move through the system by arriving at different work centers according to the parts process plan. Material handling must be flexible to accommodate many different part types, so it is usually done by manually controlled tools such as forklifts and handcarts. This is advantageous for a number of reasons:

- Each operation can be assigned to a machine that yields the best quality or best production rate.
- Machines can be evenly loaded.
- The problem of machine breakdown can be addressed easily.

Project Shop Structure

A product's position in a project shop (**Figure 7.11**) remains fixed during manufacturing because of its size and/or weight. Materials, operators, and machines are brought to the product as needed. Facilities organized as product shops can be found in aircraft and ship-building industries as well as in bridge and building construction.

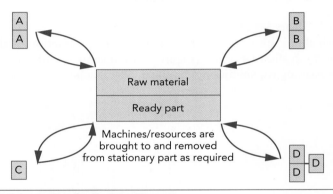

FIGURE 7.11 Schematic of a Project Structure

Cellular System Structure

Manufacturing layouts organized according to the cellular system (**Figure 7.12**) group equipment or machinery according to process combinations that have low-to-medium production volumes and lot sizes and can be clustered into a family of parts. Each cell has machines that can produce a specific family of parts, and the material flow within a cell can differ for various parts of a family. Several automotive component suppliers have adopted this layout since it is conducive to quality control. Intracellular material flow can be performed either automatically or manually.

Flow Line Structure

The flow line structure is best suited to high-volume, large lot size production. It creates a flow line in which machines and other equipment are located according to the process

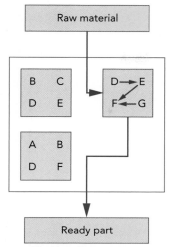

FIGURE 7.12 Schematic of a Cellular Structure

sequences in which the parts are manufactured (**Figure 7.13**). A typical example is the *transfer line*, often used in the automotive industry. This structure consists of a sequence of machines typically dedicated to fabricating only one particular part or a few very similar ones because of the dedicated machines and material-handling equipment.

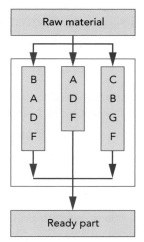

FIGURE 7.13 Schematic of a Flow Line Structure

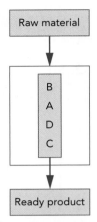

Processes are grouped
in lines according to
the process sequence
of the products

FIGURE 7.14 Schematic of a Continuous Structure

Continuous System Structure

In contrast to other manufacturing systems that produce discrete parts, the continuous structure (**Figure 7.14**) is used to produce liquid-based, gas-based, and powder-based products. As in a flow line structure, processes are arranged according to product sequencing. The continuous structure is the least flexible of the design structure types.

Production capacity model: Equations can be developed to determine the required number of workstations in a specific structure. The required hourly rate of production on a single model line to satisfy the annual demand for a product is calculated as

$$R_P = \frac{D_a}{50SH} \tag{7.1}$$

where R_P = the actual average production rate (units/hour) assuming 50 work weeks/year; S = number of shifts/week; H = hours/shift/D_a = demand rate per week.

The average production time, per minute, T_P is shown as

$$T_P = \frac{60}{R_P} \tag{7.2}$$

Production capacity is generally known as the maximum rate of output that a manufacturing facility is able to generate under normal operating conditions. Production capacity, P_C, is expressed as

$$P_C = WSHR_p \tag{7.3}$$

W = number of work centers; S = number of shifts per week; R_p = average production rate; H = number of hours/shift.

7.3.2 **Choice of Production Method**

Design, materials, and manufacture can be combined in many ways; no one route from the design stage to inspection and sales might be best. If one or even two of the three elements, design, materials, and manufacture, are fixed (such as a fixed design with fixed materials), there could still be a number of acceptable manufacturing routes to produce the desired design. The choice could depend on the number of parts to be made, the availability and quality of raw material(s), the number and capability of existing machines in the factory, and, of course, the resulting cost per part.

The design of a product can be influenced by the process of manufacturing and by the production route it takes in a manufacturing facility. In some cases, the desired properties of a product can be attained through the use of a specific process. For example, the forging process is one of the most important methods of manufacturing items for high performance uses. Forging changes the shape of a piece of material by exerting force on it. One of the characteristics associated with cold forging is that it increases the yield strength of the component.

Sometimes the use of certain materials can cause production difficulties, making the manufacture of an article nearly impossible. Materials that are extremely difficult to machine or weld can introduce severe problems. Clearly, some flexibility between designers' specifications and implementation is required for economic reasons. Required tolerances and surface finishes, for example, must be considered when different manufacturing methods are compared. The element of overall economics is linked to such considerations. Savings from one or two inexpensive manufacturing operations early in a production sequence could be lost completely if expensive finishing operations are required later.

The type of production system could vary depending on the volume of the product created, geometry of the product, and the variety of the components under production. The production system could be a transfer line for high volume and low variety components. This is also referred to as a *fixed automation manufacturing system*. The use of a manufacturing cell has become popular in recent years. The design of a manufacturing cell is based on the concepts of group technology (see the later discussion). The objective is to process some families of parts on a group of CNC machines within a cell to minimize the intercellular material handling effort. Some of the features of a manufacturing cell are the ability to use in low to mid-volume production systems, manufacturability, and flexibility in the variety of parts as shown in **Figure 7.15**.

There are special situations for using customized manufacturing systems. A special manufacturing system has machines organized to manufacture a family of parts based on the sequence of operations. The machines are linked by a fixed-path material handling system on which the parts move in a sequence from machine to machine. The system has a high production rate.

Flexible Manufacturing System (FMS) The progress in FMS systems has been documented in several references [11], [22], and [23]. Flexibility is the cornerstone of the key concepts used in the design of modern automation. It can be defined as the properties of a manufacturing system that support changes in production activities. The changes are the result of both internal and external factors. Internal changes could be caused by equipment breakdowns, software failure, operator absenteeism, and variability in processing times. To respond to uncertainties resulting from product design changes,

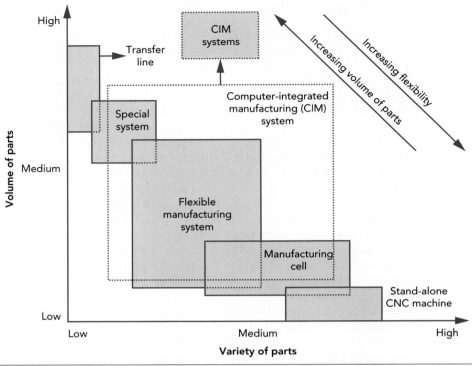

FIGURE 7.15 Volume–Variety Relationships Used to Categorize Production Systems

manufacturing systems must be versatile and able to produce the designated variety of part types with minimal cost and lead times.

An FMS lies between the two extremes of manufacturing cell and special manufacturing system. It is a true mid-volume, mid-variety, centralized computer-controlled manufacturing system providing a higher production rate than a manufacturing cell and much more flexibility than a special manufacturing system. Some FMS features follow:

- FMS covers a wide spectrum of manufacturing activities such as machining, sheet metal working, welding, fabricating, and assembly.
- It is composed of a series of flexible machines, an automated material-handling system, an automated tool changer, and equipment such as coordinate measuring machines and part washers, all under high-level centralized computer control.
- It permits both sequential and random routing of a wide variety of parts.
- It has a higher production rate than a manufacturing cell and much higher flexibility than a special manufacturing system.

Other physical components of an FMS are the following:

- Potentially independent NC machine tools capable of performing multiple functions and having automated tool interchange capabilities.

- Automated material-handling system to move parts between machine tools and fixturing stations.
- All components (machine tools, material-handling equipment, tool changers) hierarchically computer controlled.

FMS Subsystems FMS consists of two subsystems, physical and control. The FMS *physical subsystem* includes the following:

1. *Workstations* consisting of an NC machine tool, inspection equipment, part-washing devices, load and unload areas, and work area.
2. *A storage retrieval system* consisting of pallet stands at each workstation and other devices such as carousels used to store parts temporarily between the workstations or operations.
3. *Material-handling systems* consisting of powered vehicles, towline carts, conveyors, automated guided vehicles, and other systems to transport parts between workstations.

The *control subsystem*, which is required to ensure optimum performance, includes:

- Hardware consisting of computers, embedded controllers, communication networks, sensors, switching devices, and many other peripheral devices such as printers and mass storage memory equipment.
- Software of a set of files and programs used to control physical subsystems.

Types of Flexibility in an FMS System

Machine flexibility: This term refers to the capability of a machine to perform a variety of operations on diverse part types and sizes. It also represents the ease with which the parts on a machine can be changed from one type to another. The changeover time, which includes setup, tool changing, part program transfer, and part transfer times, is an important measure of machine flexibility. A machine's CNC centers are normally equipped with automatic tool changer, part buffer storage, part programs, and fixtured parts on pallets.

Routing flexibility: One way to achieve mix adaptability is through routing flexibility. That is, part(s) can be manufactured or assembled on other than normal routes if these operations can be performed on alternative machines, in alternative sequences, or with alternative resources. Routing flexibility is used primarily to manage internal changes resulting from equipment breakdowns, tool breakages, and controller failures. It can also help increase throughput in the presence of external changes such as product mix, engineering revisions, and new product introductions, each of which could alter machine workloads and cause bottlenecks.

Process flexibility: Also known as *mix flexibility*, process flexibility refers to the manufacturing capacity to absorb changes in the product mix by performing similar operations or producing similar products or parts in multipurpose, adaptable, CNC machining centers.

Product flexibility: This is the ability to add a new set of products economically and quickly in response to market or engineering changes and even on a make-to-order basis.

Production flexibility: This type of flexibility refers to the manufacturing capacity to produce a range of products without adding major capital equipment although new tooling or other resources could be required. The *product envelope* is the range of products that can be produced at a moderate cost and time.

Expansion flexibility: This refers to the manufacturing capacity to change a manufacturing system to accommodate a changed product envelope. Although production flexibility requires no change in a major investment, expansion flexibility can easily make additions as well as equipment replacements because such provisions are made in the original manufacturing system design.

7.3.3 **Value Stream Mapping**

Value stream mapping, a process from Toyota's lean manufacturing, is generally a one-page flow chart showing the current design path that a product takes beginning with the essential process flow from raw materials through delivery to the customer. It provides a clear view of information involved in manufacturing a product, including all value-added and non-value-added activities. *Value* can be defined as the capability to deliver a product to the customer at the right time and at the right price. Its benefits include enabling employees to visualize the entire process flow rather than just a single part of it and to make the sources of waste evident.

Value stream mapping is a valuable tool for identifying waste and areas for improvement. As with other lean methodologies, value stream mapping should be an ongoing effort for improvements. However, without a clear understanding of techniques and philosophies of the Toyota production system—standard work, Takt time, pull system, kaizen, kanban, one-piece flow, cellular manufacturing, and cycle time reduction—value stream mapping will not provide benefits. The goal of value streaming is to plan so that one process makes only what the next process needs and only when it is needed.

Standard work: Standard work is used to determine maximum performance with minimum waste. It is calculated to identify the best combination of operator and machine. Standard work eliminates process variability by establishing its routine. Implementation of standard work exposes problems and identifies waste. It also establishes a relationship between man, machine, and material.

Takt time: This method identifies how frequently a product must be finished to meet customer requirements. It sets the rhythm for standard work and production with sales. It is calculated by subtracting nonworking time from the available working time and dividing this value by the customer demand.

Pull system: This method controls production between the flows of two processes. Its purpose is to provide accurate production requirements to an upstream process so that it cannot make a product or part until a downstream activity signals the need for it based on a concrete customer order. Production control is visible and disciplined, typically with the use of kanban (cards with instructions sent along the production line to regulate the supply of components). Kanban is a scheduling system to control the logistical chain from a production point of view and is not an inventory control system. Kanban uses the rate of demand to control the rate of production, passing demand from the end customer up through the chain of customer-store processes. Kanbans also control material and indicate proper inventory levels, making operators' indirect work manageable. Kanbans base lead time on the customer rather than the process, eliminating the need to make predictions on demand.

Kanban cards are a key component of kanban, and they signal the need to move materials within a production facility or to move materials from an outside supplier into the production facility. The kanban card is, in effect, a message that signals depletion of product, part, or inventory that, when received, triggers the replenishment of that product, part, or inventory. Kanban cards help create a demand-driven system. In the last few years, systems sending kanban signals electronically have become more widespread. While this trend is leading to a reduction in the use of kanban cards in aggregate, it is still common in modern lean production facilities to find the use of kanban cards.

Kaizen philosophy: Kaizen (*kai* meaning "change" and *zen* meaning "for the good"), also called *Muda*, is a continuous incremental improvement to reduce waste and add value. In lean thinking, perfection is the complete elimination of any activity that uses resources but does not add value. Kaizen identifies seven forms of waste: overproducing before demand, keeping inventories at more than the absolute minimum, unnecessarily transporting materials, inefficient processing because of poor tool and product design, unnecessary employee movement during the course of their work, waiting for the next processing step, and production of defective parts.

Overproduction occurs when an item is produced and pushed forward regardless of the downstream process needs. Because this material is not yet needed, it must be handled, counted, and stored, causing waste. Inventory that is between processes can have defects that are not seen until it is used in a downstream process, making the identification of the defect's root cause difficult. Excessive inventory creates the need for more personnel, equipment, and floor space to transport and stock the inventory.

One-Piece flow: This term in lean methodology means producing one piece or part at a time. It is the progressive movement of a product to the next station in a value stream without stoppages, scrap, or backflows. One-piece flow can significantly improve quality, delivery, and cost and create flexible setups. It requires operators to be multiskilled and control several machines or complete several steps. Lean production can incorporate cellular manufacturing (discussed next); however, the production flow is determined by a customer order. Other benefits of one-piece flow include:

- Shortened lead times
- Improved product distribution
- Reduced scrap and rework
- Increased ease of scheduling
- Improved use of floor space

- Reduced material handling
- Improved labor utilization and productivity

Cellular manufacturing: This lean process produces similar products in a single cell or workstation of machines controlled by operators who work only in a cell or line. The machines are connected by a process sequence that supports an efficient production flow and determines the layout, usually a U shape that flows counterclockwise. Quality is designed into each production step rather than requiring a final inspection. Cellular manufacturing can be very beneficial depending on the volume and variety of products (**Figure 7.16**). It is ideal for one special-order product such as a space shuttle or a ship. *Product layout* is used for items with a high volume with little or no variety in products. *Process layout* is for products with low volumes but a large variety of differences between products.

Cycle time reduction: *Cycle time* refers to the amount of time required to complete one operation cycle, task, job, or function. Its reduction has become a key competitive parameter in product development. Any attempt to improve cycle time requires investigation and improvement of the development process. A company that is good at developing new products can use this advantage to gain market share. The reduction in cycle time for product development can open new market opportunities and improve a company's responsiveness to customers. Cycle time reduction also lowers market risk by reducing the time between product specification and delivery. Many companies have experienced improvement in quality and profits by reducing cycle time.

Value Stream Mapping Compared to Other Methodologies
A few techniques are comparable to value stream mapping.

Value Engineering *Value engineering* is a systemized method to analyze and improve the relationship between a component or product's function and its associated cost. It identifies and removes unnecessary costs without compromising quality or design reliability. The process studies a specific product or component to determine whether a better design, material, or manufacturing method exists. Users of value engineering

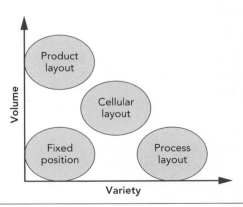

FIGURE 7.16 Layout of Volume and Variety According to Functions

STEP 1 INFORMATION	STEP 2 SPECULATION	STEP 3 ANALYSIS	STEP 4 DEVELOPMENT	STEP 5 PRESENTATION AND FOLLOW-UP
What is it? *What does it do?* *What does it cost?* *What is it worth?*				
Get all the facts	*What else might do the job?*			
Get information from best source	Seek new information	*What does that cost?*		
Get all available costs	Eliminate the function	Put money on each main idea	*What will satisfy user's needs?* *What is needed to implement?*	
Work on specifics not generalities	Simplify	Evaluate by comparison	Use specialty vendors and processes	*What is recommended?* *Who has to approve it?* *What was done?* *How much did it save?*
Define the functions	Blast and refine	Evaluate by function	Use specialty products	Use good human relations
	Use creative techniques	Use experts	Use standards	Spend the organization's money as you would your own
			Use your own judgment	Monitor progress of review and implementation
			Substantiate conclusions	
			Prepare implementation plan	

FIGURE 7.17 Steps in Value Analysis

perform five steps (**Figure 7.17**) that focus on identifying how a given design system function can be performed at minimum cost and the value that each design feature contributes to the specific function that it must fulfill. The users of value analysis question the performance of each function, cost, and the value of the contribution of each design function on the total product.

For the value engineering method to be successful, it must be performed by a team of engineers and managers from different backgrounds who have the support of top management.

The identification of the tasks and functions ensures that the all-important aspects are analyzed and itemized and have a written record of progress. The first step in value analysis is to gather the basic information. The next step is to speculate on alternative ways to accomplish the basic functions involved. The third step involves selecting the promising alternatives for further analysis and definition. A complete plan for the implementation of alternatives is developed in the fourth step. The best alternative is then selected for implementation along with several alternatives as backups. The final step is to present the best alternative for approval of the value analysis proposal.

The determination of the design function's cost should focus on its expensive elements. Total unit cost is composed of material, labor, and overhead. One method for calculating cost is to develop the cost of each design element in the manufacturing process, from raw to finished material(s) for each step in the manufacturing process. Then the value of the design or system can be determined. Value can be expressed as a ratio of cost to worth referred to as the *value index*. Large value indices signal that a part under consideration is a target for cost reduction. Pareto's law states that 80% of the total effect of any group of product comes from 20% of its components (**Figure 7.18**). It is necessary that the key components are correctly designed and manufactured.

Group Technology *Group technology* is a manufacturing technique that organizes machines into a group according to their function—for example, turning, milling, drilling, assembly, and heat treatment—according to design and manufacturing attributes, standardized part numbers, specifications of purchased parts, and process selection. These groups classify and code parts into families typically based on shapes, sizes, material types, and processing requirements. Design and manufacturing decisions must be made based on family characteristics. Arranging production equipment into cells to facilitate work flow creates manufacturing efficiencies. The use of group technology typically benefits design, tooling and setups, material(s) handling, production and

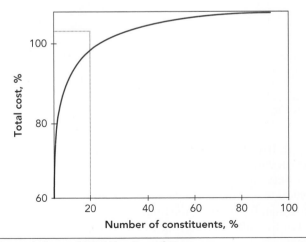

FIGURE 7.18 Pareto's Law of Distribution of Costs

inventory control, and process planning. Operators can visualize the contributions they make in their cells, leading to higher job satisfaction.

A process layout for batch production can be used (**Figure 7.19 (a)**). The product movement illustrated indicates a significant amount of material handling, a large in-process inventory, a high number of setups, and increased lead times and costs.

Refer to **Figure 7.19 (b)** for an example of a group technology layout with machines arranged in cells. Each cell manufactures a specific part family, reducing material handling, lowering setup times, and reducing work in process and lead times. The use of group technology typically benefits design, tooling, materials handling, production and inventory control, and process planning. Changeover typically requires little to no setup. Another key aspect of group technology is improvement in employee satisfaction. Operators can visualize the contributions they make in their cells, leading to improved morale and higher job satisfaction.

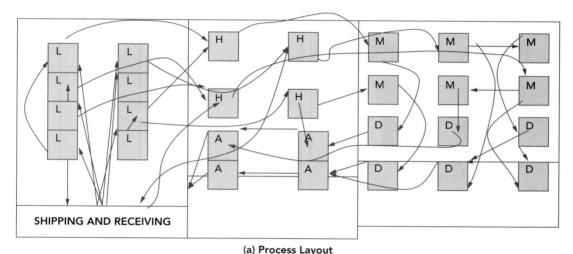

(a) Process Layout

| L = Lathe | H = Heat-treat | M = Milling | D = Drilling | A = Assembly |

(b) Group Technology Layout

FIGURE 7.19 Production Layout

Existing parts with matching codes can be easily retrieved and modified if needed instead of being redesigned. Group technology tooling is designed for group fixtures and jigs to accommodate the different members of a family of parts. Also because parts on the fixtures are similar, changeover typically requires little to no setup. Use of cells can also reduce production scheduling because there are fewer production centers to schedule.

Theory of Constraints Another methodology, the theory of constraints (TOC), is based on the theory that a chain is only as strong as its weakest link. TOC is a method based on the concept that every organization has at least one constraint that limits it from reaching a specific goal, typically to earn profit. TOC identifies the most important limiting factor or constraint (often called a *bottleneck*) that prevents the organization from achieving that goal and then eliminates or reduces it.

Eliyahu Goldratt and Jeff Fox have defined a five-step process for using a change agent to strengthen an organization's weakest link.

▸ **Step 1. Identify the System Constraint**
A constraint is anything limiting a system from higher performance. A link can be either a physical or a policy constraint.

▸ **Step 2. Determine How to Exploit This Constraint**
The process of change should maximize the capabilities of existing constraints and at the same time make sure that the changes can be implemented inexpensively.

▸ **Step 3. Subordinate Nonconstraint Components**
Doing this allows the constraint to operate at its level of maximum effectiveness. The overall system should then be reviewed to determine whether the constraint has moved to another component. If the constraint has been eliminated, the change agent skips step 4 and continues to step 5.

▸ **Step 4. Elevate the Constraint by Taking Actions to Eliminate It**
These actions can include making major changes to the existing system. This step is necessary only if steps 2 and 3 were not successful.

▸ **Step 5. Return to Step 1**
TOC is a continuous improvement process.

TOC also defines three essential measurements to drive changes in the process. The first is *throughput*, which is defined as the rate at which sales of a product or service generate income. It represents all income flowing into an organization. *Inventory* is a measurement representing income that has been invested in a product or service that the organization intends to sell. According to Goldratt and Fox, inventory includes assets such as facilities, equipment, obsolete items, raw material, work in process, and finished goods. The third measurement is *operating expense*, which is defined as the amount used to turn inventory into throughput. Operating expenses include direct labor, utilities, consumable supplies, and depreciation of assets. All improvement efforts should be prioritized by how they affect the three measures. The formula for the implementation is to maximize throughput while minimizing inventory and operating expense.

Line Balancing The objective of *line balancing* is to assign equal amounts of work to each station to remove bottlenecks by leveling work across all processes in a cell or value stream. Ideally, the goal is to balance the workload so that the time for production of a specified output is the same at all stations. If these times are unequal, the slowest station determines the production rate of the line. Line balancing promotes efficient use of labor and equipment. An unbalanced line can create morale problems for operators at the slower stations because they must work continuously to keep the flow.

A production flow typically includes several different processing and assembly operations that invariably restrict the order in which the operations can be sequenced in some way. These restrictions on the sequence of operations are called *precedence constraints*. Organizations typically set a production rate in order to meet demand. An operation's performance within the system depends on production scheduling, reliability, and line balancing.

A workstation consists of one or more individual work elements. Line balancing creates a table of work elements (**Table 7.7**) that represents a job divided into component tasks to distribute its functions among its stations. The minimum rational work element (T_{ej}) is the smallest practical task into which a job can be divided and is considered to be a constant. The total work content (T_{wc}) is the sum of all work elements to be completed. Workstation process time (T_{si}) is the sum of the time required to complete all functional elements performed at the station. Cycle time (T_c) is the length of time between receiving parts from the line.

A *precedence diagram* graphically represents the work element sequence (**Figure 7.20**), which is defined by the precedence constraints. If the work elements can be grouped so that all of the station times are exactly equal, there will be a perfect balance on the line, and smooth production flow can be expected. The goal in the balancing procedure is to distribute the total load on the assembly line as evenly as possible among workers.

Element Letter (see Figure 7.20)	Element Description	T_{ej} (seconds)	Must Be Preceded By
A	Rough drill, machine 1	10	—
B	Rough drill, machine 2	15	—
C	Finish drill	15	A, B
D	Tap	20	C
E	Tap	15	C
F	Counterbore	10	D
G	Counterbore	10	E
H	Assembly	5	F, G

TABLE 7.7 Table of Work Elements

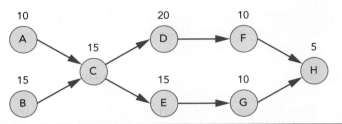

FIGURE 7.20 Precedence Diagram

Implementation of Value Stream Mapping

Value stream mapping is based on a Toyota production system (TPS) called *material and information flow mapping*. The focus at Toyota is to establish a flow, eliminate waste, and add value. Toyota identifies three flows in manufacturing: material, information, and people/process. Value stream mapping illustrates the flow of material and information and can be performed anywhere for a diverse range of products that need to be streamlined.

See **Figure 7.21** for the steps to implement value stream mapping.

‣ **Step 1. Select a Product Family**

A product that needs to be streamlined and process mapped is selected. The selection process can consider a diverse range of products with a range of benefits. For instance, the automotive industry has outlined well-established material and information flow lines resulting from the implementation of value stream mapping.

‣ **Step 2. Draw the Current State Map**

The purpose of drawing a current state map is to make a clear representation of the material and information flows in production. The map is drawn to depict the process that was present when the study began. It begins with the receipt of a shipment from the supplier and ends when the material flows through possible routes for the first few workstations.

‣ **Step 3. Plan the Value Stream**

The goal of lean manufacturing is to have one process make only what the next process needs when it needs it by linking the processes from raw material to the final customer. Areas to be addressed in planning the value stream are discussed above.

A value stream plan shows what activities are planned when and identifies measurable goals and checkpoints with deadlines and a reviewer.

The future state value stream map can be divided into sections or loops. The maps show the objectives and measurable goals for each objective. A typical pattern for

FIGURE 7.21 Value Stream Analysis Steps

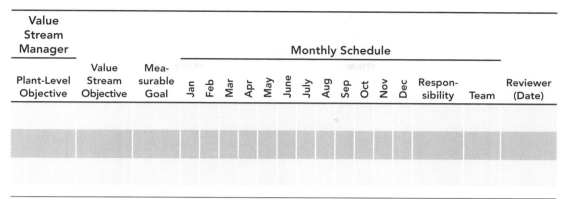

TABLE 7.8 Chart for Yearly Value Stream Plan

implementing improvements is to start by developing a continuous flow based on Takt time. Using takt time tends to be the simplest area to start and typically provides the highest value for the amount spent by eliminating Muda and shortening lead times. The next step is to implement a pull system (discussed previously). Once the order for implementation has been decided, the elements need to be recorded as the yearly value stream plan (**Table 7.8**), which looks like a variation of a Gantt chart. The key to successfully implementing the value stream plan is to incorporate it as a part of normal business practice.

‣ **Step 4. Identify Standard Work**
Identify the standard work that provides the best possible combination of operator and machine layout.

‣ **Step 5. Draw the Future Map**
The purpose of the value stream plan is to break the future state concept into reasonable steps by mapping a future value into segments or loops. Determining the material flow and information flow between pulls can identify the segments. The value stream plan then lists the objectives as well as steps planned to achieve them. The future state value stream identifies and eliminates waste. Improving the process flow addresses Takt time (previously discussed), continuous flow processing, the pacemaker production process, pull systems, production mix at the pacemaker process, increments of work at the pacemaker process, and process improvements.

Continuous flow should be developed wherever possible. This means producing one piece at a time and passing it immediately from one process step to the next without stagnation. Although this is the most efficient way to produce, sometimes continuous flow should be limited because it could merge all lead times and downtimes.

The next step is to send the customer schedule to *the pacemaker process*, which sets the pace for all upstream processes. Another requirement is to level the production mix evenly over time at the pacemaker process. By leveling the mix, the upstream can reduce storage requirements and total lead time. The material transfers from the pacemaker process to the finished goods downstream must occur as a flow. Continuous flow could

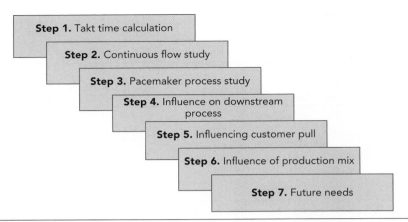

FIGURE 7.22 Basic Investigation Steps to Create a Future State Map

be difficult upstream of the pacemaker process because stations may be far away, operate at faster cycle times, require changeover for multiple value streams, or have a large lead time. A *pull system* (discussed previously) should be installed where continuous flow is interrupted and the upstream process must still operate in batch mode.

▸ **Step 6. Standardize**
The first phase of implementing a future state value stream map should ignore the inherent waste from product design, current processing machinery, and the location of some activities because changing them could require a great deal of work without immediate results. These features should be addressed in later iterations.

The basic steps to be investigated to create a future state map are shown in **Figure 7.22**.

Macro Value Stream Mapping

Macro value stream mapping extends beyond plant level maps. It is an effort taken at the cross-company level and can be extremely valuable. Lockheed Martin Aeronautics (Ft. Worth, Texas) and Northrop Grumman Electronic Systems (Baltimore, Maryland) partner on many important and valuable defense-related contracts. Lockheed Martin Aeronautics builds state-of-the-art military fighter aircraft. Northrop Grumman Electronic Systems provides highly sophisticated radars and electronics for the Lockheed Martin airplanes. Ref: "Macro Value Stream Mapping at Northrop Grumman and Lockheed Martin" (*http://www.leanadvisors.com/lean-success-stories/manufacturing/macro-value-stream-mapping-at-northrop-grumman-and-lockheed-martin*). Macro value stream mapping can be implemented after creating current and future state maps for a specific facility. These maps are created for several reasons. First, a large portion of costs can be attributed to the purchased materials. Downstream inconsistencies can threaten the leanness inside a facility. Added costs from downstream activities can also negate cost savings achieved internally, which can affect whether sales can grow. The maps also allow users to identify major asset reconfigurations by showing what is done, where, and with what tools.

The facility closest to the customer should be mapped first. Information such as frequency, distance, cost, processing time, lead time, inventories, cost/unit, daily volume, shift data, variation, frequency variation, and demand variation should be collected. The ideal macro value stream mapping locates all activities in their exact process sequence.

There has been an increase in the applications of value stream mapping over the last few years. In one example, the automotive component supplier in Connecticut, USA, had not made a profit in two-and-a-half years. Due to consistent financial losses, customer demands for a better product, and an internal need for space and growth within the plant, the company decided to implement value stream mapping. Several improvements were made including,

- implementing a kanban system,
- reducing setup times,
- implementing takt time,
- implementing U-shaped cells,
- implementing standard work, and
- creating a central market.

After implementing value stream mapping:

- The defect rate decreased from 27,000 PPM to 178 PPM.
- The paint yield increased from 60–65% to 80–85%.
- The lead time reduced from 29.4 days to 8.5 days.
- The raw inventory turns increased from 31 to 38.
- The work in progress (WIP) turns increased from 88 to 400.
- The finished goods inventory turns increased from 90 to 120.

In a second example, an aerospace component manufacturer located in New York, USA. decided to implement value stream mapping because of increased space requirements and issues with profitability and survivability. After implementing these improvements:

- The number of employees decreased from 153 to 88.
- The support factor decreased from 69% to 47%.
- The unit cost decreased from baseline to 69%.
- Quality improved from baseline to 37%.
- Takt time decreased from 9 days to the 8 days now required.
- Assembly lead time decreased from 64 days to 55 days.

7.4 **Design for Supply Chain**

The global economy has changed the way that engineering firms design, develop, and produce their products. Increasing competitive pressures are forcing companies to expand their innovation. The purpose is to shorten each product's duration in the market, thereby compressing its life cycle. Broadband networks, cost-effective global delivery services, and global business realities mandate that all firms evaluate many design, development, and comprehensive options available worldwide. Without proper management, increasing product turnover will add to design and manufacturing costs. Focusing on supply chain design in a company's operating environment is one way to combat the problems caused by increased competition and shorter product life cycles.

A supply chain is an integrated process in which suppliers, manufacturers, and distributors work together to acquire raw materials, convert these into specified final products, and deliver the final products to customers.

At its highest level, a supply chain comprises two basic, integrated processes:

1. Production planning and inventory control
2. Distribution and logistics

These processes (**Figure 7.23**) provide the basic framework for the conversion and movement of raw materials into final products.

Production planning and inventory control process: This consists of manufacturing and storage subprocesses and their interface(s). More specifically, *production planning* describes the design and management of an entire manufacturing process (including raw material scheduling and acquisition, manufacturing process scheduling, and material-handling design and control). Inventory determines the design and management of the storage policies and procedures for raw materials, work-in-process inventories, and, usually, final products.

The distribution and logistics process: This manages inventory retrieval, transportation, and final product delivery. It determines how to retrieve products from the warehouse and transport them to customers. Transportation to customers can be direct or indirect when products are first moved to distribution facilities, which in turn transfer products to customers.

These two processes interact to create an integrated supply chain. Their design and management determine the extent to which the supply chain works as a unit to meet

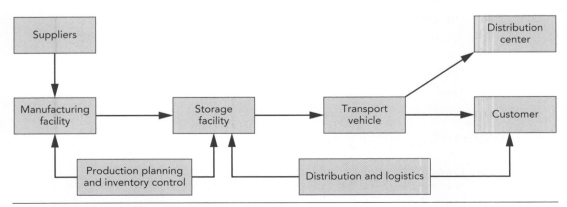

FIGURE 7.23 Supply Chain Process

required performance objectives. Optimization of the management of a number of global system suppliers at different locations must include the following:

1. Tariff, tax, and currency considerations
2. In-country content requirements
3. Lowest cost contract sites
4. Site utilization levels
5. Shipping and warehousing logistics
6. Inventory management and JIT constraints
7. Evolution of customer mix and contract constraints

Performance Measures

Qualitative performance measures must address each of the following:

Customer satisfaction: The degree to which internal and external customers are satisfied with the product and/or service received.

Flexibility: The degree to which a supply chain can respond to random fluctuations in the demand pattern.

Information and material flow integration: The extent to which all functions within the supply chain communicate information and move materials.

Effective risk management: The recognition that all relationships within the supply chain contain inherent risk and the degree to which the effects of these risks can be minimized.

Supplier performance: The consistency with which suppliers deliver raw materials to production facilities on time and in good condition.

Quantitative performance measures must include each of the following:

Cost minimization: Is the most widely used objective; typically addresses an entire supply chain (total cost) or particular business units or stages.

Sales maximization: Addresses the amount of sales dollars or units sold.

Profit maximization: Is calculated by subtracting costs from revenues.

Inventory investment minimization: Pertains to the amount of inventory costs (including product costs and holding costs).

Return on investment maximization: Addresses the ratio of net profit to capital employed to produce that profit.

Measures based on customer responsiveness include each of the following:

- **Fill rate maximization:** Pertains to the fraction of customer orders filled on time.
- **Product lateness minimization:** Addresses the amount of time between the promised and the actual product delivery dates.
- **Customer response time minimization:** Refers to the amount of time required from placing an order until the customer (usually an external one only) receives it.

- **Lead time minimization:** Pertains to the amount of time required from beginning a product's manufacture until the time it is completely processed.
- **Function duplication minimization:** Addresses the number of functions provided by more than one business entity.

CASE STUDY 7.1
Comprehensive Product Architecture

Value Stream Mapping Automotive Industry

One automotive supplier had challenges in the production line. As a result of consistent financial losses, customer demands for better product, and the need for space and growth in the plant, the company decided to implement value stream mapping. This case study demonstrates the use of the JIT concept and the TOC to develop continuous improvement and reduce non-value-adding processes (based on report of Elizabeth Cudney and D. Shetty, reference 7).

A time study of the existing production line provided the basis for reducing the cycle time, balancing the line, and designing the flow. Following the initial process of recognizing the critical problems, a production flow with kanban using an alternative layout to address the issues of optimum machine utilization was developed.

Initial process: The existing production tube line consisted of one operator and four operations separated in two stations by a large table using a push system. The table acted as a separator between the second and third operation, creating a kanban of approximately 172 parts (**Figure 7.24**).

The original process involved these steps:

▸ **Step 1.**
The first two operations consisted of boring the three inner diameters and one outer diameter and then cleaning them. The operator continued to run parts until they could not be stacked any higher. The operator would then move to the other side of the table and run the bushing press and the bearing press. Of the finished parts, 100% were inspected for accurate assembly dimensions and then stacked on a cart.

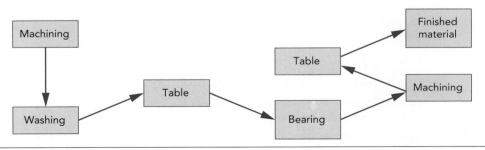

FIGURE 7.24 Initial Production Layout

▸ **Step 2.**
After approximately 30 parts, the operator packed them in a cardboard box two at a time. Double handling occurred between each grouping of operations. Also, because of small floor space with a large table, the inefficient process flow and large lot size caused a high probability of defects. The process flow also allowed possible missed operations. Missing the bearing press was a common defect that caused problems.

▸ **Step 3.**
Initially, two operating stations had been used to perform the necessary procedure for the final product. The unbalanced line created the first problem. The first station was busy approximately 70% of the time and second approximately only 30% of the time. The operators of station 2 were spending a considerable amount of their time waiting between cycle times.

▸ **Step 4.**
By combining stations 1 and 2, room for improvement became evident with respect to individual responsibility, control of inventory by the operator, and immediate feedback when a problem occurred.

Design Approach

The design of the new process had the following department objectives:

- Improve quality
- Deliver to the point of use
- Use a pull system
- Increase production
- Decrease WIP
- Decrease scrap
- Use smaller lot sizes
- Create line flexibility

The focus of the redesign was on reducing product cost by improving quality, decreasing WIP, and increasing production throughput. The most obvious change was to a U-shaped cell to develop line flexibility. The following steps were taken to determine whether this met our needs:

▸ **Step 1. Analyze the Current Process**
The results of the initial time study showed an extremely unbalanced use of time between the operations. The initial study identified the critical opportunities for time-saving steps and potential problems. The process flow chart sequenced the most effective operation of the line to reduce the amount of machine idle time and optimize throughput.

▸ **Step 2. Analyze Production and Quality Requirements**
The capacity for the next year had not been determined. Therefore, a U-shaped cell, which provided no limitation on the number of operators and reduced material

handling within the process, was selected. Capacity for the facility with the initial line was 330 pieces per shift, or 990 pieces per day. This was increased to 1,518 pieces per day. With these production expectations, the process should be able to meet all customer orders for this part.

‣ **Step 3. Evaluate Material Handling**
Day-to-day inspection of the current process revealed that a considerable amount of operators' time was spent either waiting for material-handling personnel or handling the material themselves. It became evident that delivery to the point of use would be a better system for receiving raw material.

‣ **Step 4. Analyze Costs**
The difference in cost between the current system and the proposed system was determined by investigating several elements:

- Reduced scrap
- Reduced machine and operator idle time
- Increased throughput
- Decreased inventory (WIP)
- Reduced operator material handling

Special Features of Redesign Techniques

The JIT and TOC methodologies were used to successfully achieve the goals. The concepts applied were total employee involvement (TEI), reducing lot sizes, identifying finite capacity, and employing optimal scheduling, the techniques of point of use inventory, and improved layout.

Total employee involvement: The first key element in the success of this case study was the result of employee involvement and management commitment. By illustrating the cellular layout and walking through the production flow with each operator and the production manager, the implementation phase was smooth.

Revised layout: The cellular layout of the automotive plant using value stream approach reduced WIP and increased production immediately. The redesign eliminated a work station table causing reduction in unnecessary movements. In addition, the value stream approach had other implications on supply chain, leading to reductions in material input and inventory levels, and better space utilization.

Elimination of defects: With the reduction of WIP and implementation of a pull system, defects were detected immediately, there by allowing operators to adjust the machinery tolerances or notify maintenance for timely repair.

Individual responsibility: The operators were authorized to stop the line when problems arose. The new layout eliminated the possibility of storing WIP, which had required the operator to shut down the entire line.

Flexibility: The previous layout was changed to cellular one, making the number of operators flexible depending on demand. The constraint imposed on the line is reduced.

Immediate feedback: The cellular layout provided excellent opportunities for improving communication between operators about problems and adjustments to achieve better quality.

Delivery to the point of use: With the U-shaped cell, delivery to the point of use is more convenient for the operator.

Vendor relationship: The vendor group was fully involved at every stage, resulting in speedy problem solving. The relationship with the suppliers of components and raw material improved significantly.

Setup time reduction: To reduce setup times, the tools needed for machine repair and adjustments were located in the cell.

Case Study Results

Time Studies

A comparison of the time studies of the initial arrangement (**Figure 7.24**) and the implemented layout as shown in **Figure 7.25** showed an increase in production from 300 to 506 finished products per shift. The new layout eliminated double handling between the second and third operations as well as at the pack out. The new layout reduced throughput time by making it easier to cycle all four operations in a pull system.

Machine and Operator Utilization

A simulation of the initial and implemented layouts was created to determine machine and operator utilization. Both machine utilization and operator utilization had increased.

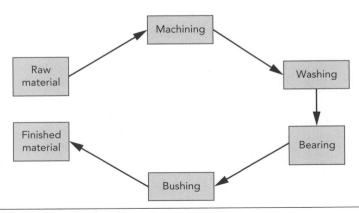

FIGURE 7.25 Final Production Layout

	Initial Production			Current Production		
Day	Daily Production	Shifts per Day	Scrap (ppm)	Daily Production	Shifts per Day	Scrap (ppm)
1	961	3	0	1,291	2	33
2	571	2	0	1,299	2	16
3	591	1	0	1,134	2	0
4	630	2	78	850	2	0
5	640	2	9	910	2	22
6	951	3	21	1,000	2	0
7	636	2	0	970	2	0
8	612	3	18	920	2	15
9	261	1	0	1,035	2	0
10	865	3	0	431	1	0
11	261	1	28	1,005	2	56
12	865	3	23	1,353	3	0
13	848	3	0	970	2	55
14	570	2	9	1,749	3	0
15	905	3	0	1,782	3	25
16	355	1	21	852	2	0
17	1,164	3	22	912	2	0
18	670	2	21	1,030	2	0
19	570	2	0	910	2	0
20	591	3	0	875	2	0

TABLE 7.9 Initial and Current Production Data

Benefits

The benefits were evaluated to determine the success of the new layout. The cellular layout benefits achieved included:

- Increased quality level
- Decreased material handling by the operator
- Increased production
- Decreased WIP
- Decreased scrap
- Improved feedback
- Unrestricted number of operators/Flexibility

	Initial	Current
Total produced	13,517	21,278
Total shifts	45	42
Total scrap ppm	250	222
Production per shift	300.4	506.6

TABLE 7.10 Comparison of Initial and Final Production Data

Conclusion

This case study reports how a company developed a process to decrease product cost and defects while increasing quality level and production throughput and improving production capacity in a design that did not limit the number of operators. Final recommendations involved evaluating the process with the various JIT concepts and performance measures. As a result, a U-shaped cell was recommended and a redesigned layout was implemented. The benefits can be summarized as:

- WIP decreased by 97%
- Production increased by 72%
- Scrap was reduced by 43%
- Machine utilization increased by 50%
- Labor utilization increased by 25%
- Labor costs reduced by 33%

The implementation of JIT and kanban concepts and TOC decreased product cost, quality rejects, and labor costs through increased labor utilization; improved process flow; and minimized finished goods delivery while improving delivery dates. The company met capacity with only two production shifts as opposed to the original three. The case study also demonstrates the way in which the process was redesigned to meet scheduled deliveries while decreasing scrap and reducing labor cost.

REFERENCES

1. M.F. Ashby, *Material Selection in Mechanical Design*, (Oxford, UK: Pergamon Press, 1992).
2. Joseph Datsko, *Material Selection for Design and Manufacturing*, (New York: Marcel Dekker, 1997).
3. M.S. Hundal, *Systematic Mechanical Designing: A Cost and Management Perspective*, (New York: ASME Press, 1997).
4. John Tanner, *Manufacturing Engineering: An Introduction to the Basic Functions*, 2nd ed. (New York: Marcel Dekker, 1991).
5. Henry Stoll, *Design for Quality and Life-cycle Manufacturing: Concurrent Product Design and Environmentally Conscious Manufacturing*, (New York: ASME, De-Vol. 94 (MED-Vol. 5), 1997).
6. J.E. Ettlie and H. Stoll, *Managing the Design Manufacturing Process* (New York: McGraw-Hill, 1990).
7. Elizabeth Cudney and Devdas Shetty, "Value Stream Mapping," *Proceedings of the 16th International Conference on CAD/CAM, Robotics and Factories of the Future, June 2000, Port of Spain, Trinidad.*

8. Eluyahu Goldratt and Jeff Fox, *The Goal: A Process of Ongoing Improvement* 2nd ed (New York: North River Press, 1992).

9. James P. Womack and Daniel T. Jones, *Lean Thinking: Banish Waste and Create Wealth in Your Corporation* (New York: Simon & Schuster, 2010).

10. James P. Womack, Daniel T. Jones, and Daniel Roos, *The Machine That Changed the World: The Story of Lean Production* (New York: Simon & Schuster, 2007).

11. M. P. Groover, *Fundamentals of Modern Manufacturing: Materials, Processes and Systems*, 5th ed. (Hoboken, New Jersey: John Wiley, 2013).

12. K. T. Ulrich and D. Eppinger, *Product Design and Development*, 5th ed. (New York: McGraw-Hill, 2011).

13. Mike Rother and John Shook, *Learning to See: Value Stream Mapping to Add Value and Eliminate Muda* (Brookline, MA: Lean Enterprise Institute, 1999).

14. Robert Hall, *Attaining Manufacturing Excellence: Just-in-Time—Total Quality, Total People Involvement* (Homewood, IL: Dow Jones-Irwin, 1987).

15. Usher John, U. Roy, and H. Parsaei, eds., *Integrated Product and Process Development: Methods, Tools and Technologies* (New York: John Wiley, 1998).

16. Karl Sabbagh, *Twenty-First Century Jet: The Making and Marketing of the Boeing 777* (New York: Scribner, 1996).

17. Z. Larsen, T. Glassey, M. Orelup, and D. Pherson "Worldwide Deployment of Design for Assembly – The Challenges of Cross-Functional and Cross Cultural Engineering", Otis Elevator, *Proceedings of the 1997 International Forum on DFMA, June 1997, Newport RI.*

18. James A. Tompkins and John A. White, *Facilities Planning* (New York: John Wiley, 2010).

19. John Janik, "A Review and Analysis in Line Balancing Methods" (thesis, University of Hartford, 1997).

20. "Lean Forum: Change Management," http://www.lean.org/community/thinkers2/htm.

21. Benita M. Beamon, "Supply Chain Design and Analysis: Models and Methods," *International Journal of Production Economics* 55, no. 3 (1998), 281–294.

22. W.W. Luggen, *Flexible Manufacturing Cells and Systems*, (Englewood Cliffs, New Jersey: Prentice Hall Inc., 1991).

23. M.Z. Mohamed, *Flexible Manufacturing Systems—Planning Issues and Solutions*, (New York: Garland Publishing Inc., 1994).

EXERCISES

7.1. Examine two products with the same function but made by different manufacturers. Suggested products are food grinder, food blender, electric can opener, hair dryer, coffee grinder, electric knife sharpener, electric razor, small electric fan, electric toothbrush, and electric tools such as a screwdriver and hand drill. Do the following:

a. List what you think the functional requirements should be.

b. Identify the performance attributes you would like it to have.

c. Identify the features you would like it to have.

7.2. In **Figure 7.1E**, identify the better design and provide the rationale for your decision.

7.3. Assume that a company decides to produce a new model of cellular telephones on a manual assembly line. The objective is to design its manual assembly line. This product has an annual demand of 100,000 units. The line operates 50 weeks per year, five shifts per week, and 7.5 hours per shift. Work units are attached to a continuously moving conveyor. Total work content time is 42.0 minutes. Assume the line efficiency is 0.97, the balancing efficiency is 0.92, and the repositioning time is 6 seconds. Find the following:

a. The hourly production rate to meet the demand

b. The number of workers required

7.4. See **Figure 7.2E** for the sectional view of two rectangular boxes that are enclosed on four sides. The boxes are made by injection molding. Assuming that their wall thickness is the same, compare the two designs. Identify the design that is less costly to produce. Explain the rationale behind your decision.

Design A	Design B	Rationale
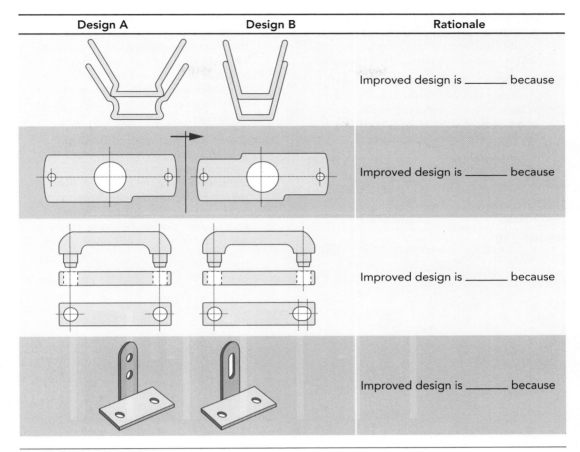		Improved design is _____ because
		Improved design is _____ because
		Improved design is _____ because
		Improved design is _____ because

FIGURE 7.1E

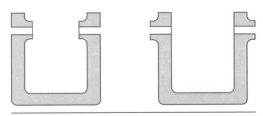

FIGURE 7.2E

7.5. How can the use of self-aligning and self-locating features facilitate automatic assembly? In what way does the method of fastening and joining the component of a product affect the feasibility of recycling?

7.6. **Figure 7.3E** shows the sectional view of two proposed alternative designs for an injection-molded box-shaped part that is enclosed on four sides. Assume that the wall thickness is the same in both designs. According to their tooling costs, which of the two designs is the least expensive?

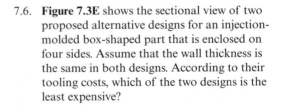

FIGURE 7.3E

7.7. As a product designer, create a product that has intricate parts that must be extruded. Identify a different way to fabricate the part without using a die. Illustrate your procedure.

7.8. See **Figure 7.4E** for two design possibilities. Assuming that the wall thickness is the same in both designs and that the part is to be injection molded, which of the two designs is less expensive to produce and why?

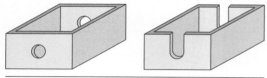

FIGURE 7.4E

7.9. A decision must be made on the layout for the production of a new product that involves a critical bearing mounting for a new escalator system. Identify the uniqueness of the three possibilities: (a) conventional layout, (b) process-based layout, and (c) cellular layout using group technology.

7.10. Describe the interrelationships among product architecture, material selection, manufacturing process assembly, maintainability, and reliability. A 1-liter jug is manufactured in large numbers as a consumer product. Identify the design alternatives for fabrication using (a) plastic and (b) metal.

CHAPTER 8
Sustainable Product Design through Reliability

OBJECTIVE

In the context of product development, products must provide a competitive solution to customer needs at a future time period. This chapter examines the basic elements that greatly affect the outcome of the development process. Sustainability in product design embraces a broad view of product development, considering a product's full life cycle and the impact of its design, manufacture, use, and retirement. Key elements of robust design, optimization, and failure mode evaluation are discussed in this chapter.

8.1 Introduction: World-Class Product Development

This section provides an initial step toward the development of a world-class product creation methodology. The effectiveness of its elements depends on both how and when they are used in the development cycle. Two lessons learned from the Japanese are the use of integrated engineering with manufacturing and developing long-term partnerships with subsidiaries and vendors. These initiatives have simplified manufacturing processes for the Japanese and enable them to understand, control, and manage their processes effectively. Japanese companies emphasize employee involvement, which has broadened workers' responsibilities and reduced the need for the overhead expenses.

World-class product development through global collaboration can address critical issues of capability and capacity at one location and how value can be added. The ability of the companies to provide service at a fraction of cost has led to the development of the outsourcing industry. With the IT revolution, global collaboration helps companies to leverage their assets both operationally and financially. Due to virtual prototyping tools and use of technologies such as CAD, CAE, and rapid prototyping, businesses can take advantage of low costs and quality product generation through offshore designers. The key benefits are reduction in manufacturing time, capability extension at a fraction of cost, and innovation through a global talent pool. Major manufacturing organizations such as United Technologies, Siemens, Microsoft, General Electric, and Motorola have used world-wide product development as an instrument in cost effective product development.

Companies can be classified into five categories based on their maturity in product creation using global partners. The first category keeps all their product engineering activities in-house. The second category of companies selectively outsources their components. They do not have a systematic policy on outsourcing. The third type of companies has well established policies on outsourcing, where certain modules are sent out and certain modules are done in-house. For example, the company might outsource CAD and support activities, but retain core design activity in-house. The fourth category outsources entire projects and co-develops products. The challenge is the risk involved in developing a product, where the entire responsibility lies with the off-shore partners. The fifth category of companies outsources all engineering activities while focusing their resources on capturing customer requirements.

8.1.1 **Product Creation: Aligning for Design and Business**

A superior product development process for an innovative product originates with product ideas and flows from left to right, ending with product's manufacture (**Figure 8.1**). Product and process technologies as well as product and process design are developed concurrently.

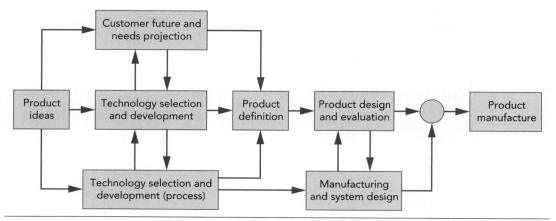

FIGURE 8.1 Superior Development Process for an Innovative Product

At the initial stage of product development process, some basic essential elements greatly affect the outcome of the development process. Some of these elements apply to the entire development process, whereas others apply specifically to particular phases.

Three of the essential product and process technology elements are so fundamental that they affect the entire product development process. These factors integrate the many diverse product development activities into a coherent, focused process. The factors are:

- Single product development team
- Customer-oriented product development
- Convergence of information at product definition

Single Product Development Team

The element that most affects superior product development is the selection of a single development and manufacturing team to control a project from initiation through the first few months of manufacturing. For example, companies such as IBM, United Technologies, and Hewlett-Packard have used product development teams (known as independent business units in some cases) to develop new and innovative products. Major corporations competing in the international market have reorganized the product development structure to become more competitive. A single development team is an essential element for world-class product development.

Because it is responsible for a product throughout the development process, a single team must contain the proper skill mix and experience to complete its job effectively. Various team members must have the design, manufacturing, marketing, testing, and other skills necessary to develop the product successfully. The most effective teams control all aspects of their project including product definition and development of specifications from technology selection through the first six months of manufacture.

Customer-Oriented Product Development

Another essential element affecting the entire development process is customer needs analysis. In the product development context, proposed products must provide a competitive solution to customer needs at a future time period beginning with product introduction and continuing at least until product development costs have been recovered. Customer-needs projections developed from this analysis then are used during the product definition stage to establish product specifications. A customer-needs analysis must project future need statements from a marketing perspective.

Encouraging product team members to participate in the analysis can enhance their creative contributions by enabling them to see opportunities that they might miss by merely reading a report from a distant market analysis group. Some companies set up a customer advisory board to provide advance notice of new products and to solicit suggestions for product changes and improvements. Design engineers attend these meetings to answer questions and to receive direct feedback from customers (Wilson and Kennedy, 1989).

Information Convergence at Product Definition

Convergence of marketing, engineering, and manufacturing information and goals is essential to create an adequate product definition. The three input arrows to the product definition box in **Figure 8.1** ensure that these three are considered simultaneously as the product is being defined to enable project leaders to agree on a common set of product goals and action plans. Doing so enables parallel product and process development to occur with minimum conflict. If common goals and plans are not developed, the simultaneous engineering of product and process is likely to cause deviations from the plan, causing major product and process reworking later in the development process.

Technology Selection and Development

Many essential elements regarding the selection of technology apply specifically to particular phases of the product development process, emphasize the use of the most appropriate engineering methods at the required time, and aid in identifying problems at the least damaging stage of the development process. The evaluation of a selected technology's ability to accomplish its intended purpose is essential. Failure to ensure technology feasibility can lead to a poor technology choice, which, in turn, could result in product failure.

8.1.2 **Seven Phases of New Product Introduction**

1. Define Customer Requirements A marketing representative who gives a detailed explanation of the present market and ways to explore sales opportunities normally leads the definition of customer requirements. The purpose of this phase is to objectively assess what the company can and cannot do. This often leads to a spreadsheet detailing the organization's entire range of capabilities.

2. Define Product Concept Next the design and marketing teams determine how to meet product requirements. They use brainstorming, storyboarding, quality functional deployment, and so on to match market opportunities with the company's capabilities.

3. Design Specification If a product concept looks feasible, the group must determine its specifications, while the marketing representative must ensure that they meet the perceived needs of the market. Creating a product specification also forces the team to define the company offering. This minimizes the chance of proposing a product that will be difficult to create and tends to preclude misunderstanding between customers and vendors about what constitutes successful delivery of the product. The team produces the first iteration of a design that can be created in the factory. Simulation techniques can avoid some trial and error, but real manufacture with real production equipment is still needed to test the process.

4. Define a Method for Manufacturing the Product (build prototype and test) The company's capabilities strongly influence the team regarding the type(s) of equipment it has and its effectiveness in using them. For example, a company that has strength in metalworking would naturally favor metals rather than plastic for a base material.

5. Evaluate the Capacity Evaluating the capacity to make this product in addition to all other products being manufactured is crucial. A serious error can occur if the company must commit capital resources to obtain capacity to manufacture the new product. This is usually done concurrently with the internal evaluation of capacity. Evaluating the capacity to make this new product in addition to all other products being manufactured is crucial. The Manufacturing engineering division leads this step, which is often overlooked.

6. Determine Product Cost The team then develops a bill of materials and routing, and then determines the total product cost. This stage of testing economic viability for full production is led by manufacturing engineering, materials, and shop operations. A prototype is structured for manufacture, and the necessary vendor supply chain is identified. A company can only remain in business if it realizes profits. Led by marketing and finance representatives, the team uses all data obtained to calculate potential profit and verifies that the new product will match the company's strategic plan. The team must take corrective action to achieve the required margin and meet customer requirements. If the calculation of profit indicates an unsatisfactory result, the team must reconsider all options to determine whether the project can be saved. Senior management should decide whether all internal and external requirements are achievable and then release the project for manufacturing or decide to stop, hold, or refine the product.

7. Product Launch The product is released for manufacturing, after all of the internal and external requirements are determined to be achievable. New products need to be introduced in a timely manner to take advantage of the market opportunity. The management team continues to monitor production and sales progress, and takes corrective actions as required. Each member needs to monitor the early progress of commercialization to ensure that everything is running smoothly.

8.2 **Sustainability Challenge in Product Development**

Sustainable product development refers to satisfying present needs without jeopardizing the ability to meet future ones. A subject that has received much attention over the past several years, sustainability in design and manufacturing involves "doing better with less" and embracing a broad view of product development. Sustainability considers a product's full life cycle and the impact that its design, manufacture, use, and retirement can have not only on business but also the environment and society.

Importance of Sustainability

Manufacturers are seriously considering sustainability as an option as a result of concerns about rising energy costs and the efficient use of resources; availability of, access to, and/or price volatility of critical materials; and the potential risks, opportunities, and costs posed by industrial and consumer waste. The application of sustainability principles, tools, and strategies during the product design phase is critical because those decisions can have the greatest impact on cost, performance, and sustainability.

8.2.1 **Financial Benefits Provided by Sustainability**

Four Winds Research, 2012, as shown in **Figure 8.2**, confirms that embracing sustainability in product design and manufacturing not only yields environmental improvements and societal benefits but also offers key financial benefits required for success. Manufacturers are increasingly finding that sustainability is about doing the right thing financially. Organizations can effectively incorporate sustainability in their business efforts, but they do not have to be expressly involved in the development of "ecofriendly" products to reap these benefits.

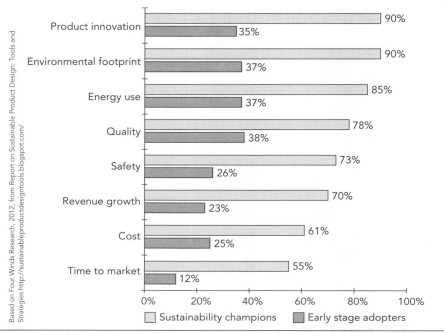

Based on Four Winds Research, 2012, from Report on Sustainable Product Design: Tools and Strategies http://sustainableproductdesigntools.blogspot.com/

FIGURE 8.2 Business Impact of Sustainable Product Development

Organizations that are *actively engaged* in sustainable product design and development cite impressive levels of improvement over their peers who have less performance success in product innovation, quality, safety, revenue growth, and anticipated environmental and energy gains. Four Winds Research, from in-depth interviews and survey responses from product designers, engineers, manufacturing executives, and sustainability experts from more than 125 organizations worldwide, found that embracing sustainability in product design and manufacturing also drives innovation, quality improvement, energy savings, and revenue growth.

Ecoefficiency efforts can result in decreased production, shipping, and transportation costs and increased operational efficiencies. Taking a more sustainable approach to product development also lowers risk (such as threats of a product recall, "hidden" presence of toxic materials in the supply chain) and reduces uncertainty (such as less exposure to potential supply chain disruptions because of resource scarcity or material shortages). Furthermore, rethinking a product's design to contain fewer parts, less packaging, or more recycled content make it more durable and more easily recycled.

8.3 **Robust Design**

Robust design is a systematic method for keeping a product's cost low and its product quality high. Its implementation can greatly improve an organization's ability to meet market opportunities. The three types of cost required to deliver a product are (1) operating, (2) manufacturing, and (3) research and development. Operating cost consists of obtaining the energy needed to operate and providing environmental control, maintenance, and inventory of spare parts. One way that manufacturers can greatly reduce their operating costs is by minimizing a product's sensitivity to environmental and manufacturing variation. Important elements of manufacturing cost are equipment, machinery, raw materials, labor, scrap, and rework. These costs can be reduced by designing not only robust parts but also a robust manufacturing process. Although research and development takes time and a substantial amount of resources, it can be shortened and its cost can be kept low by using a robust design approach that improves the efficiency of generating information needed to design products and processes. Higher quality means lower operating cost and vice versa.

Design by Quality

To fully understand robust design, the implications of quality must be considered. The ideal quality that a customer can expect from a product or service is that it will deliver the target performance each time it is used under all intended operating conditions throughout its intended life (reliability and dependability). The impact of poor product quality results in lost revenue from customer dissatisfaction, fewer future sales, and reduced market share. Reponses to these require increasing marketing and advertisement costs.

A product's specifications typically are intended to guarantee that the components are properly assembled, free from manufacturing defects, and provide the desired quality. Maintaining quality assurance is the responsibility of everyone involved in the design and production of a product.

There are established methods of sampling a product or process to determine its correspondence to its specifications. *Inspection* is the process of checking that a product conforms to its specifications.

Inspection of variables: Quantitative measurements of products and their components (dimensions, surface finish, and other physical or mechanical properties) must be inspected. Measurements are made with instruments that produce a reliable result. Inspection of a process at 100% is performed for highly critical parts with the help of an online inspection device.

Inspection for attributes: Attribute inspection is a process that determines whether the measurement of an element falls within the acceptable tolerance range.

Inspection using statistical process control methods: The systematic method of detecting assignable variability in a process is known as *statistical process control*. It involves extracting significant information from large amounts of numerical data, thus playing an important role in the quality control of large amounts of material or product. Statistical methods are also employed when dealing with variability in data such as in manufacturing processes because no two products are ever produced in the same way.

Statistical process control uses systematic methods of detecting assignable variability in a manufacturing process by sampling. Statistical quality control detects variation in the process, dimensions or properties of raw materials, and in machine and operator performance. Two types of variation, natural and assigned, can occur during a process. *Natural variability* in a manufacturing process refers to inherent, uncontrolled changes that occur in the composition of a material or in the performance of an operator or machines. These variations occur randomly with no particular pattern or trend. In contrast, *assignable variability* can be traced to a specific controllable cause. Ideally, identifying the assignable causes of variability can lead to better control and prevent defects.

8.3.1 Taguchi Principles for Cost Reduction and Quality Assessment Using Robust Design

Genichi Taguchi, a noted Japanese engineering specialist, developed methods to simultaneously reduce cost and improve quality. The Taguchi methods were first introduced to the U.S. automotive industry in 1982. Although engineers at Bell Labs and Xerox were the first in the United States to experiment with Taguchi methods, the Ford Motor company aggressively fostered its use to improve quality and reduce cost. This method has brought about fundamental changes in engineering and quality control methods that have transformed many U.S. companies.

The Taguchi design experiment involves performing a series of tests on a process when its input variables or parameters have all changed at the same time. The purpose of the experiment is to observe, identify, and isolate the variables whose interactions change the output response. Experiments using traditional methods must change variables one factor at a time (with all other factors being held constant) to find the one that most affects the response. Robust design uses ideas from statistical experiment design and addresses two major concerns:

1. How to economically reduce the variation of a product's function in the customer environment.
2. How to ensure that decisions found optimum during laboratory experiments will continue to do so in manufacturing and customer environments.

The following summarizes Taguchi's steps:

- Design quality into the product. Do not use inspection to identify poor-quality products.
- Set a target. The cost of quality is the deviation from the target values.
- Make the product insensitive to uncontrollable external factors.

The following two steps of Taguchi are involved in robust design to enhance design quality:

1. Optimizing the design of the product and process
2. Making the design robust (insensitive to the influence of uncontrollable factors)

Taguchi's Noise Factors

Two types of undesirable and uncontrollable factors—external noise and internal noise—cause deviation from the target values for a product's functional characteristics. *External noise factors* affect a product's performance. Examples include changes in operating environmental variables such as temperature, humidity, dust, and variability in human operators. *Internal noise factors* create product-to-product noise or variability. The main objective of the overall quality system is to make a product that is *robust* with respect to all noise factors. Reducing the impact of noise factors and selecting controllable factor levels ensure that the desirable quality characteristic is near target values, resulting in a robust design of the product and process. Factors that cause internal noise include:

- Manufacturing variations that include process imperfections such as variations in machine settings.
- Product deterioration that occurs over time as a result of wear and tear of parts caused by friction and the loss of spring resilience.

Taguchi Methodology for Achieving Robustness

Taguchi recommends a three-stage design methodology to determine the target values and tolerances for the respective parameters in a product and process:

1. System design
2. Parameter design
3. Tolerance design

1. **System design** uses scientific, economic, and engineering principles and experience to create a product prototype that will meet functional requirements and the process that will build it.
2. **In the parameter design** stage, the most important stage in the Taguchi method, the optimal settings of the product and process parameters are determined to minimize performance variability, that is, the effect of noise factors on a product's functional characteristics (see previous section). Taguchi defined a *performance measure* known as the *signal-to-noise* (S/N) ratio so that the parameter levels selected maximize this ratio. The term *signal* here represents the square of the mean value of the quality characteristic, and *noise* is a measure of its variability. Noise factors investigated should be those that most impact the product's performance.

However, parameter design alone does not always lead to sufficiently high quality. Additional improvement can be obtained by controlling the causes of variation when it is economically justifiable, typically by using more expensive equipment, higher-grade components, or improved environmental controls, all of which lead to higher product or operating cost, or both. The benefits of improved quality must justify the added product cost.

3. **In the tolerance design** stage, the tolerances—that is, a range of acceptable values around the target values of the control parameters identified in the design stage—are set. Narrow tolerances are specified for actual deviations in relation to the levels determined by the parameter design. Consider an example in which tolerance specification is given as 0.50 ± 0.02. In this case, it is immaterial whether the actual figure is 0.48, 0.50, or 0.52; the specifications are equally satisfied. Taguchi defines quality as product uniformity around a target value. Here the target value is 0.50, and the actual values achieved closer to the set target value are better than the values further away from the target.

Taguchi's Approach to Quality

Taguchi's approach to quality can be summarized as two fundamental concepts:

1. Product quality must be engineered or designed. It cannot be achieved economically by inspecting and screening.

2. Quality losses must be defined as deviations from the target, not conformance to arbitrary specifications. Quality losses must be measured by systemwide costs: *loss to customer and society*, not local costs at points of defect detection.

Quality Loss Function   Taguchi defines a product's *quality level* as the total loss incurred by society as a result of its failure to deliver the target performance and of its harmful side effects, including its operating costs. A product whose response is exactly on target gives the best performance. As the product's performance deviates from the target, the quality becomes progressively worse as shown in **Figure 8.3 (a)**. True quality measure is not based on the step function but as a quadratic loss function as shown in **Figure 8.3 (b)**.

The quadratic *quality loss function* is given as

$$L(y) = k(y - m)^2 \tag{8.1}$$

where

$L(y)$ = loss in dollars when quality characteristics equal y (the average quality
        loss incurred by the customers)

y = value of the quality characteristic (e.g., length, width, concentration,
     surface finish, flatness)

m = target value of y

k = constant called *quality loss coefficient*

The quality loss function is plotted in **Figures 8.3 (a)** and **(b)**.

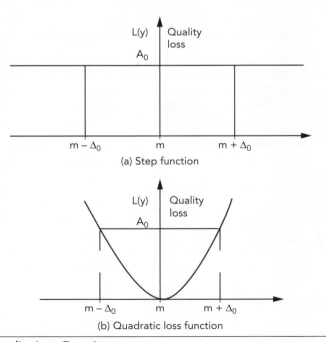

(a) Step function

(b) Quadratic loss function

FIGURE 8.3 Quality Loss Function

A convenient way to determine k is to determine first the *functional limits* for the value of y. *Functional limit* is the value of y at which a product would fail in half of its applications. Let $m \pm \Delta_0$ be the functional limits. Suppose that the loss at $m \pm \Delta_0$ is A_0, then by substitution in equation (8-1), we have

$$k = \frac{A_0}{\Delta_0^2} \qquad (8.2)$$

where Δ_0 is the cost of repair or replacement of the product and includes the loss resulting from its unavailability during the repair period.

Substituting equation (8-2) in equation (8-1) gives

$$L(y) = \frac{A_0}{\Delta_0^2}(y - m)^2 \qquad (8.3)$$

Types of Quality Loss Function:

Normal is the best:

$$L(y) = \frac{A_0}{\Delta_0^2}(y - m)^2 \qquad (8.4)$$

Smaller is better:

$$L(y) = \frac{A_0 y^2}{\Delta_0^2} \qquad (8.5)$$

Larger is better:

$$L(y) = \frac{A_0 \Delta_0^2}{y^2} \qquad (8.6)$$

Classification of Parameters Regarding Quality A schematic representation of a product/process is shown in **Figure 8.4**. The response is denoted by y and is called the *quality characteristic*. The parameters that influence it are classified as follows:

- Signal factors (M): The user sets this parameter to express the intended value for the response of the product based on the knowledge of it during development. Two or more signal factors can be used in combination to express the desired

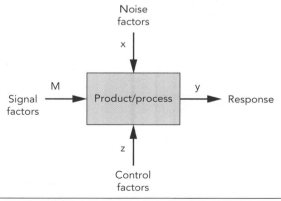

FIGURE 8.4 Block Diagram of Product/Process (P Diagram)

response. For example, speed setting of a fan is a signal factor for specifying the amount of breeze.

- Noise factors (x): Designers cannot control certain parameters known as *noise factors* (x) (see previous discussion). Parameters whose settings (also called *levels*) are difficult to control or are expensive to control are considered noise factors. The levels of the noise factors change from one unit to another, from one environment to another, and from time to time. The noise factors cause the response y to deviate from the target specified by the signal factor M and lead to quality loss.
- Control factors (z): The designer can freely specify these parameters of control factors (z), which can take multiple values or levels to result in the least sensitivity of the response to the effect of noise factors.

EXAMPLE 8.1

The critical dimension of a mechanical component made by a machining process using a turning machine must be 0.45 ± 0.005 inches. The scrapping cost is $100 per component. Samples taken from the turning machine had the following dimensions: 0.451, 0.447, 0.448, 0.452, 0.450, 0.453, 0.449, 0.447, 0.454, 0.456, 0.450, and 0.452.

Problems

1. Find the Taguchi loss equation for this operation.
2. Determine the average loss function for the parts made on this turning machine.

Solutions

1. $L(y) = k(y - m)^2 = \dfrac{A}{\Delta^2}(y - m)^2$

   $= \dfrac{\$100}{(0.005)^2}(y - m)^2 = 4 \times 10^{-6}(y - m)^2$

2. $Q = k[(\mu - m)^2 + \sigma^2]$

$$\mu = \bar{x} = \dfrac{\begin{array}{l} 0.451 + 0.447 + 0.448 + 0.452 + 0.450 + 0.453 + 0.449 + 0.447 \\ + 0.454 + 0.456 + 0.450 + 0.452 \end{array}}{12}$$

$\qquad = \dfrac{5.409}{12}$

$\qquad = 0.451$

$\sigma = \sqrt{\dfrac{1}{n-1}\sum(y_i - \mu)^2} = \sqrt{\dfrac{1}{12-1}\sqrt{(0.000090)}} = \sqrt{\dfrac{0.000090}{11}} = 0.00286$

$Q = 4 \times 10^6 [(0.451 - 0.450)^2 + (0.00286)^2]$

$\quad = 4 \times 10^6 [(0.001)^2 + (0.00286)^2]$

$\quad = 4 \times 10^6 [(1 \times 10^{-6} + 8.180 \times 10^{-6}] = 36.72$

Average quality loss is $36.72.

EXAMPLE 8.2

An auto service facility has identified two of the most important factors that help it attract and retain customers: the price of the service and the time required to perform the service. Based on the price for similar service in neighboring auto service facilities, the customer tolerance limit for price is estimated to be $0, and the associated customer loss is $60. The customer tolerance limit for the service time is 15 minutes, and the associated customer loss is $50. A random sample of 12 customers gave the following values of price: 9, 7, 8, 11, 12, 6, 9, 11, 8, 9, 7, and 12. The sample service times (in minutes) are 10, 8, 16, 9, 11, 13, 16, 14, 12, 7, 17, and 18.

Problems

1. Determine the facility's total expected loss per customer.
2. If the facility expects 3,000 customers monthly, determine its expected monthly loss.

Solutions

1. The proportionality constant for the price factor, when smaller is better, is the mean square deviation for price estimate, given by

$$\sum \frac{y_i^2}{12} = \frac{[9^2 + 7^2 + 8^2 + 11^2 + 12^2 + 6^2 + 9^2 + 11^2 + 8^2 + 9^2 + 7^2 + 12^2]}{12}$$

$$= \frac{1035}{12}$$

$$= 86.25$$

Hence, the expected loss per customer from the price factor is

$$= (0.6)(86.25) = \$51.75$$

The proportionality constant for the service time factor is

$$k_2 = \frac{50}{(15)^2} = 0.22222$$

The mean square deviation for the service time factor is given by

$$\sum \frac{y_i^2}{12} = \frac{[10^2 + 8^2 + 16^2 + 9^2 + 11^2 + 13^2 + 16^2 + 14^2 + 12^2 + 7^2 + 17^2 + 18^2]}{12}$$

$$= \frac{2049}{12}$$

$$= 170.75$$

Hence, the expected loss per customer due to the service time factor

$$= (0.22222)(170.75)$$

$$= \$37.944$$

Therefore, the total expected loss per customer due to both factors (price and service time) are $51.75 and $37.944, respectively, or $89.694.

2. The expected monthly loss $= (3,000)(\$89.694) = \$269,083.34$.

EXAMPLE 8.3

Refer to Example 8.2. If the automobile service facility employs additional service personnel to shorten the service time, the estimated additional cost is $1 per customer. A random sample of 12 customers with the added personnel gave the following waiting times (in minutes): 9, 7, 10, 12, 14, 8, 13, 9, 13, 11, 6, and 7.

Problems

1. Determine whether it is cost effective to add more service personnel.
2. Determine the total expected monthly loss.

Solutions

1. The mean square deviation for service time with added service personnel is given by

$$\sum \frac{y^2 i}{12} = \frac{\left[9^2 + 7^2 + 10^2 + 12^2 + 14^2 + 8^2 + 13^2 + 9^2 + 13^2 + 9^2 + 13^2 + 11^2 + 6^2 + 7^2\right]}{12}$$

$$= \frac{1249}{12}$$

$$= 104.916$$

The expected loss per customer because of the service time factor is

$$= (0.22222)(104.916)$$

$$= \$23.315$$

The total expected loss per customer from the factors of price and service time as well as additional cost of personnel is

$$(\$51.75 + 23.315 + \$1)$$

$$= \$76.065 < \$89.694$$

Hence, it is cost effective to add personnel.

2. The total expected monthly loss = $(3,000)(\$76.065) = \$228,193.75$

Conducting a Robust Design Experiment

The first step in conducting a robust design experiment is to identify the *factors* that influence the process, two of which are *noise* and *control*. The number of control factors is determined by the complexity of the process. After both factors have been determined, the *parameter* settings for all factors are selected. Then an orthogonal array (explained in the next section) is constructed using all factors and levels that have been predetermined. The experiment is performed according to the parameter settings of each row in the orthogonal array with or without replications.

Consideration for the replication of the experiment depends on how long it takes, sample availability, and financial aspects. After the experiment, analysis by row is

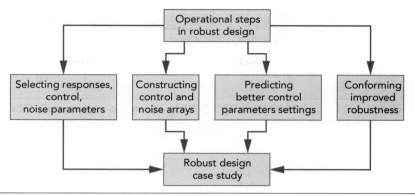

FIGURE 8.5 Steps in Robust Design Methodology

performed regarding the variability, and its bias on the data recorded. The detail in the analysis varies depending on the type of its target response: the larger the better (e.g., stiffness), the nominal the best (e.g., surface roughness), and the smaller the better (e.g., drag coefficient in cars or planes) (**Figure 8.5**).

Experimental Steps

▸ **Step 1. Determine the Degrees of Freedom**

▸ **Step 2. Select the Appropriate Orthogonal Arrays and Control Factors**

▸ **Step 3. Assign Control Factors (and Interactions) to Orthogonal Array Columns**

▸ **Step 4. Select an Outer Array for Noise Factors and Assign Factors to the Columns**

▸ **Step 5. Conduct the Experiment and Collect Data**
  - Develop a test plan.
  - Select appropriate experimental design.
  - Perform experimental set up.
  - Test and inspect as prescribed in experimental setup.

▸ **Step 6. Analyze the Data**
  - Use S/N analysis.
  - Use S/N response tables and graphs.
  - Use S/N ANOM (Analysis of Means) and ANOVA (Analysis of Variance).

▸ **Step 7. Interpret the Results**
  - Select the optimum levels of control factors.
  - Predict results for the optimal conditions.

▸ **Step 8. Confirm the Experiment Results**
  - Verify the predicted results.
  - Conduct additional experiments as needed.

Orthogonal Array	Number of Rows	Maximum Number of Factors	Maximum Number of columns at These Levels			
			2	3	4	5
L_4	4	3	3	–	–	–
L_8	8	7	7	–	–	–
L_9	9	4	–	4	–	–
L_{12}	12	11	11	–	–	–
L_{16}	16	15	15	–	–	–
L'_{16}	16	5	–	–	5	–
L_{18}	18	8	1	7	–	–
L_{25}	25	6	–	–	–	6

TABLE 8.1 Selecting a Standard Orthogonal Array (Phadke 1989)

Orthogonal Arrays

An orthogonal experiment can increase the experimenter's understanding of many factors regarding product or process quality. Although their use relieves the experimenter of the major burden of designing a fractional factorial experiment, it allows several factors to be evaluated with a minimum number of test runs.

The initial step in constructing an orthogonal array for a specific case study is to count the total degrees of freedom to identify the minimum number of experiments that must be performed to study all chosen control factors. For example, a three-level control factor counts as two degrees of freedom because of the two comparisons. In general, the number of degrees of freedom associated with a factor equals one less than the number of levels for that factor (**Tables 8.1** and **8.2**).

4 × 3 orthogonal array	9 × 4 orthogonal array	8 × 7 orthogonal array
1 1 1	1 1 1 1	1 1 1 1 1 1 1
1 2 2	1 2 2 2	1 1 1 2 2 2 2
2 1 2	1 3 3 3	1 2 2 1 1 2 2
2 2 1	2 1 2 3	1 2 2 2 2 1 1
	2 2 3 1	2 1 2 1 2 1 2
	2 3 1 2	2 1 2 2 1 2 1
	3 1 3 2	2 2 1 1 2 2 1
	3 2 1 3	2 2 1 2 1 1 2
	3 3 2 1	

TABLE 8.2 Commonly Used Orthogonal Arrays

Factors	Level 1	Level 2
A = feed rate	6.6 inches/minute	15 inches/minute
B = spindle speed	85 revolutions/minute	290 revolutions/minute
C = tool type	High-speed steel	Tungsten coated

Trial Number	A	B	C
1	1	1	1
2	1	2	2
3	2	1	2
4	2	2	1

TABLE 8.3 Control Factors of the Process and 4 × 3 Orthogonal Array

See **Table 8.3** for an example of a 4 × 3 orthogonal array for a manufacturing process involving a machining operation. The response to be monitored is surface quality. The control variables are feed rate, spindle speed, and tool type. Each of the three variables has two identified levels that can be tried experimentally. **Table 8.3** shows control factors and the construction of the orthogonal array. **Table 8.4** provides information for a molding machine with control variables of injection pressure (250 psi or 350 psi), mold temperature (150 or 200° F), and set time (6 or 9 seconds).

Variables or Factors	Level 1	Level 2
A. injection pressure	A1 = 250 psi	A2 = 350 psi
B. mold temperature	B1 = 150 ° F	B2 = 200 ° F
C. set time	C1 = 6 seconds	C2 = 9 seconds

	350 psi	200	9 sec			
	250 psi	150	6 sec			
	Injection Pressure	Mold Temperature	Set Time			
Column Exponent. no.	A	B	C	Repetitions 1	2	3
1	1	1	1	26		
2	1	2	2	25		
3	2	1	2	34		
4	2	2	1	27		

TABLE 8.4 Molding Machine Example

Steps for the Robust Design Experiment The choice of using an orthogonal array for a particular project depends on the number of factors and levels other than practical considerations.

Identify the main function, side effects, and failure modes. This requires engineering knowledge of the product or process and the customer's environment.

Identify noise factors and testing conditions for evaluating the quality loss. The testing conditions are selected to capture the effect of the more important noise factors. Testing conditions should permit a consistent estimation of the sensitivity to noise factors for any combination of control factor levels.

Identify quality characteristics to be observed and the objective function to be optimized. A robust design keeps the response mean to the target and minimizes the variation in the response. This is called the signal to noise (S/N) ratio and is different for different types of quality characteristics. So robust design is treated as an optimization process in which the objective function is S/N ratio. This is the ratio of the signal (mean) over the noise (variability). The larger the S/N ratio, the more robust the performance.

Identify the control factors and their multiple levels. The more complex a product or a process, the more control factors it has and vice versa. Six to eight control factors typically are chosen at one time for optimization. For each control factor, two or three levels are selected, one of which is usually the starting level. The levels should be chosen sufficiently far apart to cover a wide experimental region because sensitivity to noise factors does not usually change with small variations in control factor setting. Selection of a wide experimental region helps to identify good and bad regions for control factors.

Design the matrix experiment and the data analysis procedure. Using orthogonal arrays is an efficient way to study the effect of several control factors simultaneously. The factor effects thus obtained are valid over the experimental region and provide a way to test for the additivity of factor effects. The experimental effort needed is much less when compared to other methods of experimentation such as trial and error.

Conduct the matrix experiment. Levels of several control factors must be changed when going from one experiment to the next in a matrix experiment. Properly setting the levels of the various control factors is essential; that is, when a particular factor must be at level 1, it should not be set to level 2 or 3. However, small perturbations inherent in the experimental equipment should not be of concern. Any erroneous or missing experiments should be repeated to complete the matrix.

Analyze the data, determine optimum levels for control factors, and predict performance under these levels. When a product or a process has multiple characteristics, it might be necessary to make some trade-offs when choosing the optimum factor levels in robust design; however, the primary focus is on maximizing the S/N ratio. The observed factor effects with the quality loss function can be used to make a rational trade-off.

Conduct the verification experiment and plan future actions. The purpose of this final and crucial step is to verify that the optimum conditions suggested by the matrix experiments actually give the projected improvement. If the observed and the projected improvements match, the suggested optimum conditions should be adopted. If not, the conclusion should be that the additive model underlying the matrix experiment has failed and that problem should be corrected.

Corrective actions include finding better quality characteristics, S/N ratios, or different control factors and levels, or studying a few specific interactions among the control factors. Evaluating the improvement in quality loss, defining a plan for implementing the results, and deciding whether another cycle of experiments is needed are also parts of the final step of robust design. However, it is common for a product or process to require more than one cycle to achieve the desired quality and cost improvement.

EXAMPLE 8.4

The product considered for improvement is a new surface roughness analyzer. The step-by-step design aspect of this surface roughness analyzer is discussed in chapter 10. The analyzer offers a new noncontact optical method based on light diffraction principles. When applied to engineering surfaces, the analyzer rapidly provides precision surface roughness data on engineering and machined surfaces. The roughness measurement conventionally involves the use of a stylus device, which is drawn over the sample to detect and record variations in surface irregularities. The optical technique is much preferred to the method of surface roughness measurement by contact.

A laser- and microcomputer-based vision system measures the roughness of the intensity of the collimated, monochromatic light source diffracted in the spectral direction and captured by a video system. This system provides an analog signal to a digitizing system that converts the information, which is subsequently modified to display the surface roughness value. As shown in **Figure 8.6**, the intensity is measured as a function of the gray level of the image, then is processed by the digitizing circuit, and is finally compared to a previously defined calibration standard. The system's microcomputer base allows the operator to interact in the form of menu-driven steps that provide guidance through the requirements of each phase of the process: the calibration, measurement, and analysis phases.

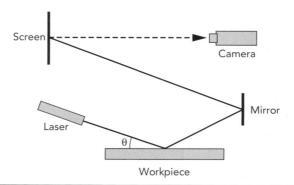

FIGURE 8.6 Schematic of the Laser-Based Roughness Analyzer

Improvement using robust design: The objective of the robust design study is to find the optimum recommended factor setting for the surface roughness analyzer to minimize the variability in the readings. This instrument relies on the spread of the laser light on the workpiece to determine surface roughness; therefore, the analyzer's reliability depends primarily on everything involved with the laser and its path.

As an example, a four-parameter setting, each at two levels, was introduced into the experiment in the form of orthogonal arrays. The parameters are as follows:

1. Laser angle
2. Background
3. Distance from the laser to workpiece
4. Background lighting

The parameter setting for the new experiment is shown in **Table 8.5**.

Parameter	Abbreviation	Level 1	Level 2
Laser angle	La	20°	30°
Background surface	Bs	Glossy	Nonglossy
Distance from laser to workpiece	Dt	4.5 in	5.5 in
Background lighting	Bl	Off	On

TABLE 8.5 Parameter Settings for Surface Analyzer

Parameter optimization: Experiments are performed to minimize the effects of the laser angle and the distance from the laser to the workpiece and the interactions between them. The parameter setting for this experiment is identified.

Refer to **Table 8.6** for the optimal setting of the parameters achieved after robust design analysis.

Many industries successfully use the robust design method to determine a system's optimum setting. The application of the method in electronics, automotive products, and photography, for example, has been an important factor in the rapid industrial growth. The robust design method is useful and easy to implement in the design process. Furthermore, the nature of the method allows the designer to investigate every possible variable and its effects on this application and reduces the time required to complete the experiment.

Parameter	Condition
Laser angle	25°
Distance from laser to workpiece	4.5 in
Background surface	Nonglossy
Background lighting	Off

TABLE 8.6 Optimal Settings for the Surface Analyzer

8.4 **Tools of Design Optimization**

Today's productive and competitive atmosphere demands that product design be optimized after considering all variables that control the design process. The designer's major aim is to find a design solution that meets performance requirements and satisfy all constraints. For a product to be successful, its design must be efficient and economical. The process of determining the best solution to the design is known as *optimization*. *Optimum design* is one that is feasible and superior to a number of alternatives.

An optimum design can be obtained by using an iterative process or solving an optimization problem. In the iterative process, the design is improved through repeated modification by which the design team changes the values of design variables based on experience. The designer is responsible for deciding which parameter to change. Solving an optimization problem provides a procedure for identifying all design parameters simultaneously to satisfy a set of constraints and optimize a set of objectives. Optimum performance can be expressed as the objective function (desired criterion), which is also a measure of the product performance and is represented as the performance index or figure of merit.

A common design objective could be reducing cost or weight. An architect designing an economic building and a comfortable surrounding could specify the objectives as minimizing the cost and maximizing the area. A jet engine designer who wants to achieve high performance can specify the objective as a power-to-weight ratio. Typical objective functions that product designers consider are maximizing the profit, production rate, and process yield while minimizing inventory and cost. The design parameters under the designer's control are called *design variables*. The objectives and constraints of a design have a direct relationship to the variables. The relationship between the input variables and the objective function can be mathematically represented.

Design Constraints

The process involved in optimization consists of considering all design constraints affecting an objective function. Defining constraint requirements for an acceptable optimization solution limits the exploration of a possible design. Represented as mathematical equations or inequalities, constraints result from physical laws involved in the design and limit individual variables. *Functional constraints*, also called *equality constraints*, specify the relations that must exist between the variables. *Regional constraints*, also called *inequality constraints*, specify the distinctions of a design. These functions are mathematical statements of the limits between which a design's parameters must lie. George Dieter and James Siddall have reviewed the development of optimal design methods and classified them into four optimization groups: those by evolution, by intuition, by trial and error, and by numerical algorithm.

8.4.1 **Numerical Optimization Techniques**

Some of numerical optimization techniques are as follows:

- Differentiation
- Linear programming
- Gradient search

- Lagrange multiplier
- Dynamic programming
- Nonlinear optimization method

The differentiation method, linear programming, and gradient search techniques are explained in detail in this chapter.

Optimization by differentiation: While using the optimization by differentiation method, the designer uses differential calculus to determine the optimum by solving simultaneous equations found by setting the derivatives of the objective function with respect to each of the parameters to zero.

$$\frac{\partial Z}{\partial x_1} = 0$$

$$\frac{\partial Z}{\partial x_2} = 0 \tag{8.7}$$

$$\frac{\partial Z}{\partial x_n} = 0$$

where Z is a function of variables $x_1, x_2, \ldots x_n$.

The optimum point in Equation 8.7 could have either a maximum or minimum value. To determine whether it is minimum or maximum, the sign of the second derivative of Z must be examined with respect to x. If the curvature is negative, the stationary point is maximum. The point is minimum if the curvature is positive.

Consider the example of a manufacturing process with two variables, x_1 and x_2. The performance Z of the process is related to the input variables x_1 and x_2 by the equation

$$Z = 25x_1 - 2x_1^2 + 41x_2 - 5x_2^2 + 4x_1x_2 \tag{8.8}$$

Find the values of x_1 and x_2 that maximize the value of Z. The differential calculus approach for two variables involves taking derivatives of the objective function Z with respect to x_1 and x_2 and setting them equal to zero. This provides two equations with two unknowns that can be solved to find the optimum set of operating conditions.

$$\frac{\partial Z}{\partial x_1} = 25 - 4x_1 + 4x_2 = 0 \tag{8.9}$$

$$x_1 = 6.25 + x_2 \tag{8.10}$$

$$\frac{\partial Z}{\partial x_2} = 41 - 10x_2 + 4x_1 = 0 \tag{8.11}$$

$$41 - 10x_2 + 4(6.25 + x_2) = 0$$

$$x_2 = 11, \ x_1 = 17.25$$

$$Z = 25(17.25) - 2(17.25_1)^2 + 41(11) - 5(11)^2 + 4(17.25)(11)x_2$$

$$Z = 441.125$$

Optimization by linear programming: *Linear programming* is a mathematical method that allocates constrained resources to attain an objective such as to minimize cost or maximize profit. It can be applied to solve problems in which the objective function and the constraints are linear functions of the variables. The objective of the problem is to maximize or minimize some linear objective function.

$$Z = a_1 x_1 + a_2 x_2 + a_3 x_3 \ldots \ldots a_n x_n \tag{8.12}$$

The objective function is constrained by resources that are shown by constraint equations.

Less-than-or-equal-to constraints

$$b_1 x_1 + b_2 x_2 + b_3 x_3 \ldots .. b_n x_n \leq b_0 \tag{8.13}$$

Greater-than-or-equal-to constraints

$$b_1 x_1 + b_2 x_2 + b_3 x_3 \ldots .. b_n x_n \geq b_0 \tag{8.14}$$

Equal to constraints

$$b_1 x_1 + b_2 x_2 + b_3 x_3 \ldots .. b_n x_n = b_0 \tag{8.15}$$

Linear programming problems can be solved analytically or graphically. The graphical approach is only suitable for solving for two variables.

EXAMPLE 8.5

The linear programming method is applied to a manufacturing engineering situation in which the input is raw material and the output consists of two products. The manufacturing process has two stages. Stage 1 is an automated rolling machine into which the raw material flows. Its output consists of two base parts for two products (A and B). The two outputs from Stage 1 are fed to the assembly line (Stage 2) where additional assembly operations take place.

The output of Stage 2 results in two products A and B (**Figure 8.7**). This result indicates that either of the products can make a profit.

Unit profit on product A = $10

Unit profit on product B = $20

Total profit can be shown as

$Z = 10x_1 + 20x_2$

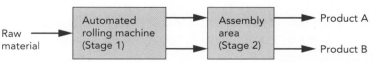

Based on Groover, Mikell P, 1980. Automation, Production Systems and Computer Aided Manufacturing. Englewood Cliffs, NJ: Prentice Hall Inc.

FIGURE 8.7 Optimization Example

where

x_1 = number of units of product 1 produced

x_2 = number of units of product 2 produced

Constraints

In addition, one unit of raw material is needed for one unit of product A and one unit of product B. The total amount of raw material that can be processed through stage 1 equals 9 units/day and is expressed as:

$$1x_1 + 1x_2 \leq 9$$

where

x_1 = units of 1; x_2 = units of 2

A designer must realize that the maximum profit cannot be realized by maximizing only product B. Instead, more labor resource is needed to make each product B. The labor required is 1 hour for each unit of product A and 3 hours for each unit of product B.

Total labor hours/day = 15

$$1x_1 + 3x_2 \leq 15$$

In the graphical problem-solving method, these two constraints are plotted as two lines. Because both constraint equations are less than or equal to each other, the useful area will be to the left region (**Figure 8.8**).

$$1x_1 + 1x_2 = 9$$
$$1x_1 + 3x_2 = 15$$

The objective is to find the combination of x_1 and x_2 to maximize the function

$$Z = 10x_1 + 20x_2$$

To meet this objective, a series of constant profit lines (shown with dotted lines) are drawn on the same graph as constraints. When the constant $-Z$ lines are superimposed on the constant region, the optimum point is at $x_1 = 6$, $x_2 = 3$, and $z = \$120$ per day. No other combination gives a higher Z value (**Figure 8.8**).

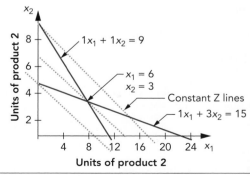

FIGURE 8.8 Graphical Solution

EXAMPLE 8.6

An assembly department producing printed circuit board outputs two types, x and y. Each board requires three operations: (1) inserting the component, (2) soldering the component, and (3) inspecting the component. See **Table 8.7** for the time required for each board in each operation, and the maximum number of work hours available per day are plotted in **Figure 8.9**.

	x Units	y Units	Operator Hours/Day
Insertion	8	4	80
Soldering	3	4	60
Inspection	1	3	24

TABLE 8.7 Example 8.6

The objective is to determine the number of x boards and y boards to produce to maximize the output. Assume that the contribution of each x board is \$120 profit and of each y board is \$60. Determine the number of each board that should be made to maximize profit.

The constraints can be rewritten as follows:

$$8x + 4y \leq 80$$
$$3x + 4y \leq 60$$
$$x + 3y \leq 24$$
$$x \geq 0; \; y \geq 0$$

The objective function in this case is the maximization of the output.

$$Z = x + y$$

Solving these equations shows that maximum output = 12.8 units ($x = 7.2; y = 5.6$). Because the units in this case are printed circuit boards, they must be rounded to whole numbers.

$$x = 7; y = 5$$

Therefore, the maximum output of 12 units is more realistic.

With the same constraints, assume that the objective function to be maximized is profit. For maximum profit, let $Z = 120x + 60y$. Therefore that maximum profit equals \$1,200 if $x = 10; y = 10$. Thus, to maximize profit, it is best to only make x boards.

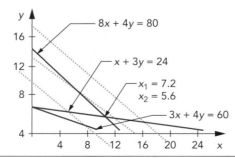

FIGURE 8.9 Graphical Solution

The *slack variable method* can be used to maximize output and profit.

- To maximize the output of x and y boards

  Output $= x + y$

- To maximize the profit of x and y boards

  Combination profit $= 120x + 60y$

As previously noted, both objective functions are subject to constraint functions.

$$8x + 4y \leq 80$$
$$3x + 4y \leq 60$$
$$x + 3y \leq 24$$
$$x \geq 0; y \geq 0$$

Introduce slack variables s_1, s_2, and s_3 into the constraint equations. The equations are modified as

$$8x + 4y + s_1 = 80$$
$$3x + 4y + s_2 = 60$$
$$x + 3y + s_3 = 24$$

In this case, there are $m = 3$ (equations) and $n = 5$ (unknowns). **Table 8.8** sets two variables to zero in each instance, and objective functions are calculated. There are 10 possible combinations for which possible combinations are calculated using the formula

$$\frac{n!}{m!(n-m)!} \tag{8.16}$$

The objective functions are optimized if at least $(n - m)$ variables are set to zero (**Table 8.8**).

Results show that for the *item no. 9*, none of the variables are negative and the output and profit are both positive.

Number	X	Y	S_1	S_2	S_3	Output	Profit	Comment
1	0	0	80	60	24	0	0	X = 0; Y = 0
2	0	20	0	−20	−36	20	1200	S_2, S_3 negative
3	0	15.0	20	0	−21	15	900	S_3 negative
4	0	8.0	48	28	0	8	480	X = 0
5	10.0	0	0	30	14	10	1200	Y = 0
6	20.0	0	−80	0	4	20	2400	S_1 negative
7	24.0	0	−112	−12	0	24	2880	S_1, S_2 negative
8	4.0	12.0	0	0	−16	16	1200	S_3 negative
9	7.2	5.6	0	16	0	12.8	**1200**	Output and profit are positive
10	16.8	2.4	−64	0	0	19.2	2160	S_1 negative

TABLE 8.8 Use of the Slack Variable Method

EXAMPLE 8.7

Product mix problems occur when several products are produced in the same production plant. Consider a refinery that uses two crude oils. Crude A costs $30 a barrel, and 20,000 barrels are available.

Crude B costs $36 a barrel, and 30,000 barrels are available. The company manufactures gasoline and lubrication oil from crude. The yield and sale price per barrel of the product and market demands are shown in **Table 8.9**.

	Yield		Sale Price per Barrel	Market Demand
Product	Crude A	Crude B		
Gasoline	0.6	0.8	$50	20,000
Lube oil	0.4	0.2	$120	10,000

TABLE 8.9 Crude Oil Data

Problem

How much crude oil should the company use to maximize its revenue? Formulate and solve this optimum design problem.

Solution

Assume that crude A = x; B = y;

The objective function is defined as:

$$(0.6x)\$50 + (0.4x)\$120 + (0.8y)\$50 + (0.2y)\$120 = Z$$
$$78x + 64y = Z$$

$$0.6x + 0.8y \le 20,000$$
$$0.4x + 0.2y \le 10,000$$
$$x \ge 0; y \ge 0$$

where

$$x = \text{Crude A}$$
$$y = \text{Crude B}$$
$$Z = \text{Revenue}$$

Solving these equations shows that $x = 20,000$ and $y = 10,000$. This maximizes revenue but not profit.

If the goal is to maximize the profit, the objective function can be written as

$$\text{Profit} = 0.6(\$50 - 30)A + 0.4(\$120 - \$30)A + 0.8(\$50 - \$36)B$$
$$+ 0.2(\$120 - \$36)B$$

$$\text{Profit} = 48A + 28B$$
$$0.6A + 0.8B \le 20,000$$
$$0.4A + 0.2B \le 10,000$$

From these equations, the values of A and B can be found.

Optimization by gradient search: Gradient search is another way to find the optimum point by using the method of steepest ascent/descent along the gradient. A *gradient* is a vector quantity whose components are along the axes of the independent variables X and Y. In other words, gradient is a vector of directional derivative of a given function. Refer to **Figure 8.10** for an illustration of the procedure, which shows some of the constant contours of an objective function $Z = f(x, y)$. The procedure requires that a starting point be chosen. From this starting point, the search moves in the direction of maximum slope (gradient). After moving a certain distance, the process stops and changes the direction to that of maximum slope from that point and so on, until the optimum is reached.

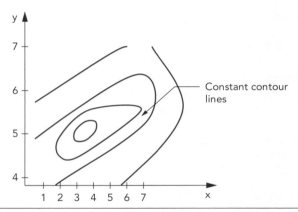

FIGURE 8.10 Gradient Vector of Directional Directive

Basic steps for optimization by gradient search include:

▸ **Step 1. Evaluate the objective function at the starting point.**

▸ **Step 2. Compute the direction of the gradient from the starting point.**

▸ **Step 3. Perform a search in the computed gradient direction for determining the minimum function along that direction.**

▸ **Step 4. Evaluate the objective function at a point $(x + 1)$ from the previous step.**

▸ **Step 5. Terminate the search procedure when optimum is reached.**

The map in **Figure 8.10** shows some of the constant contours of an objective function Z. It is generated by plotting the relationship between the objective function (also known as performance index in some applications) and the input variables, which appear similar to a geographical survey map. The figure shows the plot generated for two variables x and y. These constant contour lines are also known as *response surface*.

Z = Objective functions

x, y = Input variables on which Z is dependent

The optimum point of an objective function is the combination of x and y values at which the objective function is optimized. Many techniques searching for an optimum are based on gradient techniques. A component's magnitude equals the partial derivative of the objective function with respect to the corresponding independent variable. For two inputs, x and y, the components of the gradients are defined as

$$G_x = \frac{\partial Z}{\partial x} \tag{8.17}$$

$$G_y = \frac{\partial Z}{\partial y} \tag{8.18}$$

G_x and G_y are the components of the gradients in the x and y directions on the response surface.

$$G = iGx + jGy \tag{8.19}$$

where i and j represent unit vectors parallel to x and y axes.

The gradient proceeds in the direction of the steepest slope, which is a reasonable strategy to reach the top of the response (objective) surface. The magnitude of the gradient is a scalar quantity given by

$$M = \left[\left(\frac{\partial Z}{\partial x} \right)^2 + \left(\frac{\partial Z}{\partial y} \right)^2 \right]^{\frac{1}{2}} \tag{8.20}$$

The magnitude of the point is defined at a particular point P on the x-y surface. The direction of the gradient is defined as a unit vector

$$D = \frac{G}{M} \tag{8.21}$$

As the gradient search procedure proceeds from the starting point toward the optimum, the sequence of moves to seek the optimum is represented by a trajectory as shown in **Figure 8.11**. The definitions of gradient, magnitude, and direction already given can be extended to functions with more than two independent variables.

Gradient search technique by the method of steepest ascent: Using the method of steepest ascent, the search begins by estimating the gradient at the current operating point. The search then moves the operating point to a new position in the direction of the gradient, which is determined at the new position. The cycle of gradient determination and step-by-step movement is repeated until the optimum is reached.

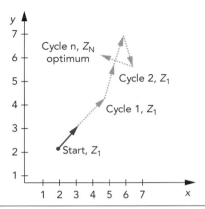

FIGURE 8.11 Trajectory Followed by the Gradient Search Technique

An analytical representation of a response surface in many industrial processes is not readily available, which limits the ability to find the gradient at each cycle. Starting from the current operating point, the slope of the response surface is found by making several exploratory moves centered around the current point. These exploratory moves are arranged in the form of a factorial experiment on neighborhood points. Four experimental points are explored around the current point. At each of those points, the objective function is calculated. Then the gradient components are estimated by means of equations:

$$G_1 = \frac{(Z_2 + Z_3) - (Z_1 + Z_4)}{2\Delta x} \tag{8.22}$$

$$G_2 = \frac{(Z_2 + Z_4) - (Z_1 + Z_3)}{2\Delta y} \tag{8.23}$$

$$\text{Gradient } G = G_1 i + G_2 j \tag{8.24}$$

where Z_1, Z_2, Z_3, and Z_4 are the values of the objective function at four experimental points (**Figure 8.12**).

Δx = Difference in the independent variable x separating the experimental points

Δy = Difference in the independent variable y separating the experimental points

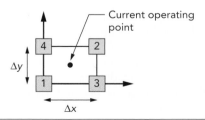

FIGURE 8.12 Objective Function at Experimental Points

Assuming the current point to be (3, 4), consider Δx and Δy to be 2 units each. The coordinates of point 1 are 2 and 3 (**Table 8.10**).

Test Point	x	y	z
1	2	3	z_1
2	4	5	z_2
3	4	3	z_3
4	2	5	z_4

TABLE 8.10 Objective Function Values

Exploratory moves are made for the purpose of determining the gradient. If the gradient is determined, a step move is made to the new operating point. The step move is taken in the direction of the gradient. The input variables x and y are incremented in proportion to the components of the direction vector.

$$\text{New } x = \text{old } x + C^{\frac{G_1}{M}} \tag{8.25}$$

$$\text{New } y = \text{old } y + C^{\frac{G_2}{M}} \tag{8.26}$$

C is a scalar quantity that determines the size of the step move. The search continues until the optimum is reached. At the optimum value of the objective function (*performance index*), the gradient has a value of zero. Quite often, the gradient changes direction abruptly, indicating values in the opposite direction.

EXAMPLE 8.8

Suppose that the response surface for a certain manufacturing process is defined by the equation

$$Z = 17x + 27y - x^2 - 0.9y^2$$

To determine the approximate optimum operating point using the method of steepest ascent, the starting point of the search should be $x = 2$, $y = 3$ and the step size should be $C = 4$.

$$\frac{dZ}{dx} = 17 - 2x \qquad \frac{dZ}{dy} = 27 - 1.8y$$

▸ At the starting point, $x = 2$ and $y = 3$

$$\frac{dZ}{dx} = 17 - 2(2) = 13 \qquad \frac{dZ}{dy} = 27 - 1.8(3) = 21.6$$

New $x = 2 + 4(0.5157) = 4.063$

$M = \sqrt{13^2 + 21.6^2} = 25.21$ $\qquad$ $\bar{D} = 0.5157i + 0.8568j$

New $y = 3 + 4(0.8568) = 6.427$

For cycle 1, calculate Z_1.

▸ **At $x = 4.063$ and $y = 6.427$**

$\dfrac{dZ}{dx} = 17 - 2(4.063) = 8.874$ $\qquad$ $\dfrac{dZ}{dy} = 27 - 1.8(6.427) = 15.431$

New $x = 4.063 + 4(0.4985) = 6.057$

$\bar{D} = .04985i + 0.867j$

$M = 17.8$

New $y = 6.427 + 4(.8670) = 9.895$

For cycle 2, calculate Z_2.

▸ **At $x = 6.057$ and $y = 9.895$**

New $x = 6.057 + 4(0.4695) = 7.935$

$M = 10.407$

$\bar{D} = 0.4695i + 0.883j$

New $y = 9.895 + 4(0.8830) = 13.427$

$\dfrac{dZ}{dy} = 27 - 1.8(13.427) = 2.831$

$\dfrac{dZ}{dx} = 17 - 2(7.935) = 1.13$

For cycle 3, calculate Z_3.

▸ **At $x = 7.935$ and $y = 13.427$**

$\bar{D} = 0.4695i + 0.8830j$

$M = 10.407$

New $x = 7.935 + 4(0.371) = 9.419$

New $y = 13.427 + 4(0.929) = 17.143$

For cycle 4, calculate Z_4.

$$\frac{dZ}{dy} = 27 - 1.8(17.143) = -3.859$$

$$\frac{dZ}{dx} = 17 - 2(9.419) = -1.838$$

▸ **At $x = 9.419$ and $y = 17.143$**

There is a change in the slope. Reduce step size to $C = 2$

New $x = 9.419 + 2(1.43) = 8.56$

$$Z = 17(8.56) + 27(15.34) - (8.56)^2 - 0.9(15.34)^2$$
$$= 274.64$$

New $y = 17.143 + 2(-0.903) = 15.34$

For cycle 5, calculate Z_5.

When the vicinity of the optimum is reached, the step size should be reduced. The point at which the gradient direction changes and the objective function does not change appreciably indicates that the vicinity of optimum point has been located. The objective function tends to fluctuate.

Repeat the last step until no further improvement of the objective function results.

8.4.2 Learning Curve Analysis for Decision Making

When a new model of a product is introduced, learning curve analysis is a useful method for studying its economics, especially if the new product has work content similar to others. As an organization gains experience in manufacturing a product, the resource inputs required per unit of output diminish over the product's life. The labor times to manufacture the first unit of a new automobile is typically much higher than that needed for the 100th unit. As the cumulative output of the model increases, the labor input continues to decline. The performance time drops rather dramatically at first, and it continues to fall at some slower rate until it reaches a constant. This learning curve pattern also applies to groups and individuals.

Consider the general shape of this curve, called the *learning curve* (**Figure 8.13**). This exponential curve becomes a straight line when plotted on the logarithmic coordinates as opposed to arithmetic coordinates. In this example, the initial unit requires 60 labor hours to manufacture. As output and experience continue, labor hours diminish to about 23 hours for the 20th unit.

By convention, the learning rate is specified as a percentage. A 90% curve means that each time the cumulative output doubles, the most recent unit of output requires 90% of the labor input of the reference unit. If unit 1 requires 100 hours, unit 2 requires 90% of 100 labor hours, or 90 hours; unit 4 requires 90% of 90 hours, or 81 hours, and so on.

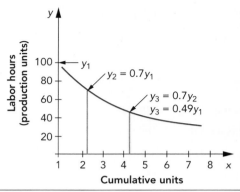

FIGURE 8.13 Learning Curve

The general equation for this curve is

$$y = Ki^b \tag{8.27}$$

where

Y = labor hours required to produce the ith units (or the production time per unit after producing a number of units equal to i)

K = labor hours required to produce the first unit

i = ordinal number of unit produced, that is, first, second, third, and so on

b = index of learning (a constant that depends on the constant percentage reduction characterizing a learning curve, for example, -0.5146 for a 70% curve, -0.322 for an 80% curve, and -0.152 for a 90% curve).

The value of b in Equation 8.27 for a given learning curve ($P\%$) can be found as follows in **Table 8.11**. Each time the production quality doubles, average cost is reduced by a constant percentage that defines the slope. Slope percentage is given as:

$$P = 2^b \times 100 \tag{8.28}$$

Assume that initially, $Y = Ki^b$ for unit $i = i_a$.

Learning curve %, (P)	Exponent, (b)
65	−0.624
70	−0.515
75	−0.415
80	−0.322
85	−0.234
90	−0.074

TABLE 8.11 Exponent Values for Typical Learning Curve Percentages

For double the unit number, that is, $i = 2i_a$, the time is reduced to $Y = PY_a/100$. Substituting this in the equation 8.27,

$$Y_a = Ki_a{}^b \tag{8.29}$$

$$\frac{PY_a}{100} = K(2i_a)^b \tag{8.30}$$

Dividing the second equation by the first provides

$$\frac{P}{100} = 2^b \text{ or}$$

$$\frac{PY_a}{100} = K(2i_a)^b \tag{8.31}$$

The exponent b is a negative number with an absolute value of less than 1, which defines the rate at which the average cost decreases as quantity increases.

For an 80% learning curve,

$$y_2 = 0.8 y_1 \text{ for } i_2 = 2i_1$$

$$\frac{y_2}{y_1} = (\frac{i_2}{i_1})^b;$$

$$\frac{0.8 y_i}{y_1} = (\frac{2i_1}{i_1})^b$$

$$b \log 2 = \log 0.8, \quad b = \frac{-0.0969}{0.3010} = -0.322$$

EXAMPLE 8.9

For a product being evaluated, the first unit cost in terms of operator hours is shown as 1,200. In this case, experience has shown that an 88% learning curve can be anticipated.

Problem
Calculate the projected costs for the first 50 units and for 100 units following the first 50 units.

Solution

$$Y = Ki^b$$

Taking logarithm on both sides,

$$\log(Y) = \log(K) + b\log(i)$$

Determine b, given $P = 88\%$

$$b = Log(0.88)/Log 2 = -0.1844.$$

Substituting 1,200 for the first unit K in the main equation, we get

$$Y = 1,200(i^{-0.1844})$$

Calculate Y at $i = 50$ and $i = 150$

   Cumulative average cost for 50 units $= 1,200(50^{-0.1844}) = 583.2$

   Cumulative average cost for 150 units $= 1,200(150^{-0.1844}) = 496.3$

This shows that total cost for 50 units is 50(583.2), or $29160.

   Total cost for 150 units $= 150(496.3) = \$71445$.

Cost for 100 units following first 50 units equals $42,285

EXAMPLE 8.10

The first of the group of eight machines costs $100,000.

Problem

If a learning rate of 80% is expected, calculate how much time would it take to complete the eighth machine.

Solution

$$Y = Ki^b$$

For $P = 80\%, \quad b = -0.322, \text{ and } k = 100,000$

$$y = 100,000\,(8)^{-322}$$

$$y = \$51,200$$

Rate of learning: The rate of learning is not the same for all manufacturing applications. Learning occurs at a higher rate in some applications than others and is reflected by a more rapid descent of the curve. The use of the learning curve concept as an estimating tool involves more than inserting variables into an equation. The actual behavior of the manufacturing cost trend is influenced by a number of key factors that must be considered in terms of both their effect on the actual database and influences on the cost of new or follow-up work.

8.5 **Failure Modes and Effects Analysis (FMEA)**

Failure modes and effects analysis (FMEA) is an iterative technique used in product design and development. FMEA is used to detail the effects of each individual possible failure mode of a new product or system and to prioritize the relative importance of identified failures. The method addresses how bad a failure could be and what needs to be done to prevent it.

FMEA was first developed by the U.S. Armed Forces in 1949 by the introduction of Mil-P 1629 *Procedure for performing a failure mode effect and criticality analysis* and later was adopted by nuclear power and automotive industries. It has been incorporated as a design standard by many world-class manufacturers and is now a critical step in the design process. FMEA can be performed at the system, subsystem, assembly, and component levels.

The standard FMEA format evaluates single-point failures only. Because a failure in one component can influence failure on an interconnected component, FMEA analysis can be modified to encompass common cause failures (CCFs) in redundant and interdependent pieces of equipment that are likely to fail within a short period because of similar reactions in the operating environment. The FMEA methodology provides a practical approach well suited for products and systems for which little human interface exists, as well as for software-driven operations.

FMEA should be completed with inputs from several disciplines that have knowledge of or experience with a system to be analyzed. Some functional disciplines normally involved are design engineering, quality control, manufacturing, and service. The main point is that the analysis should be performed to account for as many different perspectives as is practically possible. These are analyzed generally because of a desire to improve the reliability of a component or system, to reduce or eliminate failures, to document failure modes for future use, or to use as input for other analyses. FMEA usually is performed early in the design phase of a product, continues throughout development, and is completed prior to the final release of engineering production drawings. An active FMEA worksheet identifies and records the design process and, when used properly, can greatly reduce the need for late design changes that can cause significant schedule delays and can be very costly.

FMEA is also commonly used on existing systems to provide comparative data for a new development or to capture all possible failure mode causes of systems that have quality problems. One drawback is that FMEA analysis requires a lengthy and in-depth study of the system and could be a major concern if schedule is an issue. Other types of analysis for these issues, although not as comprehensive, can be done with relative ease and speed. One such tool is fault tree analysis (FTA), used when a system to be analyzed is complex with many interconnected functions and human interface situations.

FMEA analysis provides a vehicle for tracking known risk items and for prioritizing those risks appropriately using the experience and knowledge of the design and service team. Another advantage is that it can be manipulated and adapted for a wide variety of applications.

8.5.1 **FMEA Process**

The FMEA process is discussed extensively in quality improvement literature and varies according to its application. See the major steps for completing an FMEA analysis in **Table 8.12**.

Step 1	Define the system including subsystems as seen in assembly. Specify the system level at which the analysis will take place.
Step 2	Identify all operational system characteristics.
Step 3	Detail what the expected functions and outputs of the design are. Include expectations and limiting parameters.
Step 4	Determine environmental profiles. Clarify the type of environment in which the design will be operating, what an operation cycle is, and any external factors that could influence the design's functionality.
Step 5	Develop a functional block diagram to clarify the inputs, functions, and outputs of the system.
Step 6	Define the possible failure mode(s) for each hardware item. Determine the effect of each mode on the rest of the system.
Step 7	Classify the failure effect in an initial severity ranking (SR).
Step 8	Determine the cause(s) of each potential failure mode. (Use fault tree analysis (FTA).)
Step 9	Determine the initial occurrence ranking (OR) based on the knowledge available with respect to frequency of the particular causes.
Step 10	List the design or test verification tasks that can detect the failure cause during development.
Step 11	Determine the failure mode initial detection ranking (DR).
Step 12	Calculate the failure mode/effect initial risk priority number (RPN) $RPN = SR \times OR \times DR$
Step 13	Develop an action plan and responsibilities to reduce the severity and occurrence of the failure mode effect and/or change design verification tasks.
Step 14	Implement the actions on the product or system.
Step 15	Determine new SR, OR, DR, and RPN after corrective measures have been taken.
Step 16	Document conclusions and recommendations.

TABLE 8.12 FMEA Steps

An FMEA worksheet is a form in which all analysis information and evaluations are recorded (**Table 8.13**). Most of the information gathered from the steps listed in **Table 8.12** is recorded on the FMEA worksheet. Organizations can generate their own ranking scales.

The following are examples of FMEA activities:

- Identify the system's scope and function.
- Identify potential failures by brainstorming.
- Determine the effects of potential failures using a fishbone (cause-and-effect) diagram.
- Determine the cause of each failure type.
- Prioritize potential failures according to safety, quality, and cost.
- Monitor the plan and document it.

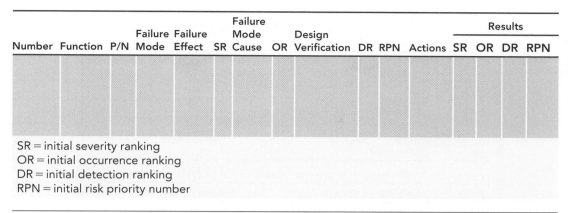

Number	Function	P/N	Failure Mode	Failure Effect	SR	Failure Mode Cause	OR	Design Verification	DR	RPN	Actions	Results SR	OR	DR	RPN

SR = initial severity ranking
OR = initial occurrence ranking
DR = initial detection ranking
RPN = initial risk priority number

TABLE 8.13 Information Recorded on an FMEA Worksheet

EXAMPLE 8.11

A company that produces travel mugs and other food and beverage containment products is designing a new travel coffee mug. The company has incorporated FMEA in its design process to help improve quality and reduce rework and scrap, which have been problems in past products. The worksheet in **Table 8.13** is a typical one that must be completed shortly prior to production start-up. It focuses specifically on the mug's cover in addition to the handle, base, and cup components, either individually or as an assembly as shown in **Table 8.15**.

As the FMEA worksheet is filled out, it soon becomes obvious that completing this cascade of detail can be a somewhat major effort. This is one reason that the FMEA process is usually time consuming. The advantage, however, is that this worksheet helps to quickly establish priorities for corrective actions in order to obtain the maximum benefits from early-stage corrections and improvements.

Function: In this example, after completing the worksheet information and identifying the item(s) to be evaluated (example of travel mug cover), the next step is to list all important cover functions. The analysis team must understand the product to ensure completeness. If a critical function is omitted, its failure mode(s) will not be considered, and no conscious effort will be made to address possible design flaws until after the item has been designed. At that point, the cost to correct the situation will be at least on the order of one magnitude higher.

Failure modes: Once the functions have been listed, all possible failure modes and their effects for each expected function can be added. A single item can have multiple functions, which can have multiple failure modes that could cause multiple failure effects.

Failure effects and severity rating: One failure mode is identified as "Cover comes off unexpectedly." This failure has two similar but significantly different effects. The first, "Total loss of fluid," will irritate the user, especially if it occurs frequently. Over time,

(a) Severity Ranking Scale (SR)		
Rank	Severity Effect	Explanation
1	Minor	Has no effect on system performance
2, 3	Low	Has light effect on system performance
4, 5, 6	Moderate	Causes some system performance deterioration
7, 8	High	Causes system malfunctions
9, 10	Very high	Compromises system safety or does not comply with codes

(b) Occurrence Ranking (OR) Classifications		
Rank	Probability of Failure/Event	(Quantitative Probability)
1	Remote	Failure unlikely (0.000001)
2, 3, 4	Low	Relatively few failures (0.00005, 0.00025, 0.001, respectively))
5, 6	Moderate	Occasional failures (0.00025, 0.0125, respectively))
7, 8	High	Repeated failures (0.025, 0.05, respectively)
9, 10	Very high	Almost certain failures (0.125, 0.5, respectively)

(c) Detection Ranking (DR) Classifications		
Rank	Detection Ranking Scale	Likelihood of Detection by Verification Program
1, 2	Very high	Will almost certainly detect potential failure cause
3, 4	High	Has a good chance of detecting potential failure cause
5, 6	Moderate	Might detect failure cause
7, 8	Low	Is not likely to detect failure cause
9	Very low	Will probably not detect failure cause
10	None	Cannot detect failure cause

TABLE 8.14 FMEA ranking scales

this could result in the loss of the customer, so the SR for this effect is 8, fairly high. The user could consider the second effect, "Major hot fluid spill on user," as a safety hazard and sue the company. Therefore, the highest severity rating of 10 should be assigned.

Failure mode cause: After all failure effects have been documented and their severity ratings assigned, it is helpful to determine the causes of the failure modes on the worksheet. This information can be gathered from prior experience, test data, other methods such as root cause analysis (RCA) or FFA, or a combination of these as the particular

Item	Function	Failure Mode	Failure Effect	SR	Failure Mode Cause	OR	Design Verification	DR	RPN	Recommended Action	Action(s)	SR	OR	DR	RPN
												Actions/Results			
Cover	To restrain fluid from unintentional spillage	Cover leaks	Small spills on user	5	Poor fit due to tolerance or design	5	SPC on cup and cover dimensions	3	75	Analyze design and process to improve fit	Temperature control module on molding machine updated for better control	5	3	3	45
					Uneven expansion due to difference in materials	7	Reliability testing of assembly	4	140	Review material selection compatibility	Materials with same thermal expansion coefficient selected		1	4	20
					Wrong material used	5	No incoming material inspection	10	250	Review suppliers, QC, and alternative suppliers	New supplier certifies material is correct; Has SPC in place.		1	1	5
		Cover comes off unexpectedly	Total loss of fluid	8	Poor fit due to tolerances or design	4	SPC on cup and cover dimensions	3	96	Analyze design and process to improve it	Temperature control module on molding machine updated for better control	8	2	3	48
					Uneven expansion due to difference in material	6	Reliability testing of assembly	4	192	Review material selection compatibility	Materials with same thermal expansion coefficient selected		1	4	32

(continued)

Item	Function	Failure Mode	Failure Effect	SR	Failure Mode Cause	OR	Design Verification	DR	RPN	Recommended Action	Action(s)	SR	OR	DR	RPN
			Major spill on user of hot fluid	10	Wrong material used	5	No incoming material inspection	10	400	Review suppliers, QC, and alternative suppliers	New supplier certifies material is correct; Has SPC in place	1	1	1	8
					Poor fit due to tolerances or design	4	SPC on cup and cover dimensions	3	120	Analyze design and process to improve it	Temperature control module on molding machine updated for better control	10	2	3	60
					Uneven expansion due to difference in material	6	Reliability testing of assembly	4	240	Review material selection compatibility	Materials with same thermal expansion coefficient selected		1	4	40
					Wrong material used	5	No Incoming material inspection	10	500	Review suppliers, QC, and alternative suppliers	New supplier certifies material is correct; Has SPC in place.		1	1	10
	To provide smooth, controlled fluid flow for drinking	Air vent or fluid opening blocked or insufficient	Poor fluid flow	4	Mold flashing in hole	6	SPC on cover dimensions	3	72	Analyzing molding process and equipment	Molding cavity & process refined with minimal flashing. Added meshing removal step as needed	4	2	1	8

368

Function	Potential Failure Mode	Potential Effect(s)	SEV	Potential Cause	OCC	Current Design Controls	DET	RPN	Recommended Action	Action Taken / Results	SEV	OCC	DET	RPN
To slow fluid heat loss	Does not retain heat effectively	Fluid cools too quickly	5	Too small by Design	1	calculation and prototype testing	2	8	No action	No action		1	2	8
				Poor material design	3	Reliability testing of assembly	4	60	Review material selection compatibility	Material with acceptable thermal insulation properties	5	2	4	40
				Wrong material used	5	No incoming material inspection	10	260	Review suppliers, QC, and alternative suppliers	New supplier certifies material is correct; Has SPC in place		1	1	5
				Fluid opening and air vent too large	1	Calculation and prototype testing	2	10	No action	No action		1	2	10

TABLE 8.15 FMEA example of a coffee mug

situation requires. These results are then fed into the FMEA process in the "Failure Mode Cause" column. Each cause is rated based on its frequency or probability of incidence in an occurrence rating table similar to that on the FMEA worksheet in **Table 8.15**. In the present example, three failure mode causes are listed twice. The cause of each is listed next to each of the two failure mode effects for calculation purpose because the two effects have different severity ratings. In this particular case, the failure has two possible design-related causes; one involves fits and tolerances, and the other is material selection. The third cause is process related; the wrong material is being used in the cover's molding.

Design verification: The third function of FMEA, to determine the effects of potential failure, involves uncovering the failure causes of the specific design or process before the product reaches the end user. The causes are recorded in the column "Design Verification" and are then rated according to how successfully the flaws in question can be detected. At this point, the ratings are multiplied to find the initial RPN. The focus is then on the failure mode cause with the highest RPN; improvements in these areas will provide the greatest return on investment.

At this point, the four highest-ranking RPN issues on the worksheet involve the use of the wrong material. The team reviewed the supplier's quality, researched other suppliers, and developed plans to address any other issues that were within the control of the design or process. The causes were then investigated in order of the assigned priorities.

Action and results: The results of the review regarding the wrong material indicated that this was a process problem traced to the combination of poor supplier and lack of material inspection upon receipt. The company changed to another supplier with good quality control methods in place and designated an inspector to certify the material when each shipment arrived. With these and changes from the other efforts, all RPNs were recalculated. Nothing changed the severity ratings of any of the failure modes because none of the failure modes had been eliminated, but it was determined that the changes in the occurrence and detection ratings had been reduced significantly. This resulted in a considerable decrease in the new RPN ratings for wrong material, dropping this category to the bottom of the list, essentially implying that this failure mode was no longer a concern. The team then focused on the next problem.

When this process has been completed for all failure mode causes and their RPNs have been recalculated, it can be repeated for each category until all RPNs are at an acceptable level, or the analysis can be considered complete and documented for future reference. What constitutes an "acceptable level" is solely the determination of those conducting the analysis.

FMEA can be a highly effective tool for reliability assurance. It is extremely efficient when it analyzes elements that cause the failure of an entire system. The flexibility allowed by the FMEA format also makes it especially easy to use in a wide variety of situations. The drawback of this analysis is that it quickly becomes very cumbersome when applied to complex systems with many components or multiple operating modes.

8.5.2 **Assessing Reliability by Root Cause Analysis (RCA)**

RCA is the persistent pursuit of the initial controllable event in a chain of events in order to identify a defect or failure. Correcting a problem will prevent its recurrence. The key to this definition is the word *controllable*. Circumstances are often identified when performing an RCA that, for one reason or another, cannot be controlled or corrected. Acts of nature, preexisting conditions, or even circumstances that are impractical to consider changing are examples of uncontrollable root causes. In that case, other steps can be taken to reduce the severity or likelihood of the event causing a failure.

RCA is a logical approach to eliminating problems, usually significant ones involving defects, process errors, or any other undesired element to major property or causing personal damage. This analysis seeks to find a solution to prevent the problem from recurring. RCA requires a persistent, patient, and open-minded search for the problem's initial cause, not a perceived cause that is in fact only a symptom or result of an earlier issue.

The driving force behind RCA is the idea that the cost of solving a problem increases by an order of magnitude for each process that is allowed to proceed beyond the problem's origin. It is used extensively as a risk management tool by a wide spectrum of organizations. Regardless of the field or industry, RCA operates on the premise that there are no isolated incidents, and that every incident indicates a problem experienced earlier. Each incident is viewed as being the tip of the iceberg with many underlying circumstances that must be identified and understood. Some refer to RCA as making order out of chaos and assert that everything happens for a reason, and that many errors occur before an undesirable outcome is identified. If such an outcome can be eliminated, failures are often regarded as opportunities to improve quality and profitability, not as problems decreasing the bottom line.

RCA is deeply ingrained in quality management systems. It can be used in virtually any context in which there is a desire to prevent a problem's recurrence. In many industries, human safety and/or financial impact are the primary motivation of its use. In the manufacturing industry, the major benefits of RCA come from the reduction of scrap, rework, repair, and warranty expenses.

RCA is primarily considered to be a reactive, or *after-the-fact*, tool. In other words, it is used after a failure has occurred. When coupled with other tools, RCA can successfully be used to improve a company's profitability by identifying root causes to problems that have commonly been accepted as the cost of doing business.

For instance, if a piece of manufacturing equipment requires a 10- to 20-minute resetting or adjustment five or six times a day, it could go unnoticed because each individual occurrence seems to be a minor inconvenience. But the number of occurrences over a year's time can have a significant cost. Using RCA to analyze such events can increase a company's profits considerably by reducing downtime. Another benefit from the routine use of RCA is that its analysis can call attention to trends. By recording the types of root causes found, seemingly unrelated failures could be identified as being from a common theme.

For example, failure in manufacturing a part, filling out required forms, performing a maintenance task, and shipping a product could all result from poor instructions.

This highlights the need for more focus on creating clear, understandable documentation for these and other tasks.

Six basic steps to performing a successful RCA are commonly found in RCA literature.

▸ **Step 1. Identify and Define the Problem**
This requires making a clear and complete statement of the precise nature and scope of the failure being analyzed. It is especially important at the beginning that everyone on the team have the same understanding of what occurred and where the focus lies.

▸ **Step 2. Preserve and Collect Data**
This is the single most important step of the entire analysis. If a poor job is done in this step, problem resolution is nearly impossible. A failure event usually leaves parts, data, paperwork, or some other physical evidence behind. Preserving them at the onset of an investigation is crucial. Interviews with people involved and reviews of procedures used are key elements of good data collection.

▸ **Step 3. Analyze the Data**
The type and importance of the failure, as well as the level of effort required for the analysis, determine the specific analysis tools used. If the problem considered is a hydraulic power transmission system, for example, the analysts use a fishbone diagram. Digging until the root cause is found is critical to the success of this step.

Fishbone Diagram:
The Fishbone diagram is a formal technique used for brainstorming problems that affect the customer. The main objective of the Fishbone diagram as shown in **Figure 8.14** is to assist in the brainstorming process, to ensure that all possible problems are addressed and categorized in an organized manner.

▸ **Step 4. Identify and Verify the Root Cause**
This is the main point of the analysis although its first part is really the conclusion of the third step. When the root cause has been identified, it should be tested and verified. The failure can be recreated when the cost allows. When safety and/or cost is an issue, use a model, calculation, or other way to verify the root cause.

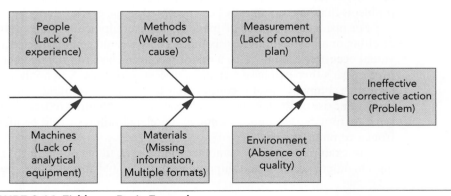

FIGURE 8.14 Fishbone Basic Example

▸ **Step 5. Communicate Findings and Recommendations**

This step is necessary to enable those in positions of power to make informed decisions. Recommendations for preventing the root cause from recurring should be specific, clear, concise, actionable, and verifiable and avoid unacceptable risk. General recommendations such as "operator should use caution when performing this task" are impossible to confirm or enforce. A more useful one is "operator should confirm that the locks are engaged and the safety screen is in place before performing this task."

▸ **Step 6. Implement Corrective Actions**

It may or may not be within the authority of an RCA team to take corrective actions, which is why it is important to build a strong case for implementing the team's recommendations. Occasionally, the cost of the recommendation is higher than the cost of living with the failure. This becomes a business decision, and ideally should never outweigh the cost of human life or injury.

CASE STUDY 8.1
Use of Root Cause Analysis to Monitor Elevator Service Quality

This case study presents an approach to monitoring the service quality of general elevators using root cause analysis. This example highlights the importance of recognizing structural changes in the value stream that have the greatest impact on service quality. Original manufacturers can use the information discussed here to demonstrate to customers the benefits of proper maintenance and the disadvantages of insufficient maintenance. This case study also examines ways to identify the most influential drivers of service quality in the elevator industry with the ultimate goal of creating a tool that allows the supply chain to simulate changes in the supply chain and to identify their effect(s) on service rate.

Background

A design challenge for elevator companies is to integrate the many components and subsystems within elevators to produce the most cost-efficient and reliable product. The scenario in the elevator industry shows that elevator companies may choose to produce some components and outsource others.

In most cases, the elevator industry provides the following functions to qualified customers:

- Installing new equipment
- Modernizing existing elevators
- Outsourcing service aspects to supply chain and qualified customers
- Providing repairs or upgrades

The basic functional codes for an elevator system are governed by elevator standards such as ANSI A17.1-2007. Their purpose is to ensure that a product is reliable and meets riders' safety expectations. The value of reliability and maintenance is relative to the environment in which it is installed and its location and height. See **Table 8.16** for an overview of the most common maintenance procedures performed on major elevator systems.

	Elevator Subsystems	Maintenance Functions	Repair Functions
Mechanical	Fluid power	Lubricate	Use seals, cylinders, control valve, piping
	Frame of elevator carriage	Lubricate sliding guide, roller guide bearing	Lubricate roller guides, limit switch actuation mechanism such as cams
	Rail brackets	N/A	
	Mechanism for hoisting	Lubricate	Inspect drive sheave, liner, gearbox, motor brushes, contact points, and isolation pads
	Hoist-way material	N/A	Limit switches
	Counterweight	Lubricate sliding guide, roller guide bearing	Ensure 2:1 sheave counter-weight safeties
	Overspeed governor	Control speed	Replace governor, rope, switches, tripping mechanism
	Suspension	Lubricate ropes	Replace ropes
	Critical components and contact devices	Lubricate pivot points	Replace knurled rollers, worn pads
Electrical	Motion controller	Clean relay contacts	Check connectors, dynamic braking resistors, wiring harnesses
	Operation controller	Clean relay contacts	N/A
	Actuator drive	Clean contacts on relays or other high-current capacity components	Inspect dynamic braking resistors, SCRs
	Hoist-way wiring	N/A	N/A
	Traveling cable	N/A	Replace traveling cable
Both	Carriage and doors	Adjust door rollers, lubricate pivot points for linkage-driven operators	Replace operator motor, brushes, or closer
	Position reference system	Clean magnets, lubricate pivot points	Replace magnets, switches, cams
	Car operating panel	N/A	Replace buttons, circuit boards, wiring harnesses
	Exit/Entrance buttons	N/A	Replace buttons, circuit boards, wiring harnesses

TABLE 8.16 Elevator Subsystem Maintenance and Repair Functions

The RCA process in this case study examines action items that can be used to resolve the root cause of declining service rate and profitability. *Service rate* is the average number of service hours spent maintaining or repairing an elevator.

An RCA tool can be employed to help identify the root cause of issues that negatively affect the service rate of elevators. Because of the very large number of possible

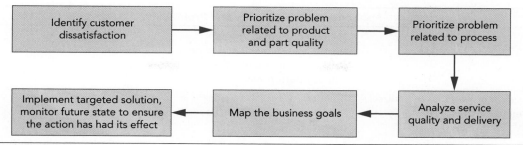

FIGURE 8.15 Overview of Value Stream Optimization

Based on United Technologies Corporation, NAA Otis ACE Community (Osowiski, Otis Elevator Company)

causes that can affect an elevator's service rate, this tool helps to focus decision making on factors that significantly affect service rate (see **Figure 8.15**).

- Customer dissatisfaction
- Prioritized problems related to service, product, component, part quality, and reliability according to market analysis
- Prioritized process problems
- Problems with supplier quality and delivery problems
- Shortfalls in business goals (related to revenue growth, inventory, customer and employee satisfaction, and health and safety)

General Steps to Perform RCA

▸ **Step 1. Identify Customer Requirements by Using an SIPOC Diagram**
   SIPOC is a type of process map typically used in lean Six Sigma projects to identify the primary elements of a process. An SIPOC (Suppliers, Input, Outputs, Customers) diagram is a tool that identifies all relevant elements of a process improvement project before work begins. It helps define a complex project that may not be well scoped and is typically employed at the early stage of Six Sigma implementation. The stages of Six Sigma implementation are: define, measure, analyze, improve, and control. The SIPOC tool is particularly useful when it is not clear:

   - Who supplies inputs to the process?
   - What specifications are placed on the inputs?
   - Who are the true customers of the process?
   - What are the requirements of the customers?

   The tool name prompts the team to consider the suppliers of your process (S), the inputs to the process (I), the process the team is improving (P), the outputs of the process (O), and the customers that receive the process outputs (C). The map identifies every relevant element of a process and refines the scope of complex projects.

▸ **Step 2. Make a Fishbone Diagram**
   Table 8.17 is the result of a completed Fishbone diagram for the problem "Service quality and profitability are decreasing." A group of process stakeholders are

Fishbone—Service Quality and Profitability

Customer	Systems	Communication
	Maintenance management system allows mechanic to be paid for maintenance that is not actually performed.	
Customer wants to minimize maintenance costs.	Maintenance support system is Category A without backup in place.	
Customer expectations are not defined.	New elevator equipment being installed is harder to maintain and repair than legacy equipment	Communication process is unclear between the branch and the customer.
Customer does not understand the costs associated with low reliability.	New elevator equipment being installed is less reliable than legacy equipment.	Miscommunication between the local branch and the customer.
Customer desires least expensive investment in new elevator systems.	Maintenance management system does not schedule resources efficiently.	Unachievable promises have been made to the customer.
Customer expectations too high.	Resources are not available to assist with maintenance management system upkeep.	Customer is not aware that their elevator's reliability may be decreasing.

Problem

Service quality and profitability are decreasing

Process	Resources/Materials/People	Training
Standard processes are not defined for maintenance of new elevator equipment.	Service support (supervisor/branch role) resources are stretched thin.	Mechanics are not properly trained to perform maintenance efficiently.
Standard processes are not defined for maintenance of elevator equipment installed by another company.	Mechanics are not given enough time to perform the maintenance that is required for each particular elevator.	Supervisors are not trained to provide sufficient support to their mechanics.
Standard process is not defined for establishing priorities of which elevators are critical and which are not.	Lack of senior management involvement in service quality.	Sales reps are not properly trained to establish clear requirements and expectations from the customer.
	Management focus is not on service quality or profitability.	
	Proper maintenance tools are not provided to mechanics.	

Mathew Morrison, "RRCA Study of the Elevator Service Industry: Elevator Maintenance and Condition Monitoring," University of Hartford Masters Thesis, May 2014, (Supervisor: Dr. Devdas Shetty).

TABLE 8.17 Fishbone Analysis—Service Quality and Profitability

queried for potential root causes. The objective is to list all of the possible root causes of the problem, when taking into account the many different perspectives of all the process stakeholders.

▸ **Step 3. Organize Results**

After confirming all possible root causes, they are arranged as:

- Customer
- Systems
- Communication
- Process
- Resources/Material/People
- Training

Based on this information, it is possible to come to many conclusions. The research by Morrison and Shetty identifies a method known as Elevator Condition Index (ECI). ECI allows efficient measurement and communication of the current condition of a customer's elevator. Based on the available up-front information on the reliability of the original equipment and projected average life cycles of components, proper maintenance intervals can be identified and recommended.

This case study uses an SIPOC worksheet (**Table 8.18**) to identify the requirements of customers in a value stream and those processes and stakeholders that facilitate meeting the requirements. Important customers and their requirements are clearly identified and quantified. Customer surveys can be used for SIPOC data, but in their absence, business financial performance or other indirect customer feedback can be used to assist in building the SIPOC.

The root cause investigation identifies many significant factors contributing to any decrease in service quality and profitability. The preferred strategy focuses on root causes that have the greatest impact with the least amount of resources invested.

Use of the root cause method helps to identify the following:

- The process of identifying critical priorities.
- The presence or absence of a standard process for establishing priorities for elevators.
- State of training of sales representatives related to customer expectations.
- Customer awareness of the elevators' reliability.
- Customer education on the costs associated with low reliability.

The processes identified in this case study that significantly contribute to the problems as shown in **Table 8.18** are:

1. Creating an efficient maintenance schedule
2. Defining acceptable service rate.

	Inputs			Process	Outputs					
Suppliers	Description	Quantified measure		Process	Description	Quantified measure Delivery	Quantified measure Quality	Quantified measure Cost	Customer	Impact (1–10)*
Mechanic	Work schedule	Supplied accurately and before the start of the work week.		Perform maintenance.	System or component level maintenance performed.	Per contract schedule (varies based on individual contract).	Components renewed.	Maintenance performed within the time allotted by the contract.	Building management, building tenants	8
Maintenance management system	Maintenance contract	Contract complete and accurate.		Create efficient maintenance schedule.	Schedule of all mechanics based on contract commitments.	Provided before the start of the work week.	Schedule provides for most efficient use of mechanic resources.	Travel time is minimized between jobs.	Mechanic	9
Supervisor	Maintenance schedule	Provided before the beginning of the work week.		Support mechanics, resolve conflicts.	Resolve conflicts, provide required tools/materials to complete the job.	Resolve any scheduling conflicts or lack of resources within 15 minutes.	Correct tools and parts provided 100% of the time.	Mechanic downtime (waste) minimized.	Mechanic	7
Customer	Desired equipment reliability	Percent downtime acceptable.		Provide definition of acceptable downtime (service rate).	Specific information about which elevators are critical and which are not.	Information supplied during contract negotiation.	Critical elevators identified correctly.	Maintenance is performed on those elevators which are most critical more often and with higher priority than less-critical elevators.	Sales rep	9

| Sales rep | Basic information on elevator equipment in question. | Information is complete and accurate. | Communicate with customer during/after contract negotiation. | Customer requirements and branch commitments are in-line. | Before start of maintenance contract, during update of maintenance contract. | Customer requirements are clearly communicated to supervisor; maintenance program is created to meet/exceed customer expectations. | Information is provided to branch supervisor in a complete/accurate manner at the start of the maintenance contract or immediately following any change in customer requirements. | Supervisor | 8 |

TABLE 8.18 SIPOC data for Elevator Maintenance

Elevator Condition Index (Morrison and Shetty)

A more efficient system for maintaining elevators based on their current condition can be implemented using Elevator Condition Index (ECI). Data collection and processing is key to driving this advanced maintenance model. A wide array of data is required in order to drive the model. The ECI of a brand new elevator that is specified within its operational limits for the application, manufactured per the design, and installed correctly is defined as 100%. In the real world, there are variables that exist in the design, specification, manufacturing, and installation of an elevator, and it may be difficult to achieve an ECI of 100%. ECI is calculated based on original equipment reliability, projected average life cycle of key wear components, number of run cycles since maintenance was last performed on each component, cost of emergency repair (parts and labor at local rate) versus cost of maintenance versus likelihood of failure. One of the key drivers toward implementing such a maintenance protocol is the desire for the corporation to capitalize on this information. Profit can be increased by implementing an ECI-based maintenance model in several ways:

1. Charge the customer a specific rate for maintenance based on their elevators' specific ECI, as an alternative to flat-rate maintenance fees. A company would likely see higher signup rates from elevators with a better ECI, as the maintenance pricing should decrease as a result.
2. Using ECI will result in preventive maintenance as the problems are fixed before the breakdown. This has the potential for increase in customer satisfaction.

Case Study Conclusions

The RCA methodology is a useful tool for addressing the root cause of problems not only in elevator industry but in any industry. It allows for drawing the most appropriate conclusions based on available resources, business, and technical requirements. Targeted actions can be taken based on the results of this type of study, so that resources expended have the greatest benefit. In this example, the problem is identified as being directly related to poor communication of the actual elevator condition with the customer. The ECI tool can be used to discuss the reliability of the elevator system with the customer over time. The example with elevator shows that electro mechanical components in the elevator system are not designed to last forever. However, if the customers are provided with a tool that clearly communicates the value of proper maintenance, as well as original equipment reliability, they can then make more informed decisions that will result in significant cost savings over time.

Appendix

Software Packages

With the advent of the personal microcomputers, a range of statistical software packages for forecasting statistical process control (SPC) and experimental design are available for product and process development and optimization. Several different types

Software	Contact Address
NCSS	NCSS, Inc., 865 E 400, North Kayville, UT, 84037
Experimental Design	Statistical Programs, 9941 Rowlett, Suite 6, Houston, TX, 77075
Simplex-V	Statistical Programs, 9941 Rowlett, Suite 6, Houston, TX, 77075
Screen	Statistical Programs, 9941 Rowlett, Suite 6, Houston, TX, 77075
C-Comp	Statistical Programs, 9941 Rowlett, Suite 6, Houston, TX, 77075
The Bass System	Bass Institute, P. O. Box 349 Chapel Hill, NC 27514
Turbo Spring-Stat	Spring Systems, P. O. Box 10073, Chicago, IL, 60610
CSS	Stat Soft, Inc., 2325 E. 13th St., Tulsa, OK 74104
Sigstat	Significant Statistics, 3336 N. Canyon Rd., Provo, UT 84604
STATA	Computing Resource Center, 1080 National Blvd., Los Angeles, CA 90064
Statpac Gold	Walonick Associates, Inc., 6500 Nicollet Ave. S., Minneapolis, MN 55423
BMDP/PC	BMDP Statistical Software, 1440 Sepulveda Blvd., #316, Los Angeles, CA 90025
P-Stat	P-STAT, Inc., P. O. Box AH, Princeton, NJ 08542
SAS	SAS Institute, SAS Circle, Box 8000, Cary, NC 27512-8000
SPSS/PC	SPSS, Inc., 444 N. Michigan Dr., Chicago, IL 60611
Systat (With Sygraph)	SYSTAT, Inc., 1800 Sherman Ave., Evanston, IL 60201
Statgraphics	STSC, Inc., 2115 E. Jefferson St., Rockville, MD 20852
Prodas	Conceptual Softward, Inc., P. O. Box 56627, Houston, TX 77256-6627
RS/1	BBN Software Products Corp., 10 Fawcett St., Cambridge, MA 02138

TABLE 8.19 Statistical Software Packages

of experimental designs are commonly used, and software is available to engineers. See **Table 8.19** for a list of some commercial statistical software packages. The ease of use and the number of features available with each package varies. The packages range in price from $100 to $5,000 and in complexity. Most of them can be used to develop Simplex, screening, and response surface experiments. These packages offer a wide variety of other statistical features including factor analysis, multiple regression analysis, time series analysis, and SPC routines and can produce graphical output. We strongly suggest that researchers unfamiliar with statistical designs begin on a small scale and gradually build confidence with the programs and techniques.

REFERENCES

1. George Chryssolouris, *Manufacturing Systems—Theory and Practice* 2nd edition (New York: Springer Verlag, 2006).
2. Ibrahim Zeid, *CAD/CAM Theory and Practice* (New York: McGraw-Hill, 1991).
3. James P. Womack and Daniel T. Jones, *Lean Thinking: Banish Waste and Create Wealth in Your Corporation* (New York: Simon & Schuster, 2010).
4. James Womack, Daniel T. Jones, and Daniel Roos, *The Machine That Changed the World: The Story of Lean Production* (New York: Free Press, 2007).
5. Madhav S. Phadke, *Quality Engineering Using Robust Design, AT&T Bell Laboratories* (Englewood Cliffs New Jersey: Prentice Hall, 1989).
6. S. M. Yoo, D. A Dornfeld, and R. L. Lemaster, "Analysis and Modeling of Laser Measurement System Performance for Wood Surface," *Journal of Engineering for Industry* 112 (1990), 69–76.
7. D. Shetty and H. Neault, Method and apparatus for surface roughness measurement using laser diffraction pattern. U. S. Patent 5,189,490; filed on Sep 27, 1991 and issued February 23, 1993.
8. G. E. P. Box, S. Bisgaard, and C. A. Fund, "An Explanation and Critique of Taguchi's Contributions to Quality Engineering," *Quality and Reliability Engineering International*, 4(1988), 123–131.
9. D. M. Byrne and S. Taguchi, "The Taguchi Approach to Parameter Design," *Quality Progress* 2, no. 12 (1987), 19–26.
10. K. Dehand, *Quality Control, Robust Design and the Taguchi Method* (Pacific Grove, CA: Wadsworth & Brooks/Cole, 1989).
11. J. S. Hunter, "Statistical Design Applied to Product Design," *Journal of the Quality Technology* 17, no. 4 (1985), 210–221.
12. R. N. Kackar, "Taguchi's Quality Philosophy: Analysis and Commentary," *Quality Progress* 19, no. 12 (1986), 21–29.
13. J. S. Lawson and J. L. Madrigal, "Robust Design Through Optimization Techniques," *Quality Engineering* 6, no. 4 (1994), 593–608.
14. R. L. Mason, R. F. Gunst, and J. L. Hess, *Statistical Design and Analysis of Experiments* (New York: Wiley, 1989).
15. A. Mitra, *Fundamentals of Quality Control and Improvement* (Upper Saddle River, NJ: Prentice Hall, 2012).
16. D. C. Montgomery, *Introduction to Statistical Quality Control*, 7th ed. (New York: Wiley, 2013).
17. G. Taguchi, *Introduction to Quality Engineering: Designing Quality into Products and Processes* (White Plains, NY: Asian Productivity Organization, UNIPUB, 1986).
18. K. L. Tsyi, "An Overview of Taguchi Method and Newly Developed Statistical Methods for Robust Design," *IIE Transactions*, 24, no. 5 (1992), 44–57.
19. Troy Chicoine, "Failure Modes and Effects Analysis and Root Cause Analysis of Hydraulic Elevator Power Transmission Systems" (master's thesis, University of Hartford, 2000).
20. M.P. Groover, *Automation, Production Systems and Computer Integrated Manufacturing*, 3rd edition (Englewood Cliffs, New Jersey: Prentice Hall, 2008).
21. C. Wilson and M. Kennedy, "Some Essential Elements of Superior Product Development," Paper 89-WA/WE-7 (San Francisco, CA: American Society of Mechanical Engineers (ASME), Winter Annual meeting, 1989).
22. J. Usher, U. Roy, and H. Parsae, *Integrated Product and Process Development* (New York: John Wiley and Sons, 1998).
23. Karl Ulrich and Steven Eppinger, *Product Design and Development*, 5th ed. (New York: McGraw-Hill, 2011).
24. George E. Dieter and Linda C. Schmidt, *Engineering Design*, 5th ed.) (New York: McGraw Hill, 2012).
25. James N. Siddal, *Expert Systems for Engineers* (New York: Marcel Dekker, 1990).
26. Report on Sustainable Product Design: Tools and Strategies http://sustainableproductdesigntools.blogspot.com/.

EXERCISES

8.1. Consider a popular consumer product such as hand soap dispensers currently used in the market. Assuming you will improve the quality, identify how you would go about redesigning the product. Indicate how you would develop robust design methodology to create the best product.

8.2. Regarding learning curve analysis,
   a. Under what circumstances is it most applicable?
   b. What determines the degree of slope for a given learning curve?
   c. A company producing defensive missiles spent 125,000 hours to

complete the first unit. The second and third units were produced with an 86% learning factor. Assuming the same learning factor and the rate of $40 per hour, what would be the cost of the fourth unit?

8.3. In response to a consumer inquiry, a manufacturing industry is estimating the cost of a new product, which is similar to the one it has now. It is estimated that 400 labor hours will be required to produce the first unit. Draw graphs of labor requirements for units 1 through 25 for 80% and 90% learning curves.

8.4. A department in a small manufacturing organization produces two types of parts, A and B. Assume that it produces y_1 units of type A and y_2 units of type B. Each part is manufactured using a two-step process involving two robots, M and N. The manufacturing times for y_1 units of A on robots M and N are 4 hours and 6 hours, respectively. Similarly, the manufacturing times for y_2 units of B on robots M and N are 8 hours and 4 hours, respectively. For a three-week period, the availability of robot M is 120 hours and of robot N is 100 hours. The estimated profit generated per unit by type A is $40 and by type B is $30. The department can sell the parts as soon as they are manufactured. Estimate graphically how many units of each part type should be manufactured to maximize profit. (*Suggestion:* In this case, the objective function is the profit function expressed as $Z = 40y_1 + 30y_2$, where Z is the total profit. The availability

time constraints for robots M and N, respectively, are

$$4y_1 + 8y_2 \leq 120$$
$$6y_1 + 4y_2 \leq 100$$

8.5. A manufacturing firm has the capability to produce three different products in its factory. Fabricating each product requires several manufacturing operations. See **Table 8.1E** for the time required for each operation, the maximum operation capacity per day, and the unit profit for each operation.

Set up the relevant constraint equations to select the appropriate product mix to maximize profit.

8.6. Briefly explain what a product's robust design implies. Construct an orthogonal array for a robust design experiment on a newly designed surface roughness instrument. Roughness is measured in gray scale values by using a laser- and microcomputer-based vision system.

Control parameters are

Laser distance	10	15
Ambient light	Present	Not present
Laser power	500 miliwatt	1000 miliwatt
Gray-scale setting	190	200

8.7. One of a company's production shops makes two products, each of which requires three manufacturing operations. The company can sell all it can make of both. In fact, it

Operation	Product 1 Time/ Unit (minutes)	Product 2 Time/ Unit (minutes)	Product 3 Time/ Unit (minutes)	Operation Capacity (minutes)
Machining	1	2	1	430
Welding	3	0	2	460
Casting	1	4	0	420
Profit/Unit ($)	3	2	5	

TABLE 8.1E Time versus Operation

Product	Production Time Required per Unit (hour/unit)			Cost ($)	Selling Price ($)
	Operator 1	Operator 2	Operator 3		
1	1.2	2.3	4.5	$80	$95
2	2.3	6.8	1.9	110	130
Hours available	24.0	32.0	24.0		

TABLE 8.2E Operation's constraints

is considering an increase its production capacity. See **Table 8.2E** for the operation's constraints and other data.

a. Determine the product mix that maximizes profit.

b. If production capacity were to be increased, to which operation should the additional capacity be allocated?

8.8. Automatic cameras can have a number of associated variables, for example, film speed, flash, and focus. Each independent variable affects the dependent variable, which is the picture quality. The levels of these variables follow:

Film speed, A	400 ASA	100 ASA
Flash, B	On	Off
Focus, C	In	Out

a. Construct a suitable orthogonal array and substitute appropriate control variable values.

b. Will 100% interchangeability or statistical interchangeability generate wider tolerances?

8.9. The Taguchi quadratic loss function for a part in snow blowing equipment is $L(y) = 4000(y - m)^2$ where y = actual value of critical dimension and m is the nominal value. If $m = 100.00$ mm, determine the value of loss function for tolerances (a) ±0.15 mm and (b) ±0.10 mm.

8.10. What role does research play in product design? Indicate four sources of research and the desired information that could be expected from each source. Using web-based research, identify five corporations and their product lines. In what way do these products support the corporate strategy?

CHAPTER 9
Virtual Product Prototyping and Additive Manufacturing

OBJECTIVES

Sustained business success depends on continuous improvement in products and processes. The emerging emphasis on product reliability and the desire to reduce product development time have focused on the use of software tools for design and production. Interactive modeling is crucial to the design process, and it can occur in an environment that combines real and virtual objects. This chapter with examples focuses on virtual prototyping tools that help companies get new products to market. This chapter also examines the role of additive manufacturing that emphasizes quickly creating output in the form of a prototype.

9.1 Introduction: Virtual Reality

Product innovation is the primary instrument that companies use to drive competition. At the same time, global economic pressure has caused product manufacturers to slash budgets and cut costs wherever possible. To compress new product development time, most major companies have deployed sophisticated product life cycle management solutions to streamline and speed product development.

Companies, particularly cost-conscious industries, use digital manufacturing tools to get new products to market faster and less expensively. With new digital factory layout tools and improved three-dimensional visualizations, companies can digitally design and make full factories operational in a fraction of the time previously required. They use product life cycle management systems that allow original manufacturers to simulate the behavior of logic controllers and other automation equipment on the factory floor before making any physical layout.

Sustained success depends on continuous improvement in products and processes. The emphasis on product reliability and the desire to reduce product development time led companies to investigate and use software tools for design and manufacturing. For example, a manufacturer can have a real robotic arm do end plate laser welding on a virtual assembly line if the hand on the robotic arm can reorient an object. The key to such use is the interaction between the virtual objects and their control counterparts with the real objects by having program codes to operate robotic devices.

9.1.1 Customer Design Using Virtual Prototyping

The inclusion of the customer early in the product creation process has increased. Virtual reality enables customers to increase their involvement in the design and early stages of product development by using an interactive web-based platform for customizing a product family design. Customers can experience the product from a combination of sensory inputs such as vision, touch, and hearing.

Virtual prototyping is rapidly gaining importance as the engineering practice of choice for product development to shorten the design cycle. Designers are

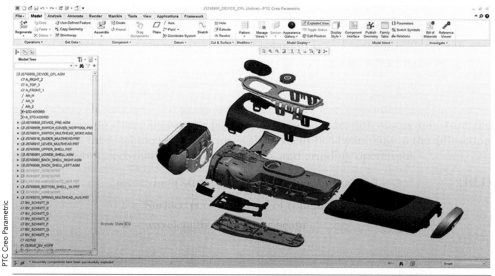

FIGURE 9.1 Modeling of Jet Engine Component

trained to introduce simulations of human interaction with objects within a CAD environment.

Central to virtual prototyping to enable product customization is the fact that the Internet can enhance communication and collaboration between customers, product developers, and manufacturers located in geographically dispersed places. For example, customers and developers can review an actual three-dimensional (3D) model of a product such as the one in **Figure 9.1**.

9.2 **Digital Tools**

In the product development process, designers use computer-aided design (CAD) to communicate and test their ideas from the initial concept phase to the product life cycle management phase. This optimizes the overall manufacturing process from product design to inspection by integrating all related information into a common database. The CAD system's knowledge of part geometry function can be used to determine the reference values of manufacturing process variables.

CAD tools can create a geometrical modeling environment and perform all simulations and testing activities based on enriching the geometry with additional information. The information from various process-related sensors can be used to achieve on-line optimization of the process and can be integrated to improve the reliability and quality of sensing. This shared information, such as the geometric data of a part from the CAD/CAM (computer-aided manufacturing) database, can be used to select the optimum combination of tools, manufacturing processes, and finish quality.

9.2.1 **Computer-Aided Design Environment for Product Development**

Selecting the appropriate CAD/CAM software for an industry requires extensive investigation of its accuracy, consistency, and adherence to product standards. This software is a very powerful tool for conducting product simulation. Designers use this tool to simulate products' physical characteristics and view realistic pictures of them; they also can use virtual modeling to create exploded views, part details, and manufacturing drawings. As an individual component is designed, the software can be used to specify desired product characteristics and identify its physical dimensions and physical properties. This information can be documented and then used during production.

Design is an iterative process that includes steps such as need identification, problem definition, problem synthesis, analysis and optimization, and evaluation of the particular design for the market.

See **Figure 9.2** for the design activity in a computer-aided design environment. The system and the designer are in direct communication, as the designer uses the commands for actions and responds to the computer system's questions. The use of CAD systems increases the design engineer's productivity by enabling him or her to view the product, part components, and subassemblies, thereby reducing the time required for synthesis, analysis, and documentation.

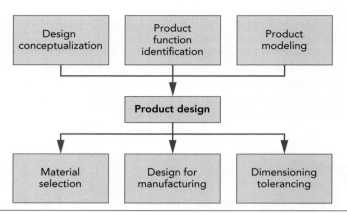

FIGURE 9.2 Design Activity in a Computer-Aided Design Environment

Design-related tasks assigned to CAD systems include:

- Geometrical modeling
- Engineering analysis and optimization
- Review of design and presentation
- Synthesis
- Presentation
- Creation of a database to use in manufacturing and to improve communication

For example, aerospace designers must consider the behavior of the aircraft wings in their plans. The design process requires not only geometric data related to the wing but also details such as material properties, material volume, welding behavior of the structure, optimum layout of the materials for minimum scrap, and so on. The geometric data on the wings can be used to analyze fatigue and crack, aerodynamic computation, and so forth. Furthermore, the geometric model can be integrated with the actual machining and assembly manufacturing processes. Any design project requires some engineering analysis, such as stress and strain calculations, heat transfer computation, analysis of properties, and optimization. Individual parts are grouped into discrete physical elements that can be analyzed independently with regard to stress and displacement. This process helps to optimize the design and keep part failure to a minimum. A CAD system has the ability to review and evaluate design and perform tasks such as automatic dimensioning, tolerancing, identifying interference, and providing views from various angles (**Figure 9.3**).

9.2.2 **Feature-Based Design and Modeling**

Feature-based design and modeling involves creating a product on the basis of features and the relationships among them. A *feature* is an object's perceived geometric or functional element or property that is useful in understanding its function, behavior, and performance; it represents the engineering meaning of the geometry of a part or assembly. A feature can map an existing generic shape.

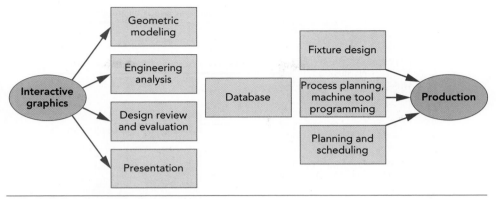

FIGURE 9.3 CAD/CAM System Activities

A feature-based design provides a method for building a complete CAD database with mechanical features from the start of design using basic shapes known as primitives. A *feature model* is a data structure that represents a part or an assembly in terms of its component features. Features are categorized according to (1) form, (2) tolerance, (3) assembly, (4) function, and (5) material. A feature-based design and modeling system can group entities into forms such as ribs, bosses, flanges, and pockets. The procedure considers the design attributes and examines the product from the point of view of the way it will be fabricated.

Parametric systems represent 3D model dimensions, lengths, and angles as drivers that control model shape. Parametric solid modeling originated in design modeling that uses the principles of Boolean algebra and relationships. Some parametric modeling products help the user define and remember nondimensional geometric constraints, such as those that are parallel and perpendicular. Recognition of geometric constraints add overall ease of use while capturing important design considerations.

An associated programming link can connect a two-dimensional representation to the 3D model's geometry, primarily solid geometry. With this link, drawings reflect 3D model changes. Changing a dimension value updates the entire 3D model shape.

The solid modeling system can calculate full mass properties, check for interference between models, interface with e-applications such as finite element analysis, and generate machining codes for machine tools that the computer numerically controls. *Surface modeling* is used to define a product model's outer skin but not its interior. Except for high-end styling products, surface modeling typically exists as an adjunct to either wire frame or solid products.

9.3 **Virtual Prototyping (VP) Technology**

Virtual prototyping (VP) technology as shown in (**Figure 9.4**) is the natural extension of 3D computer graphics with advanced input and output devices. VP simulation uses a digital model (virtual prototype) to identify the presence of problems or bottlenecks during manufacturing, assembly, and disassembly processes. VP techniques offer the

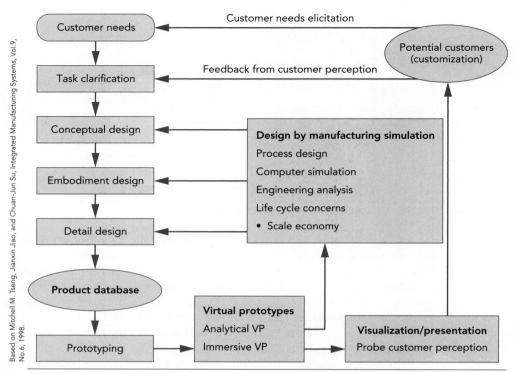

Based on Mitchell M. Tseng, Jianxin Jiao, and Chuan-Jun Su, Integrated Manufacturing Systems, Vol. 9, No. 6, 1998.

FIGURE 9.4 Virtual Prototyping Design Environment

possibility of experiencing highly realistic virtual worlds. A product's data model can be used as a virtual prototype, which avoids creating a real one to model and analyze geometry, functionality, and manufacturability. VP users can interact with a simulated immersive 3D environment in real time through multiple sensors. The "virtual assembly" process identifies any areas where itemized cataloged components do not meet requirements and therefore must be custom fabricated. The model's drawings, specifications, and instructions can be sent to the fabricator with the 3D drawings as a communications aid to ensure that custom or modified items are fabricated correctly.

Before launching a product, physical prototypes are required for making an iterative evaluation that can provide feedback for design modification such as selection of design alternatives, engineering analysis, manufacturing planning, and visualizing a product.

A VP system consists of three parts: (1) VP engine, (2) VP database and model base, and (3) input/output (I/O) devices.

VP engine: This graphic modeling and processing system is used for object modeling, texturing, mapping, lighting, rendering, and finally displaying 3D scenes in a real time environment.

VP database and model: This base stores virtual objects that are ready to be loaded into the scene whenever required.

I/O devices: These devices provide the human–machine interface to control the VP system and for communication.

9.3.1 Virtual Prototyping Design Environment (VPDE)

The interaction between the design world and the virtual prototyping design environment (VPDE) (**Figure 9.5**) creates a virtual world that harmonizes product definition, design, and manufacturing.

Process models include all physical processes required for representing product behavior and manufacturing processes. *Activity models* represent various engineering activities, whether by humans or computer, for product engineering and production management, such as costing and scheduling. A well-established *product model* and its related process model not only capture and use information generated during the design phase but also

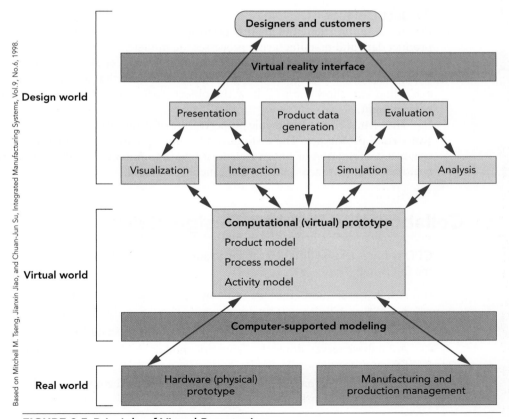

Based on Mitchell M. Tseng, Jianxin Jiao, and Chuan-Jun Su, Integrated Manufacturing Systems, Vol.9, No.6, 1998.

FIGURE 9.5 Principle of Virtual Prototyping

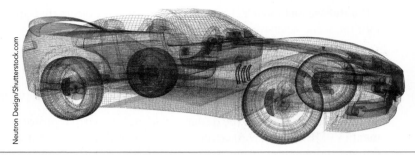

Neutron Design/Shutterstock.com

FIGURE 9.6 Virtual Prototype of a Ferrari F1

perform as a virtual analytical prototype as shown in **Figure 9.6** for engineering analysis based on computer simulation. VPDE enables the design, evaluation, and improvement of products by bridging different stages of product development via virtual prototypes and enabling the customer to be in the loop. VPDE is also referred to as collaborative product design (CPD).

Product Definition

Product definition activities are first performed in the virtual world where all necessary product data and manufacturing processes are modeled.

With the help of new human computer interfaces such as data gloves and headsets, product definition can make the user believe that a virtual object actually exists and can show the user how the object performs and behaves in its intended environment.

The virtual effects and tactile properties are of primary importance in these immersive virtual prototypes. The distinguishing characteristic of product definition is the use of a combination of hardware and software technologies that allow a designer or customer to experience virtual sight, touch, and sound.

9.4 **Collaborative Product Design (CPD)**

CPD requires several different servers to support its various functions. **Figure 9.7** shows the following:

- A web server allows the client to download the CPD interface and other utilities.
- A solid modeling server allows the designer to modify the geometric models.
- A visualization server produces a polygonized representation of a model.
- A database server stores the geometric models produced from the collaborative design process.
- A catalog server provides a directory of predesigned components that can be integrated in an assembly during a process.

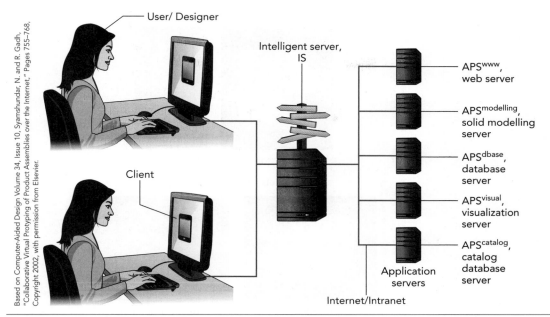

Based on Computer-Aided Design Volume 34, Issue 10, Syamshundar, N. and R. Gadh, "Collaborative Virtual Protyping of Product Assemblies over the Internet," Pages 755–768, Copyright 2002, with permission from Elsevier.

FIGURE 9.7 **Architecture of Collaborative Product Design (CPD)**

9.4.1 **CPD Session**

CPD enables the improvement of products by examining stages of development via virtual prototypes with the customer in the loop. A CPD session begins when the user opens a web browser and links to a virtual prototyping system. The designer installs the client's interface on the local workstation computer system, which establishes a connection. An appropriate virtual environment identifies, structures, and maps the customer requirements and develops a real-time graphical representation of the product so that the designer gains direct interaction with the customers as shown in **Figure 9.8**.

The designer requests a part list from the application server. A product frame describing the significant parameters and features of the product and its functional structure and model appears. The customer and designer can share new models, feedback on other models, and adjustments needed instantly through the web-based network as shown in **Figure 9.9**.

Designers can use virtual prototyping software tools such as *Cartona3D*, *Flux Player*, *Cosmo Player*, *Instant Player*, *View3dscene, Xj3D,* and *SwirlX3D* to view and manage 3D virtual reality drawings as shown in **Figure 9.10**.

Schematic models created in CPD

(a) Piston   **(b)** Shock absorber

Models imported into CPD

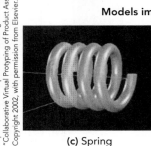

(c) Spring   **(d)** Gear box part

Reprinted from Computer-Aided Design Volume 34, Issue 10, Syamshundar, N. and R. Gadh, "Collaborative Virtual Protyping of Product Assemblies over the Internet," Pages 755–768, Copyright 2002, with permission from Elsevier.

FIGURE 9.8 Schematic Models of Machine Components Created with CPD

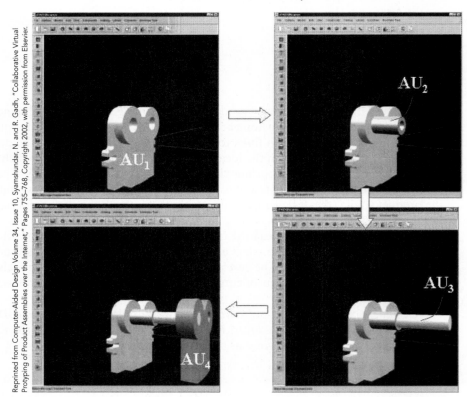

Reprinted from Computer-Aided Design Volume 34, Issue 10, Syamshundar, N. and R. Gadh, "Collaborative Virtual Protyping of Product Assemblies over the Internet, " Pages 755–768, Copyright 2002, with permission from Elsevier.

FIGURE 9.9 Automotive Components Assembled

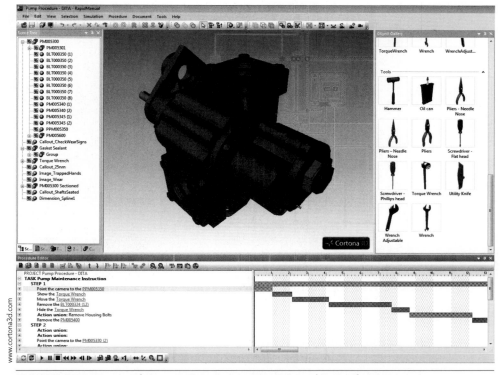

www.cortona3d.com

FIGURE 9.10 Text and Supporting Animation Created in Real time

9.5 Virtual Prototyping Technologies for Medical Applications

VP tools enable users to develop and optimize medical devices specifically for orthopedic applications and biomechanical simulation.

VP techniques are employed in medical training and education, surgical planning, diagnosis, reconstruction of digital human organs, design of medical devices, physical rehabilitation and telemedicine/telesurgery. VP tools can be used to study how rehabilitation procedures affect movement in post stroke and healthy subjects and to develop customized therapies and therapy devices for people with diminished motor capabilities.

Virtual Simulation Prototypes

A complete virtual engineering prototype is generally composed of a 3D solid model, a 3D CAD model, a human product interaction model, and model-related perspective tests (such as for structural and motion analyses).

Computer-based biomechanical models can be classified as virtual or computational biomodels. *Virtual biomodels* are used to visualize biological structures and human organs as shown in **Figure 9.11**. *Computational biomodels* are used for

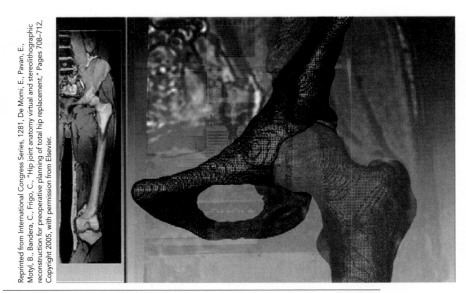

FIGURE 9.11 Reconstructed Hip and Femur Bones Shown with MRI

biomechanical analysis, for example, of finite element models (FEM) of anatomical structures and joints (knees, hips, mandibles, feet) (**Figure 9.12**) to determine stress and strain distributions. These models can be static, dynamic, or kinematic. *Static models* with the help of simulation are used to investigate the deformations of the anatomical structures or bones under load and muscle forces. *Dynamic models* of the human body are used to study muscle force system dynamics. *Kinematic models* are used to analyze movement mechanisms.

9.5.1 Simulation Tools for Orthopedic Implant Design

Manufacturers of orthopedic devices have provided additional resources to VP technologies, in particular in biomechanical simulations and finite element analysis (FEA) used to conduct full body simulations of the musculoskeletal system. Better understanding of joint kinematics and the human body can assist in the creation of advanced biomedical software. Implant companies have extensively used physical and rapid prototypes for testing the reliability of hip joints, shoulder joints, knee replacements, and spinal implants as well as for obtaining regulatory approval.

Developing modeling and simulation activities requires considerable time to complete, but the information the models and activities provide is invaluable and can be used to analyze other aspects of a device, such as its strength, durability, and wear.

Modern designers use VP and tests to reduce the number of design iterations by replicating laboratory tests within a virtual environment that allows the simulation of real-life events (such as climbing stairs, sitting, standing, and jumping) to assess the device's performance.

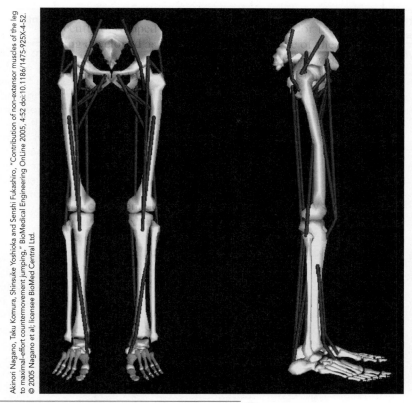

Akinori Nagano, Taku Komura, Shinsuke Yoshioka and Senshi Fukashiro, "Contribution of non-extensor muscles of the leg to maximal-effort countermovement jumping," BioMedical Engineering OnLine 2005, 4:52 doi:10.1186/1475-925X-4-52. © 2005 Nagano et al; licensee BioMed Central Ltd.

FIGURE 9.12 Full Body Musculoskeletal Model

See **Table 9.1** for an overview of the three most commonly used software packages for biomechanical-musculoskeletal simulation. The three software packages considered here represent the state of the art in kinematics and dynamic musculoskeletal modeling and simulation. Because of their specific characteristics, a thorough analysis of the application context of each must precede their selection and adoption.

In the recent orthopedic device market, the combination of VP, FEA, and multibody dynamic simulation represents an effective tool for maintaining a competitive advantage and enforcing it.

By undertaking studies on various groups of anatomical models within a virtual environment, designers can obtain better answers related to device performances and optimize the design to fit populations according to size, shape, and weight.

Remarks on Biomedical Prototyping

The use of VP for medical purposes is increasing because of the high quality of the software applications with continuous upgrades of computational resources and the refinement of modeling and simulation methods.

With implant costs rising and difficulty running in vivo testing, the orthopedic segment continues its search for new methods to gain competitive advantage.

Software Package		Comments
OpenSim	https://simtk.org/home/opensim	OpenSim by SIMtk. org is an open-source platform for modeling, simulating, and analyzing the neuromusculoskeletal system. It provides a platform on which biomechanical researchers can build and exchange their models and create a shared library of simulations.
AnyBody	http://www.anybodytech.com	The AnyBody Modeling System is a musculoskeletal simulation software used for modeling and analyzing multibody dynamic models. This software suite now consists of two main sections, one for model and one for testing and analysis. This system defines the biomechanical model by text-based input declared in a specific object-oriented language (AnyScript).
LifeMOD	http://www.lifemodeler.com	LifeMOD is a commercial musculoskeletal modeling and simulation environment based on the rigid body dynamics solver MSC.ADAMS. It computes biomechanical system kinematics and forces (such as knees, hip joints, spinal column) using forward and inverse kinematic algorithms.

TABLE 9.1 Software Packages for Biomechanical-Musculoskeletal Simulation

Biomechanical simulation technology helps designers to create better performing products. Moreover, the application of VP techniques to the production of medical devices leads to greater innovation in products, which can be developed rapidly and in close collaboration with medical staff.

9.6 **Additive Manufacturing**

This section examines the role of additive manufacturing (AM) in advanced manufacturing and its impact on product prototyping. AM is also known as *additive fabrication, additive layer manufacturing, layered manufacturing,* and *free-form fabrication.* AM is a basic technology that initially generates a model using a 3D model drawing created by a CAD system. The model is used to build functional, efficient, and effective components without the need for process planning. AM is a form of direct manufacturing that involves rapid prototyping. The output is a prototype or basic model that is modified to eventually create the final product.

Many companies started using AM for prototyping rather than for production. Aerospace and automotive companies create prototypes for experimenting the use of AM parts in their final products. Addressing the challenges of AM parts in the final product has emerged as one of the recent trends. The potential cost saving by the reduction of fuel consumption in the jet engine is a major motivating factor.

9.6.1 **Additive Process**

Generic Processing Steps

STEP 1. Creation of the CAD Model The piece must be CAD designed using 3D or surface modeling.

STEP 2. Conversion of the CAD Model into the STL Format The Standard Tessellation Language format implies a mesh of the inside and outside surfaces of the piece in triangular elements.

STEP 3. Geometric Layout and Orientation The direction of product's optimal growth greatly influences dimensional accuracy, surface finishing, and production time.

STEP 4. Dissection The Standard Tessellation Language model is sliced by a series of planes perpendicular to the growth direction. Slicing can be *uniform*, resulting in layers of uniform thickness, or *adaptive* in which case the surface curvature determines the thickness to minimize the step appearance (staircase effect).

STEP 5. Fabrication of Layered Sections The physical construction of the piece is formed by overlapping layers.

STEP 6. Removal, Cleaning, and Finishing of the Supports Eventually, fixtures are taken away from the piece, which is cleaned by removing the excess material attached to its inner and outer surfaces. Depending on the equipment used, some parts are polished and others are washed. Finishing can involve sanding, filing, or painting.

In addition to a CAD system, the AM process requires the use of a printer and a special finisher. The system makes 3D drawings by using parametric modeling to create feature-based geometries. In preparation to be printed, these drawings are translated into stereolithography (STL) files. Slicing an STL file generates each individual layer that consists of a cross-section of the part.

Liquid-based system: A system using a liquid organic resin that cures or solidifies under exposure to laser radiation, usually in the UV range. The formed layer is lowered and the next cross-section is processed. This process continues until the part has been completed. The main advantages of rapid prototyping systems are continuous operation with little or no monitoring, good accuracy, and material flexibility. Some disadvantages include the requirements for support structures for design overhangs, postprocessing for cleaning, and postcuring to fully secure the finished part.

Solid-based system: Other rapid prototyping machines use solids as the primary medium to create the part. Fused deposition modeling (FDM) uses a heated filament that is extruded to create the part. After the layer hardens, a new layer is deposited. This process is repeated until the part has been completed. Some advantages of FDM include minimal waste and ease of removal of supports and of material change. The main challenges of FDM include its limited accuracy as the result of filament size, slow processing, and unpredictable shrinkage. Current technological developments have been in the study of thermal effects and their relationship with stresses and geometry.

Powder-based system: Some 3D rapid prototyping systems use powder-based rapid prototyping built on injection printing technology. The printing starts by spreading a layer of powder over a piston-driven table. The printer applies binder solution to bond and create the layer. The table is lowered and the process is repeated until the part has been completed. Excess powder is vacuumed away from the part. Major advantages of power-based 3D printing include high speed, simple operation, and creation of no material waste. However, some of these developments are still at the early stages of innovation and growth.

9.6.2 **New Generation of Aerospace Materials for AM**

The aerospace industry is constantly looking for lightweight materials and new fabrication techniques to reduce costs by reducing engine weight and fuel consumption. Some critical components that the aerospace industry has experimented with involve engines. Some gas turbine components are candidates for AM construction. See the following list and **Figure 9.13.**

- Engine components such as augmenters, combustors, compressor stators, gearboxes, drive and turbine shafts, ducts, fan and turbine frames, fan stator, and diffuser cases
- High-value components such as casings and vanes as well as rotating parts (blades, rotors, and blade-integrated disks)

The aerospace industry is interested in materials such as Inconel, titanium, and aluminum. The AM process can produce parts from Inconel quickly and affordably. This material is ideal for many high temperature components used in gas turbine parts, instrumentation parts, and power and process industry parts. Inconel also has excellent

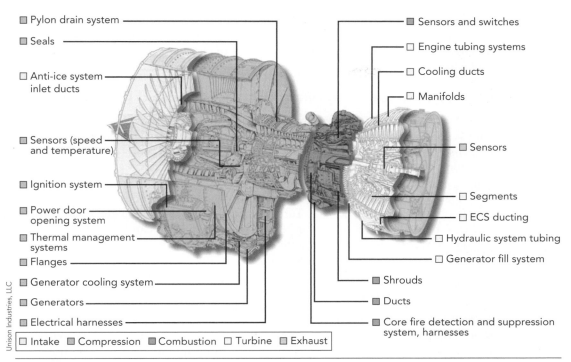

FIGURE 9.13 Aerospace Gas Turbine Components

very low temperature properties and potential for use in some applications. It is typically used in:

- Aero- and land-based turbine engine parts
- Rocket and components used in space research

Some characteristics that make titanium ideal for aerospace applications also make it difficult to machine. Its hardness and low heat conductivity reduce tool speeds and life; it requires a great deal of liquid cooling during machining; and it limits the productivity of certain shapes, such as thin walls. Laser-sintered titanium, however, retains the metal's beneficial properties and has no tool-wear or coolant costs. In addition, nearly any geometry, including thin sections, can be created with laser sintering. Titanium components made by AM are used in the direct manufacture of functional prototypes, structural and engine components, spare parts, and parts requiring a combination of high mechanical properties and low specific weight.

Alternative materials for the aerospace industry are metal matrix composites such as ceramic matrices, carbon fiber, and carbon epoxy. Weight reduction is the greatest advantage of composite material use and is a key factor in making decisions regarding its selection. Other advantages of metal matrix composites include their resistance to high corrosion and damage from fatigue and their ability to be formed into almost any shape using molding processes. These factors play a role in reducing the operating costs of an aircraft in the long term, further improving its efficiency. A major disadvantage of the use of composites is that they are relatively new to the field and as such have a high cost. The high cost is also attributed to the labor-intensive and often complex fabrication process.

9.6.3 Aerospace Component Requirements and Challenges

The aerospace industry desires parts that are lightweight, strong, and sometimes electrically conductive. Furthermore, the industry requires material and process standards that ensure part quality and consistency across machines and prototypes. The challenge to leading aerospace manufacturers is to develop AM plastics and metals as well as testing standards to ensure quality and consistency that meet industry needs. Successfully meeting these challenges requires more aggressive use of AM for testing new airfoil and turbine blade designs. Turbine blades made by direct metal laser sintering (DMLS) have already been used in testing facilities. DMLS can be used not only for flight hardware but also for jet-powered boats, land-based power generators, and other applications of gas turbine engines.

The aerospace industry produces ducts for military and commercial jets. Aerospace companies such as Boeing use laser-sintering technology extensively to manufacture these ducts for fighter jets and new 787 commercial jets. Previously, Boeing assembled up to 20 parts or more to produce one air duct assembly, which is very complex in shape. Many air ducts require vanes and other internal features to direct the flow of air. Each individual part that makes up the duct requires tooling of some type and welding, and fasteners are often needed. Today, Boeing manufactures many of its air ducts in one piece using AM. This practice has eliminated the need for identifying parts by number (and the inspections and documentation required for each), tooling, inventory, labor, entire assembly lines, and maintenance. The laser-sintered parts also weigh less than the assemblies they replaced, contributing to a savings in fuel.

Rib-Web Structural Components

The majority of aircraft substructure is composed of rib-web components (**Figure 9.14**) such as bulkheads, spars, and ribs. These components (typically made from aluminum- or titanium-based alloys) mainly feature a web-reinforced shape spaced across the structure as shown in **Figure 9.13**. These large, major structures are designed to last

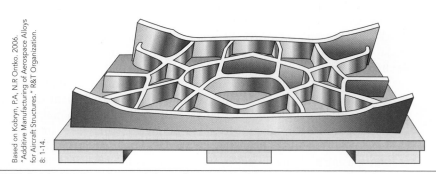

Based on Kobryn, P.A., N.R Ontko. 2006. "Additive Manufacturing of Aerospace Alloys for Aircraft Structures." R&T Organization. 8: 1-14.

FIGURE 9.14 AM Process on Rib-Web Structural Component

the entire lifetime of the airframe. They are typically machined from die forgings, hand forgings, or plate stock, which are extremely costly and hard-to-machine components. The benefits of using the AM process for making these components are that they reduce raw material use, raw material stock size, machining operations, and lead times, and when compared to forgings, reduce hard tooling requirements. These rib-web parts can be difficult to procure in a timely manner, particularly in small quantities, because the lead time required for making thick-section plate and forgings can be significant. Use of AM for repair is determined by a number of factors, including the ability to inspect the repair and resulting material for current and possible future defects, to restore the full mechanical capability of a part, to repair it in a timely manner, and to use faster or less expensive repair techniques.

Turbine Engine Cases

Turbine engine cases such as that in **Figure 9.15** are major structural components that form the outer surface of an engine and generally are made of titanium- or nickel-based alloys. These cases are typically composed of thick, cylindrical sections with a small number of low-volume, asymmetric protuberances. Thus, advantages expected from the use of AM technology to make these components are reductions in procurement lead time and cost and also the possibility of using the process in repair work.

Engine Blades and Vanes

Engine blades and vanes in **Figure 9.16** typically comprise complex airfoil-shaped sections that contain internal cooling passages to reduce the overall internal temperature of the blades and vanes. Engine blades are often made of titanium- or nickel-based

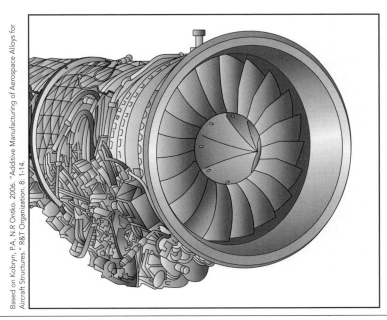

Based on Kobryn, P.A., N.R Ontko. 2006. "Additive Manufacturing of Aerospace Alloys for Aircraft Structures." R&T Organization. 8: 1-14.

FIGURE 9.15 Engine Case (A candidate for AM)

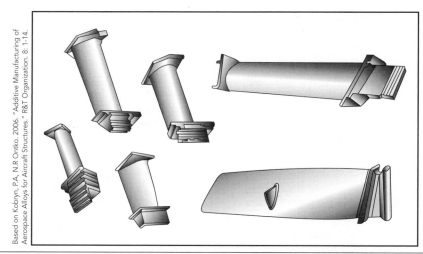

Based on Kobryn, P.A, N.R Ontko. 2006. "Additive Manufacturing of Aerospace Alloys for Aircraft Structures." R&T Organization. 8: 1-14.

FIGURE 9.16 AM-Processed Engine Blades and Vanes

alloys using precision die forging or casting, can be quite expensive to obtain, and typically experience significant wear and/or damage during their service life, resulting in very high repair cost and reduced operation time. The use of AM to repair blades and vanes is potentially significant because they often must be repaired or refurbished as part of a typical engine overhaul. AM has the potential to compete with existing repair methods and to enable previously impossible repairs.

9.6.4 **Factors Influencing the Adaptation of AM**

Actions to Prevent or Reduce AM Process Errors

Almost all AM processes can result in poor accuracy, which can affect a part's functioning. Thus, it is desirable to develop some method to reduce such errors. This can be done by processing fabricated parts using a computer numerical control (CNC) machine or other machining tool to remove the excess material or to fill the gaps and voids with processed material. Another way to reduce errors is to keep the minimum allowable layer thickness to ensure that the part is relatively free from them, but this can excessively increase *build time*, which refers to the time required to fabricate the complete part. It is important to minimize errors while minimizing build time.

More research is to be done before AM becomes a mainstream process for part production. Corporations, government agencies, and research institutions are working to solve these problems because they recognize the vast opportunities this technology can offer. Research focuses on both the current limitations of the technology and its use for new product development. The main AM limitations currently are cost and volume. In the future, technical developments are expected to increase process throughputs, making AM systems faster and more productive.

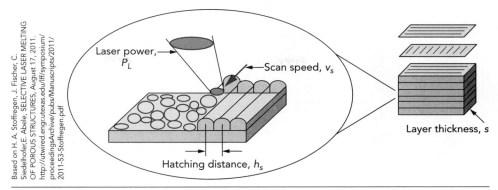

Based on H. A. Stoffregen, J. Fischer, C. Siedelhofer, E. Abele, SELECTIVE LASER MELTING OF POROUS STRUCTURES, August 17, 2011. http://utwired.engr.utexas.edu/lff/symposium/proceedingsArchive/pubs/Manuscripts/2011/2011-53-Stoffregen.pdf

FIGURE 9.17 Process and Varied Parameters

Porosity

Porosity, the amount of empty space in a material, is a factor that can negatively impact or affect material properties such as strength, hardness, and surface finish. Research on additive manufacturing emphasizes the importance of porosity reduction. **Figure 9.17** shows variables that influence porosity. All four parameters are defined by

$$E = \frac{P_L}{v_s \times h_s \times s} \tag{9.1}$$

where

P_L = laser power

v_s = scan speed

h_s = hatching distance

s = layer thickness

Because the existence of a fully melted pool is an important factor in AM part fabrication, a lower energy density should result in porous materials. The relationship between volume energy density and porosity leads to an inversely proportional relation, which is defined by:

$$\varepsilon = \frac{1}{E} = \frac{v_s \times h_s \times s}{P_L} \tag{9.2}$$

For optimum porosity analysis and best possible combinations, one parameter should be varied while others remain constant at their standard values.

Geometric Inspection

A sample test piece in **Figure 9.18** is an hourglass shape made from H13 tool steel using an AM machine. The hourglass shape requires a maximum outside diameter about 44 mm and a height of about 76 mm. A dimensional analysis of the test pieces indicates

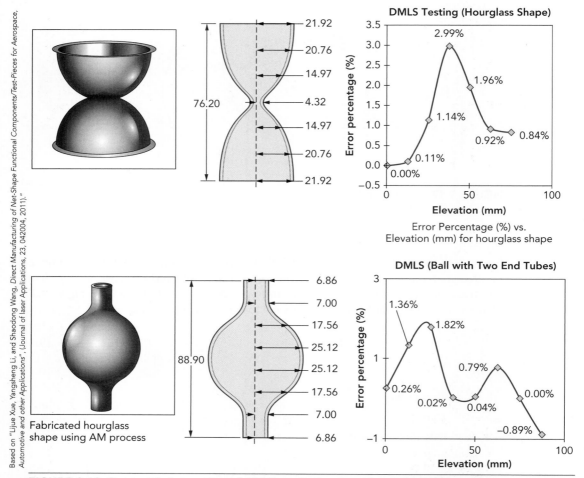

Fabricated hourglass shape using AM process

FIGURE 9.18 Geometric Inspection of AM Sample Compared to CAD Profile

geometric accuracies and limits. The maximum error percentage of the AM sample compared to that of the CAD model is about 3%, and the minimum error percentage is 0% at the base. In the case of a sample with two end tubes that is about 89 mm in height and about 52 mm in maximum diameter, the maximum error compared to the CAD model is about 2% while the minimal error percentage is 0% at 76.2 mm elevation (shown in **Figure 9.17**).

The prototypes of the housing g-hardened electronic components of the unmanned microvehicle systems in **Figure 9.19** were created with the use of additive manufacturing process. Experimenting with different design versions of them using AM before finalizing the design is cost-effective.

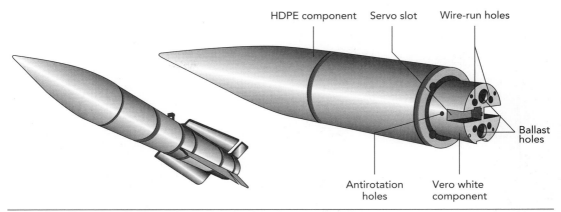

FIGURE 9.19 Models of Unmanned Aerial Vehicles Created by AM

9.6.5 **Comparison of AM to Conventional Manufacturing**

Conventional processes, including casting, rolling, forging, extruding, machining, and welding typically produce metal parts. To obtain the finished pieces usually requires performing multiple steps and using heavy equipment and tools. The use of this equipment is reasonable and an advantage in mass and large production batches. But when the part must be created in an unusual form, has complex internal cavities, or must be made in small batches, the costs and time required to prepare the production increase rapidly, and the use of conventional technologies is of little value.

In contrast, the AM process fabricates required components in a single step with material in its final stage and mechanical properties that are close to or in some cases higher than components fabricated with conventional processes.

A conventional process requires a careful and detailed analysis of part geometry to determine information such as the order in which different features can be fabricated, what tools and processes must be used, and what additional fixtures could be required to complete the part. AM can fabricate parts with extremely complex geometries, usually oversized by 0.025 mm to obtain the tolerances and the surface finishes required without the need for process planning.

AM machines essentially break a complex 3D part into a series of simple 2D cross-sections with a nominal thickness. This removes the connection of surfaces in 3D and enables continuity to be determined by the proximity of one cross-section to an adjacent one. Geometry features such as undercuts, enclosures, sharp internal corners, and other features are relatively easy for AM to process.

9.6.6 **Growth Trends in Additive Manufacturing**

Organizations in the additive-manufacturing business are optimistic about the future growth of AM for part production applications. Companies responding to a survey on the subject believe that AM part production represents 35.9% of their business in 2013. Statistics shows that in 10 years, the same companies believe that it will represent more

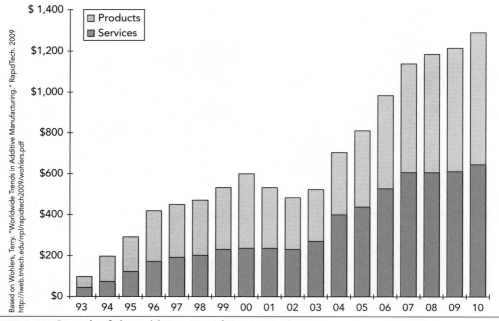

Based on Wohlers, Terry. "Worldwide Trends in Additive Manufacturing." RapidTech. 2009
http://iweb.tntech.edu/rrpl/rapidtech2009/wohlers.pdf

FIGURE 9.20 Growth of the Additive Manufacturing Industry

than half (50.5%) of their business. See **Figure 9.20** for estimated revenues (in millions of dollars) for the AM industry worldwide. The lower portion of the bars represents AM products that include AM systems, system upgrades, materials, and aftermarket products such as software, lasers, and print heads. The upper portion represents services revenue generated from parts produced on AM systems by designers/engineers, from system maintenance, and from training, seminars, conferences, shows, advertising, publications, contract research, and consulting related to AM processes.

Figure 9.21 graphically identifies a broad use of the technology across multiple industries worldwide. As the additive manufacturing industry has grown, more and more companies have begun to use this technology to produce finished products instead of prototypes.

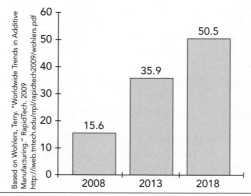

Based on Wohlers, Terry. "Worldwide Trends in Additive Manufacturing." RapidTech. 2009
http://iweb.tntech.edu/rrpl/rapidtech2009/wohlers.pdf

FIGURE 9.21 Growth Potential of Using AM for Part Production

Opportunities and Forecasts

As demand for metal AM parts increases, the number of raw material suppliers also should increase. As these materials are tested and qualified by industrial standards organizations, the range of AM applications will grow. New materials and processes will be developed, including those that produce metal powders directly. Meanwhile, AM systems will become faster and larger, increasing throughput. Doubling throughput would be similar to having two machines instead of one, thus reducing the cost of ownership.

Significant resources are being used to develop AM systems, materials, and applications for metal parts. Series manufacturing of metal parts from AM systems in production volumes higher than those of conventional processes is expected.

A major leap for the industry is predicted when aerospace industries test, accept, and expand the use of AM metal powders and the metal parts produced by AM systems. They will play an increasing role in the aerospace industry as a result of cost and environmental issues. When AM machines can build much larger structural parts than they currently can, the industry will see rapid expansion in the aerospace, defense, automotive, chemical, and marine sectors (**Figure 9.22**). The availability of metal alloys that meet mechanical property specifications for structural aerospace applications will have a significant impact on industry growth. The creation of technical standards to govern the AM industry in general will be a key step in its growth and acceptance. That work is currently underway by ASTM International Committee on AM Technologies.

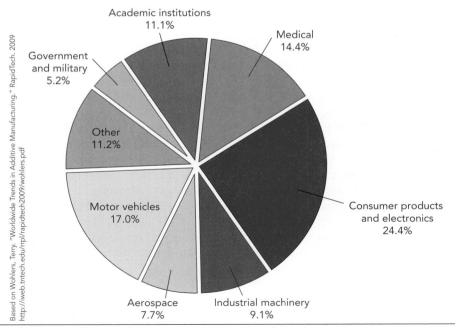

Based on Wohlers, Terry. "Worldwide Trends in Additive Manufacturing." RapidTech. 2009
http://iweb.tntech.edu/rrpl/rapidtech2009/wohlers.pdf

FIGURE 9.22 Industries Embracing Additive Manufacturing

CASE STUDY
9.1 Transformable Driver Seat/Wheelchair for Automatic Lifting

Introduction

This case study reviews the step-by-step sequences of virtual product development for a transformable driver seat/wheelchair automatic lifting system. The product is for use by people who are disabled and rely on wheelchairs for mobility. Most wheelchairs are designed for short distance movement only. When users need to travel longer distances, they must rely on other people to transfer them from the wheelchair to an automobile because automobiles have no lifting device. This situation not only limits users' independence but also requires assistance. The transfer might lead to an injury if not handled properly. Storage of the wheelchair in the car can also be a problem. To address the transfer problem, this case study discusses the design and development of a transformable driver seat/wheelchair automatic lifting system.

This design includes:

- Driver seat/wheelchair rotation
- Driver seat/wheelchair lift
- Folding driver seat/wheelchair roller
- Driver seat/wheelchair rotation function
- Driver seat/wheelchair latching

Design Methodology

The design in **Figure 9.23** includes a single actuator with dual parallel four-bar linkages for up and down movement between ground level and a car seat, a mechanism for a driver seat/wheelchair lift, and a six-linkage mechanism for wheelchair roller folding and unfolding.

The basic conceptual is that of an eccentric rotational mechanism driven by an electric actuator to reorient the driver seat before moving the user to and from the automobile. Interactive programs such as *Techoptimizer* and *Knowledgist* can assist in identifying existing design ideas and patent information. The diagrams and pictures in **Figure 9.23** show the model creation using basic shapes (a) and (b), the steps involved in the kinematic motion simulation of this wheelchair-lifting device (c), structural analyses using a finite element method (d), assembly analysis (e), and testing (f). As an example, an animated drawing is created in the software *Cortana3D* using a 3D viewer. It assists the designer in navigating the parts and assemblies and in viewing the assemblies from various points.

In conclusion, the example of "Transformable Driver Seat/Wheelchair for Automatic Lifting" demonstrates hands-on design and virtual product development concepts while simultaneously using various design software applications to develop a

Based on Lee, Ming-Yih and Chung-Hsein KUO. 2005. Hands-on Design and Virtue Experiment Training in Medical Mechatronics. Taiwan: Institute of Medical Mechatronics. http://www.ineer.org/Events/ICEER2004/Proceedings/papers/0255.pdf

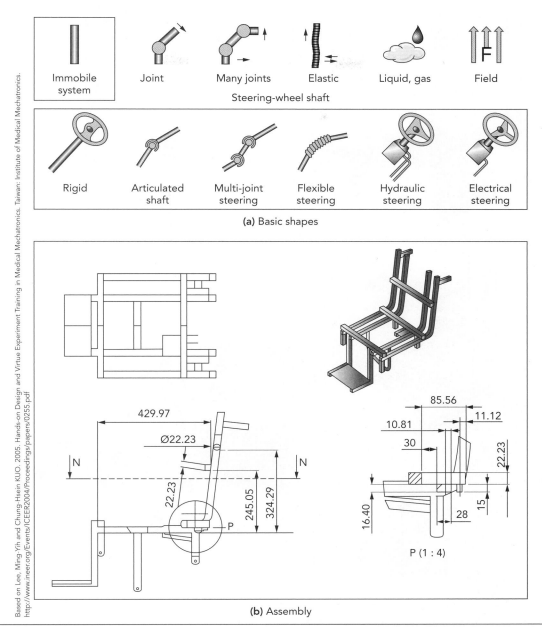

(b) Assembly

FIGURE 9.23 Step-by-Step Virtual Prototype of Wheelchair Automatic Lifting System (*continued*)

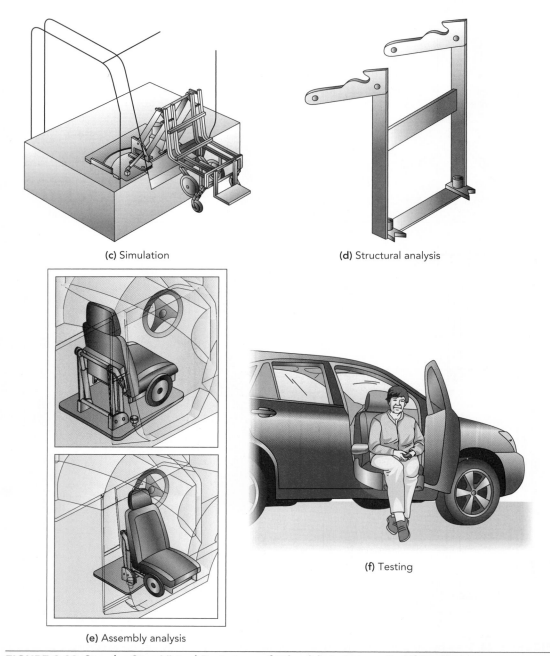

(c) Simulation

(d) Structural analysis

(e) Assembly analysis

(f) Testing

FIGURE 9.23 Step-by-Step Virtual Prototype of Wheelchair Automatic Lifting System

medical mechatronics product. The design methodology included sharing of distributed product development resources to construct the virtual product development experiments. Virtual network computing was used to connect the remote computers that have installed the corresponding product design and development software such as the *QFD, TechOptimizer, Knowledgist, SolidWorks, AutoCAD,* and *PowerMILL.*

The product design consists of system requirement analysis using quality functional deployment techniques, establishing design specifications based on house of quality, performing conceptual mechanisms utilizing TechOptimizer, constructing 3D computer models using the SolidWorks package, simulating motion mechanism with Working Model software, performing structural analysis using a finite element analysis package, and finally constructing a prototype for functional verification using rapid prototyping software. In this manner, the client user can use the online interface to utilize the product design software components through the network.

REFERENCES

1. Ian Gibson, David W. Rosen, and Brent Stucker, *Additive Manufacturing Technologies* (New York: Springer, 2010).
2. Monica Bordegoni and Caterina Rizzi, *"Innovation in Product Design: From CAD to Virtual Prototyping,"* *Proceedings of the ASME 2011 World Conference on Innovative Virtual Reality, June 2011, Milan, Italy.*
3. Gabriele Guidi and Laura Loredana Micoli, *"Modeling System for Quick Generation of Reality Models,"* *Proceedings of the ASME 2011 World Conference on Innovative Virtual Reality, June 2011, Milan, Italy.*
4. Devdas Shetty, *"Virtual Product Design Using Innovative Mechatronics Techniques for Global Supply Chain," Proceedings of the ASME 2011 World Conference on Innovative Virtual Reality, June 2011, Milan, Italy.*
5. E. De Momi, E. Pavan, B. Motyl, C. Bandera, and C. Frigo, *"Hip Joint Anatomy Virtual and Stereolithographic Reconstruction for Preoperative Planning of Total Hip Replacement" International Congress Series* vol. 1281 (2005), 708–712.
6. AnyBody: http://www.anybodytech.com/; LifeMOD: http://www.lifemodeler.com/; OpenSIM: https://simtk.org/project/xml/downloads.xml?group_id=91.
7. Lijue Xue, Yangsheng Li and Shaodong Wang, *"Direct Manufacturing of Net-Shape Functional Components/Test-Pieces for Aerospace, Automotive and other Applications"*, (Journal of Laser Applications, 23, 04 2004, 2011).
8. Mary E. Kinsella, "Additive Manufacturing of Superalloys for Aerospace Applications", AFRL-RX-WP-TP-2008-4318, Air Force Research Laboratory Materials And Manufacturing Directorate, Wright-Patterson Air Force Base, OH.
9. Kobryn, P.A.; Ontko, N.R.; Perkins, L.P.; Tiley, J.S. (2006) Additive Manufacturing of Aerospace Alloys for Aircraft Structures. In *Cost Effective Manufacture via Net-Shape Processing* (pp. 3-1 – 3-14). Meeting Proceedings RTO-MP-AVT-139, Paper 3. Neuilly-sur-Seine, France: RTO. Available from: http://www.rto.nato.int/abstracts.asp.

EXERCISES

9.1. Explain how you would select a team to design a product for a multicomponent product. What strategies would you use to have an effective design as speedily as possible? How would you use the prototype design process? Describe two rapid prototyping methods, identify the benefits of each, and provide sketches.

9.2. Design communication is an important aspect of product design. How does a design engineer communicate product-related ideas with multiple suppliers around the world?

9.3. Describe the role of integration of information technology in bringing together technological and business aspects of product creation. Give a suitable example, and explain the concept of designing a process for manufacturing integration.

9.4. Simulation and modeling are used in a wide variety of design configurations that examine the capabilities of components and systems. Consider a product such as a food processor and identify CAD/CAM roles for making major components.

9.5. Compile a list of different AM technological machines that are currently on the market today. What are the distinguishing features of each? What different materials can these machines process?

9.6. A manufacturing company introduces rapid prototyping techniques in the creation of biomedical products. What extra considerations might be needed when producing medical models using AM in comparison to conventional manufacturing?

9.7. Describe why surface modeling software is not ideal for describing models to be created using AM. What problems might occur when using surface modeling only?

9.8. Various types of models and prototypes are created before a product is fully configured. These can be models for proof of concepts, mathematical models, physical prototypes, model kinematic models, and virtual models. Describe the way in which these models differ in complexity, cost, and completeness.

9.9. Why is the interchangeability of parts important in the production of consumer products? How have digital manufacturing and additive manufacturing impacted it?

9.10. Imagine that you must choose between a plastic or a metal for a part you are designing. What are some reasons for selecting plastic? What are some reasons for selecting metal?

CHAPTER 10
Additional Case Studies

CHAPTER OUTLINE

OBJECTIVE: PUTTING IT TOGETHER

Successful companies who are at the forefront of product development adopt modern techniques and use concurrent engineering to improve the processes. In this chapter, we have included two additional case studies that address some of the design methodologies outlined in the previous nine chapters. Case study 1 describes the design approach for developing a surface inspection instrument with major usage in aerospace industry. Case study 2 investigates the design, simulation, and prototyping of a transformable hybrid projectile.

10.1 Case Study 1: Product for Inspection and Measurement of Surface Roughness

Background

Surface texture is a fingerprint of the production process. Monitoring surface texture, especially surface roughness, enhances the understanding of the production process. An essential aspect of the quality in the manufacture of precision parts is the status of the surface finish, and more particularly, the roughness of the surface. Within the aerospace industry, the need to produce high precision machined parts has increased because of the impact of surface quality on aerodynamics. The aerospace industry needs accurate measurement devices for examining the root attachment areas of jet engine turbine blades. The root of the turbine blade can exhibit a series of scratches

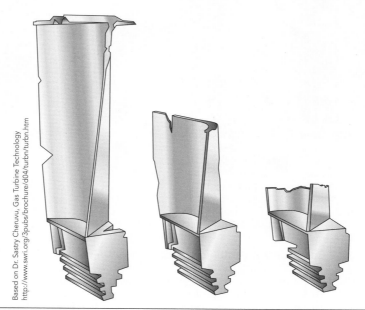

Based on Dr. Sastry Cheruvu, Gas Turbine Technology
http://www.swri.org/3pubs/brochure/d04/turbn/turbn.htm

FIGURE 10.1 Turbine Blades

that seem harmless until it is in use. As the blade flexes under stress, the forces trans-mit to the root which causes the scratches to enlarge and extend, then eventually snap (See **Figure 10.1**). In the automotive industry, micro-scale differences in R_a impact the performance and useful life of automotive engine and drivetrain components. Thus, there is a need to develop a reliable method for measuring these values accurately. Reliable measurement of the average surface roughness (R_a) is critical for quality control in a range of manufacturing processes.

10.1.1 **Surface Roughness Measurement Devices in Industry**

There are different types of surface roughness measuring devices available in the mar-ket. Two major types are (i) conventional measurement, such as contact type using a stylus, and (ii) noncontact measurement. Stylus-type, or contact measurement, devices (**Figure 10.2**) are popular and they operate by moving a small probe across a test surface to detect variations in its height. A primary limitation of this technique is that the stylus always has to traverse perpendicular to the lay on the surface.

For accuracy, the measuring probe must remain in contact with the sample surface. The load exerted by the probe on the surface can cause destructive metal deformation, and if the speed at which the stylus moves across the surface is too fast, the probe can lose contact with the surface and produce erroneous readings. Limitations of a stylus instrument are listed below.

- Excessive usage results in wear and tear on the probe tip.
- Time consuming measurement due to physical contact with the surface.
- Vibrations can cause inaccuracy.
- Contact with the surface potentially damages the surface or creates scratch marks.

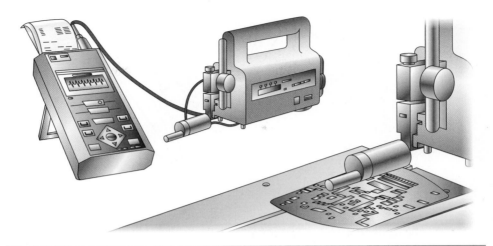

FIGURE 10.2 Stylus Contact Measurement Instruments

The use of noncontact laser-based measurement instrument has gained interest in recent years (**Figure 10.3**). Initially, a light source (laser beam) is cast on the surface workpiece, and the light then is allowed to bounce off a mirror and finally is captured and shown on a computer screen. A video system converts the data from an analog signal to a digitizing system. The captured image is analyzed and compared to the previously generated images stored as pixel values that identify two extreme limits of surface finished standards. After comparison, the real roughness value of the surface is displayed graphically and numerically. The major shortcoming of this device include

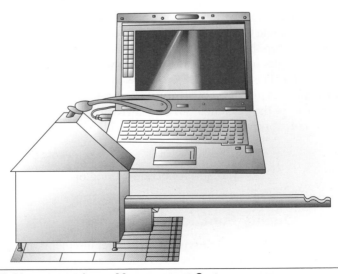

FIGURE 10.3 Noncontact Laser Measurement System

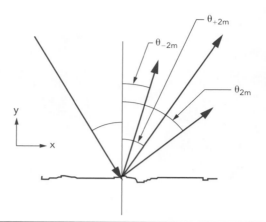

FIGURE 10.4 Light Scattering Principles

constraints imposed by the working distance to the sample and the slowness in measuring large areas.

10.1.2 **Theoretical Background**

The following section explains the steps for the instrument using laser reflection and computer coding to show the roughness of a surface.

The method developed by Shetty and Neault (1993) takes advantage of the fact that a light source reflected off the surface of a workpiece provides a signature pattern based on the roughness of the surface. Initially, a light source (laser beam) is cast upon the surface workpiece, and then the reflected light is allowed to bounce off a mirror and finally captured onto a screen. By a video system the data is converted from an analog signal to a digital signal. By computer programming the average roughness (R_a) is displayed both graphically and numerically.

An electromagnetic wave of known wavelength is incident upon the rough surface at an angle θ_1. For the case of periodic roughness, the scattering is made up of a specular component, at an angle predicted by ray tracing optics, and discrete components at angles predicted by the relationship as shown in Equation 1 and illustrated in **Figure 10.4**. The angle of diffuse scatter, θ_{2m}, is related to the period of the roughness. Since surfaces produced by various processes exhibit distinct differences in texture, the specimens of the machined surfaces can easily be identified by looking at the diffraction pattern of workpiece made by turning, milling, and grinding operations.

$$2m = \mathrm{Sin}^{-1}(\mathrm{Sin}\,\theta_i + m\,\lambda/T), \quad \text{where } m = 0, \pm 1, \pm 2, \tag{10.1}$$

T = Surface period (period of roughness)

λ = wavelength of the laser beam

10.1.3 **Quality Function Deployment**

Quality is created by identifying customer requirements and providing them by linking them to engineering characteristics. Each requirement is given an importance rating of 1 to 5 on a QFD (**Figure 10.5**) to rank the various engineering parameters identified.

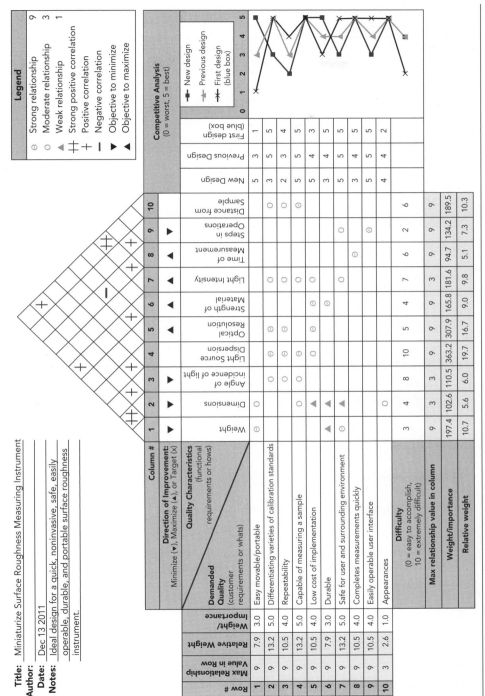

FIGURE 10.5 Quality Functional Deployment (QFD) Matrix

Customer Requirements	Importance Ratings (scale from 1–5)
Calibration standards for varying manufacturing processes	5
Capability to measure the specimen	5
Safety for the user and surrounding environment	5
Repeatability	4
Low-cost product implementation	4
Quick measurements	4
User-friendly interface	4
Portability	3
Durability	3
Acceptable aesthetics	1

TABLE 10.1 Customer Ratings

Functional Requirements	Relative Importance Weight (%)	Rank
Light source dispersion	19.7	1
Optical resolution	16.7	2
Weight	10.7	3
Distance from sample	10.3	4
Light intensity	9.8	5
Strength of materials	9.0	6
Number of steps in operation	7.3	7
Angle of incidence of light	6.0	8
Dimensions	5.6	9
Time requirement	5.1	10

TABLE 10.2 Specifications for Engineering Parameters

Use of the correlation matrix in the QFD quantifies the engineering parameter related to each customer requirement for the device being designed (**Table 10.1**). By multiplying these correlation coefficients (value of 1, 3, or 9) and using the importance rating of each customer requirement, the rank of each engineering parameter for the new design can be determined (**Table 10.2**).

10.1.4 **Functional Layout**

Figure 10.6 shows the breakdown of the system into components. The final product is created by having an optical path of the product, where the laser beam diffracts from a surface and is captured by the camera and configured as a compact product as shown in **Figure 10.7**. The instrument is configured as a box-type measuring instrument. The image is recorded and analyzed by the algorithm created using the software (LabVIEW in this

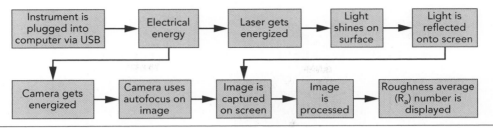

FIGURE 10.6 Functional Layout

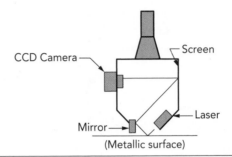

FIGURE 10.7 Schematic of the Product (Surface Roughness Measuring Instrument)

example). The measured value of the roughness is displayed. The key part of the concept is the evaluation of the image of the diffraction pattern from the surface, digitizing it and comparing it to a calibrated curve. The calibrated curve is initially created by using a set of perfect reference standards for ideal surfaces (for example, ground machined surface or milled surface or turned surface). The methodology described in this case study provides a robust approach for evaluating the surface finish of engineering surfaces, regardless of the workpiece orientation. Combined with a simple computer algorithm, the engineering surfaces in question can be classified by means of the measurement of the scattering intensities of the diffracted image. The accuracy of the instrument can be improved by taking multiple measurements on the specimen under test.

10.1.5 **Concept Generation and Selection**

At this point the detail design of the instrument is done by considering the overall size, location of the mounting brackets, portability, and the need for miniaturization. The ability to use the instrument in real-time environment, especially for online surface measurement while machining process is in progress, is an additional feature for the product. **Table 10.3** considers three different layout configurations based on the usage and adaptability of the instrument. The product designed using concept A is useful for vertical mounting on a machine tool and can be used for a machining operation such as a milling operation. The product designed using concept B can be used as a hand-held roughness measuring unit in a measurement and metrology laboratory. The product designed using concept C can be used as table top measurement unit.

In addition, Table 10.3 identifies three different possible applications of the product. A product developed using concept A can be useful for on-line measurement while machining processes take place. A product developed under concept B could be a

Concept A

Concept B

Concept C

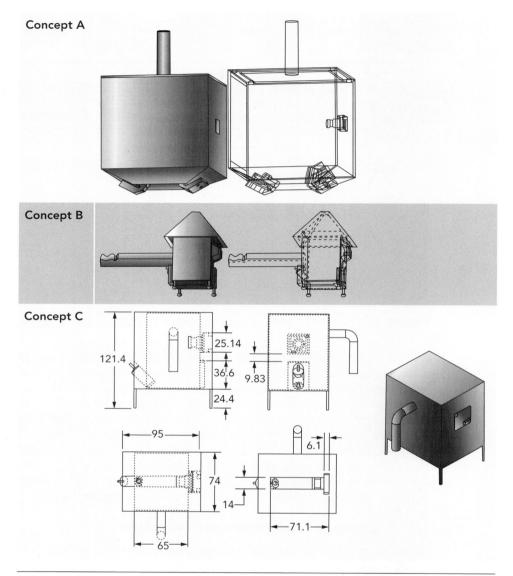

TABLE 10.3 Geometric Layout

portable inspection instrument while a product developed by concept C could be used as an off-line instrument in an inspection laboratory.

A *Pugh chart* (**Table 10.4**) is a concept selection methodology that is used to resolve design problems by evaluating, rating, and comparing alternatives based on the selected criteria. Each option is ranked 0 to 4 to indicate its importance in the selection criteria; 4 is the most important option.

+ Indicates that the alternative is better than the datum based on the criteria.

0 Indicates that the alternative is identical to the datum based on the criteria.

Selection Criteria	Weight	Concept		
		A	B (datum)	C
Cost	4	−	0	0
Availability of materials	3	0	0	−
Simplicity of design	3	−	0	+
Ease of fabrication	2	−	0	+
Camera focus	4	−	0	−
Functionality	3	0	0	0
Safety	2	0	0	0
Aesthetics	1	−	0	0
Size	2	−	0	+
Portability	4	−	0	0
Weight	2	−	0	+
Robust	3	+	0	−
Sum +		3	0	9
Sum 0		3	12	5
Sum −		22	0	10
Net score		−19	0	−1

TABLE 10.4 Pugh Chart

− Indicates that the alternative is worse than the datum based on the criteria.

By using the data in the Pugh chart, concept B is selected in this special example.

10.1.6 DFA for Chosen Design

The next step in the design process is to analyze the product assembly procedure so as to determine whether the number of parts in the design could be reduced or any could be combined with another. The following questions are generally considered while identifying whether the component is a critical part of the product.

- Does the part move relative to all other moving parts?
- Must the part absolutely be made of a material different from that of other parts?
- Must the part be a different design that makes disassembly difficult?

The designer evaluates the geometry of each component in the product and determines the degree of difficulty in handling and inserting each part. The analysis can provide an estimated assembly cost and a direction for redesign to improve the design. Design improvement focuses on reducing the number of parts and shortening the associated process times. The best way to achieve this goal is to reduce the number of components used and then to ensure that those actually used are easy to install and assemble.

See **Table 10.5** for the selected design's tabulated data.

Number	Part Description	Number of Identical Operations	$\alpha + \beta$ Symmetry	Manual Handling Code	Manual Handling Time	Manual Insertion Code	Manual Insertion Time	Total Assembly Time	Theoretical Minimum Number of Parts
1	Laser	1	360	10	1.5	01	2.5	4	1
2	Mirror	1	540	20	1.8	02	2.5	4.3	1
3	Hinge	1	540	20	1.8	00	1.5	3.3	0
4	Screen	1	360	18	3	00	1.5	4.5	0
5	Camera	1	540	20	1.8	08	6.5	8.3	1
6	Camera support	1	540	20	1.8	10	4	5.8	0
7	Screws	2	360	10	1.5	39	8	19	0
8	Component housing/ casing	1	720	30	1.95	08	6.5	8.45	1
9	Handle	1	360	10	1.5	01	2.5	4	0
10	Legs	4	360	10	1.5	00	1.5	12	0
11	Computer	1	–	–	–	–	–	–	1

TABLE 10.5 DFA of Chosen Design

Current design efficiency = $3 \times Nm / Tm$ = 53% (not including part 11)

10.1.7 **Algorithm Development and User Interface**

The interface for an operator of a surface roughness device must be user friendly. See **Figures 10.8** and **10.9** for features of the LabVIEW interface.

The interface for concept B has three requirements:

1. The user must be able to physically position the device directly above the surface whose roughness is being measured. The user can do this by either moving the surface while keeping the device stationary or moving the device while keeping the surface stationary, depending on the nature of use.

2. The user must be able to capture an image; this can be done by selecting the RUN button on the interface software. The Reflection Pattern window shows the potential image to be captured so that the user can ensure that the device captures an image of the desired surface region.

3. The user must be able to correct the standard calibration for the workpiece by adjusting the appropriate switches. Finally, the user must be able to view

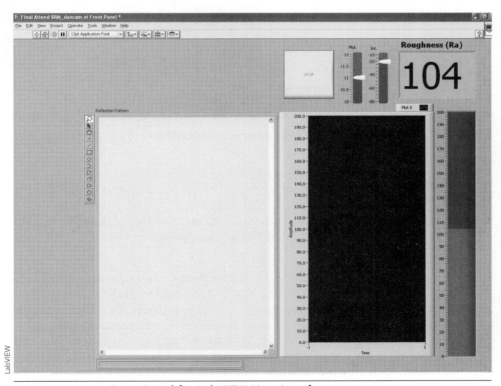

FIGURE 10.8 (a) Front Panel for LabVIEW User Interface

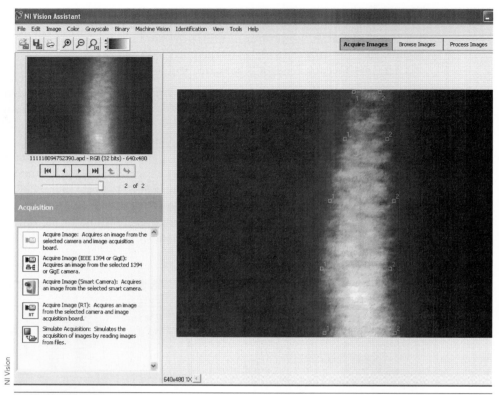

FIGURE 10.8 (b) NI Vision Assistant (IMAQ) Working Parallel to LabVIEW

the plot, bar graph, and displayed screen of the measured R_a value. The software can be programmed to measure workpieces manufactured using different processes.

Figure 10.10 shows the surface roughness measuring instrument (using concept A) mounted on the CNC Milling Machine and used for real-time measurement of surface finish while manufacturing process is under progress. The instrument described in this case study is capable of successfully displaying surface roughness values for a variety of manufacturing operations. With the increasing need in aerospace and automotive industry for high-quality precision surfaces, manufacturers constantly look for new surface quality products that provide surface finish requirements.

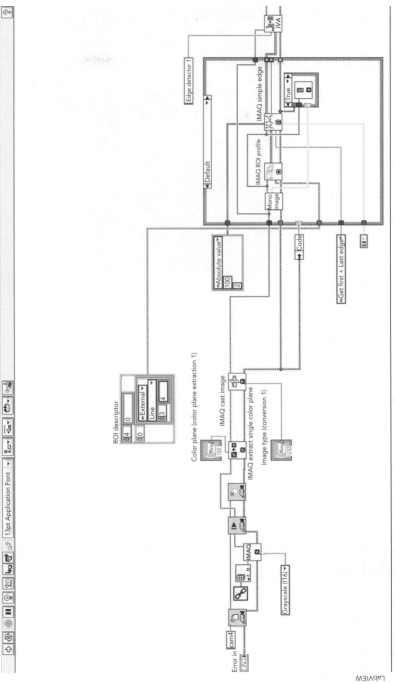

FIGURE 10.9 LabVIEW Block Diagram

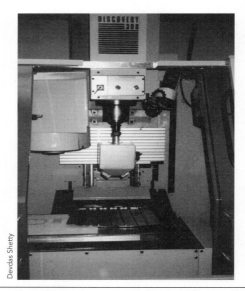

FIGURE 10.10 Surface Roughness Instrument Mounted on a CNC Milling Machine

REFERENCES

1. D. Shetty and J. Hill, "Precision Measurement Method of Misalignment, Cracks, Contours and Gaps in Aerospace Industry," *Proceedings of American Society of Engineering Education Conference, San Antonio, Texas, June 2012.*

2. D. Shetty, C. Campana, and S. Moslehpour, "Standalone Surface Roughness Analyzer," *IEEE Journal of Instrumentation and Measurement* 58, no. 3 (2009), 698–706.

3. D. Shetty, S. Ramasamy, and S. Choi, "A Non-Contact Visual Measurement System Integrating Labview with Matlab," *International Journal of Engineering Education* 21 (2004).

4. D. Shetty and Q. Han, "Methodology and Modeling of a Laser-Based In-process Surface Finish Inspection Probe," *Proceedings of the Thirty-Second International MATADOR Conference, UMIST, Manchester, UK. July 1997,* 597–602.

5. D. Shetty and H. Neault, "A technique and methodology for surface roughness evaluation using computer vision" *U.S. Patent, February 23, 1993. No. 5,189,490.* Product delivered to Pratt and Whitney Aircraft Co. (United Technologies, CT) for the evaluation of the surface finish of the turbine blades.

10.2 Case Study 2: Design of Transformable Unmanned Aerial Vehicle (UAV)

Background

The focus of this case study is the guided hybrid projectile, referred to as a *transformable unmanned aerial vehicle* (UAV), which is a powered aerial vehicle that operates without a human operator onboard and uses aerodynamic forces to provide lift. It has both civilian and military applications. A UAV used by the military can be launched from a gun and can be transformed into a projectile that is visually guided to a specific location. Guided mortar projectiles have been used for military applications since late 1970s. More recently developed projectiles are still used for targeting tanks and bunkers, but are being developed as low-cost products. Some of them have grown in complexity of the guidance and control systems, allowing more widespread use of guided rounds of ammunition. Many of these projectiles are equipped with a mechatronic system that can perform multiple tasks and a micro camera and transmitter for target recognition that can capture and relay image signals to the ground base where they are processed. The major factors in determining the relative merit of the different concepts are their design integrity including complexity, structural viability, and overall system survivability from gravitational forces.

10.2.1 Guidance Systems for Transformable UAVs

The current guidance system designs for transformable UAVs use a multidisciplinary approach that integrates expertise across key areas such as fluid dynamics, aerodynamics, guidance laws, control theory, flight dynamics, and microelectronics. The projectile design provides a person-in-the-loop guidance of a mortar round using real-time video (optical) feedback for precision position guidance while extending the interaction range by using gliding projectile designs. Optical guidance provides high accuracy, improves the projectile's effectiveness, and reduces the likelihood of damage in the gun's barrel during launching and on impact. The current mortar canister for guided projectiles has a diameter from 40 mm to 120 mm. The projectiles are specified by their outside diameter, and these dimensions match the diameter of the barrel of the gun. The smart mortar has target acquisition techniques that include camera-based feedback systems, laser tracking, and global positioning systems (GPS).

GPS Guidance

GPS-guided systems for transformable UAVs have certain advantages, primarily because they do not depend on real-time line-of-sight targets, and they cannot be influenced by poor visibility, cloud cover, or smoke screen countermeasures. These factors

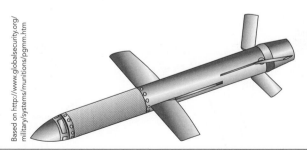

Based on http://www.globalsecurity.org/military/systems/munitions/pgmm.htm

FIGURE 10.11 Diehl Bussard/Lockheed XM395 PGMM

give GPS-guided mortar projectiles true fire-and-forget (self-guidance to its target once it is fire launched) capability. Because the target coordinates are programmed before the round is fired, GPS guidance allows override and mid-course corrections to be initiated during the gliding stage. The disadvantages of GPS-guided systems are that they are less accurate than some advanced targeting systems and are incapable of striking moving targets.

10.2.2 **Background of the Wireless, Hybrid Transformable UAV**

Various methods are used for creating and testing a wireless, hybrid surveillance projectile launched from a gun that gives users real-time video feedback. The design process establishes a framework of component selection for the projectile of certain diameter D and functionality. The projectile's specifications are to be gun launchable and to withstand a setback force of 5,000 G. It should have deployable wings and should reach an apogee of about 500 m with a nominal flight range of 1,500 m that includes tracking and gliding to a destination. Transformable UAV will have the ability to change the track while it is in the process of gliding (**Figure 10.12**).

The design also must do the following:

- Convert a gun-launched transformable UAV into a projectile that can be optically guided to a specific location.
- Have a wireless link to control the projectile's track.
- Use an electromechanical system to control the projectile's wings.
- Incorporate software to send and receive control signals from the projectile to its ground base.

The stages and components of a wireless transformable UAV are provided in **Figure 10.13**. The system includes a wireless micro camera for relaying video stream captured during the flight, and the ground base that can control the UAV's path after the wings get deployed and the projectile transforms itself.

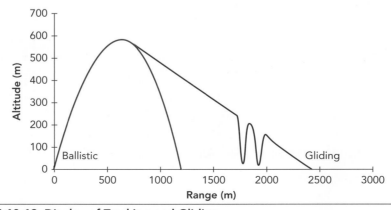

FIGURE 10.12 Display of Tracking and Gliding

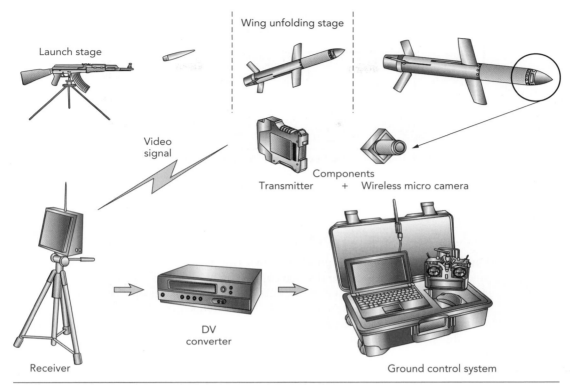

FIGURE 10.13 Wireless Transformable UAV Stages and Components

Based on Tarek Sobh, Khaled Elleithy, Jeongkyu Lee, Ali El-Rashidi, Jovin Joy, and Leon Manole, "Design and Implementation of the Wireless Camera and Control Components for a Transformable Unmanned Aerial Vehicle," Journal of Intelligent & Robotic Systems 66 (2012), 401–414.

The basic specifications of a transformable UAV for this case study are assumed to include:

- Launchability by gun with identified diameter D and projectile compatibility with the launching system normally used in the institution.
- 60–120 m/s service velocity
- Weight of <1.5 kg
- Expected setback force of 5,000 G
- Ability to operate in wind conditions of 8 m/s
- Cost of <$500 at 10,000 volume of production
- Use of mechatronics system and off-the-shelf components
- Mass and center of gravity identical to those of a conventional projectile that has no path control capability.
- Projectile with a minimum of 30 seconds of flight time after reaching apogee
- Video feedback regarding functionality
- Real-time data communication for course correction

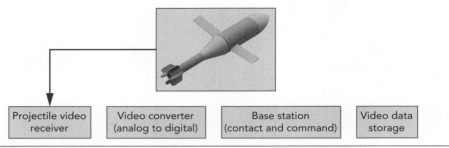

FIGURE 10.14 UAV Wireless Feedback and Control Components

10.2.3 **Functional Layout**

The initial mechanical design of a transformable UAV consists of packaging the precision components optimally in a compactly held casing. In addition to the wireless camera, other components include wings, wing actuation mechanism, servo motors, battery pack, antenna, and IR-integrated sensors. Some of these components modify their shape and transform into a new shape that has the ability to glide during the flight when a command signal from the wing actuation mechanism deploys the wings and fins. Some projectiles deploy the wings by a servo mechanism after the components are commanded to move from the initial folded and packaged position. The transformable UAVs can be launched in many ways. The use of an aerodynamic gun to launch the UAV enables it to carry a heavier payload of high explosives and to reach farther targets than conventional guns. Sometimes the use of cannon with its great force launches the projectile so that it can travel a long distance and glide to deliver the payload to the target. The transformable UAV's major focus is the maximization of available wing area for extended range and minimization of storage volume to house all the components. In contrast to transformable UAVs, conventional projectiles cannot be steered during flight or recalled once launched, nor can they reach their targets as far away as the UAVs can.

10.2.4 **Concept Selection**

Aerodynamic Concept

The aerodynamic design of the basic shape and wing profile should show the capability of at least a 3:1 glide slope when the UAV is carrying a predetermined payload. The initial concept rests on the aerodynamic viability required to send the vehicle to its target. Initial design is initially limited to verifying basic size/fit and balance with provisions for a control mechanism. The mechanical design follows a computational fluid dynamics (CFD) analysis to confirm that the design has lift/drag and neutral or positive aerodynamic stability in all three axes.

Based on the aerodynamic requirements identified during initial modeling and assigned during path planning, the feasibility of several preliminary configurations are evaluated by using a spreadsheet with the standard equations for lift and drag (refer to Shevell, 1989). The UAV's basic maneuverability is also evaluated to estimate maximum roll angle (φ) and the corresponding minimum sustained turning radius for the configurations based on typical $C_{L\text{-max}}$ values; velocity is calculated using the following equations.

$$L = C_L \frac{1}{2} \rho A_{ref} V_\infty^2$$

$$D = C_D \frac{1}{2} \rho A_{ref} V_\infty^2 + D_{(skin\ friction)}$$

$$C = C_{D(profile)} + C_{D(induced)}$$

$$C_{D\,(induced)} = \frac{C_L^2}{\pi(AR)e}$$

$$D_{(skin\ friction)} = \left(\frac{1.33}{\sqrt{Re}}\right)\frac{1}{2}\rho A_{wetted} V_\infty^2$$

$$\phi_{max} = cos^{-1}\left(\frac{2\,mg}{C_{L-max}\,\rho A V_\infty^2}\right)$$

$$R_{min} = \frac{m}{C_{L-max}\,\rho A \cdot sin(\phi_{max})}$$

where

ρ = atmospheric density

A_{ref} = wing area

A_{wetted} = total surface area

V_∞ = indicated airspeed

e = span efficiency factor

When these requirements have been satisfied based on modeling, fabrication, and testing, a wind tunnel test of the baseline design should be conducted to verify the computational fluid dynamic results. A morphological chart is described in the earlier chapters of this book. The chart helps the designer to identify preliminary concepts and the means to achieve them. See **Table 10.6** for a morphological chart that identifies main features of a UAV in the first column, the methods used to achieve them, and other information in the remaining columns.

10.2.5 **Conceptual Ideas for the Hybrid Projectile**

Features	Methods				
Launching system	Gun	Shoulder	Grenade	Rifle	Aircraft
Guidance system	Video (from camera)	RF emission	Laser homing	GPS support	
Battery power	Lithium ion	Nickel metal hydride (NiMH)	Nickel cadmium (NiCd)	Lead acid	Solar
Wing actuation	None		Cruciform shape		
Setback force	2,000 G	3,000 G	5,000 G	10,000 G	15,000 G
Casing material	Aluminum	Stainless steel	Titanium	Hard plastic	Carbon fiber
Testing method	Gun launch	Pneumatic	Drop	Wind tunnel	Drop

TABLE 10.6 Features of Hybrid Projectiles

10.2.6 **Prototyping**

Based on the results of the wind tunnel testing and CFD modeling, the prototype's external shape is created before finalizing which internal components will be included (**Table 10.7 (a)**). Concurrent finite element analysis must be performed during the design process to verify the size and material choices for the components. To facilitate mechanical functionality and size/fit of various components, a rapid prototyping machine is used to create the parts as shown in **Table 10.7 (b)**. Prototypes can be used to identify mechanical problem areas and validate assembly and fabrication procedures. The photopolymer's mechanical properties are adequate for a number of small internal components used to position, isolate, and mount the electronics inside the metallic frame. Creation of a prototype by rapid prototyping machine allows components that would otherwise require extensive tooling to be produced quickly in-house, simplifying logistics and reducing costs.

Additional CFD modeling is conducted to develop a dynamic aircraft model that can determine pitch, roll, and yaw rates and the control response (**Table 10.7 (c)**). A baseline model for the wind tunnel test should include functional actuators to verify the CFD results. Iterative design procedures optimize the aerodynamic performance of the model's design.

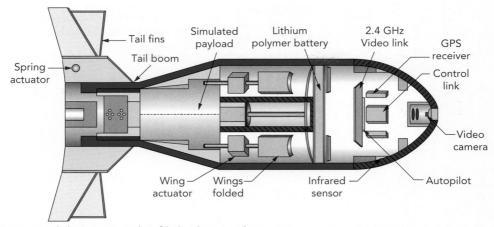

(a) Conceptual design example of hybrid projectile

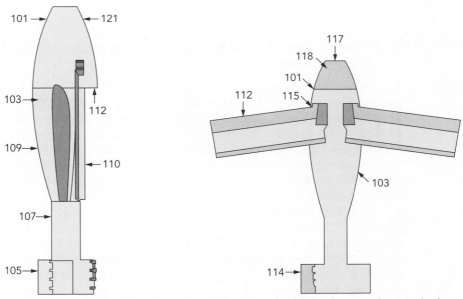

(b) Gun-launched hybrid projectile (Manole et al.)
The figures (a) and (b) illustrate the initial shape of the projectile launched and its changed configuration, respectively

(c) Gun-launched hybrid projectile as it deploys its wings. The projectile is under transformation

(continued)

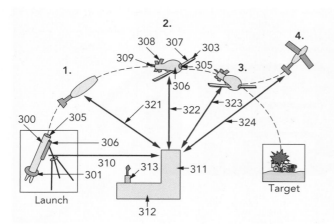

(d) The projectile undergoes transformation and glides to a target location.

TABLE 10.7 Conceptual Ideas for the Hybrid Projectile.

Based on gun-launched hybrid projectile US 8115149 B1

10.2.7 **Algorithm Development and Hardware-in-the-Loop Simulation**

By using visual tools in a real-time environment, major portions of the mechatronic product and the hardware-in-the-loop (HIL) can be simulated. The electronics can be simulated in the design simulation model while considering the mechanical systems, actuators, and sensors as hardware. HIL simulation and experimental validation are required. Simulation-based testing is used because conducting flight tests is expensive. In the prototyping step, many of the noncomputer subsystems in the projectile model are replaced with actual hardware. Sensors and actuators provide the interface signals necessary to connect the hardware subsystems to the model.

Tarek Sobh et al. (2012) summarize some typical components used in their experimental project. Typical hardware includes a wireless micro camera with a 2.4 gigahertz (GHz) operating frequency, audio-video receiver in the 2.4–2.5 GHz frequency range, video device for converting analog to digital signals, connectors to the computer's USB ports, desktop data storage device, micro stepper motors for wing actuation and control interface, and a wireless sensor device to identify barometric pressure, GPS, magnetic field, sound, photosensitive light, humidity, and temperature. The control interface software should be capable of displaying video captured in real time so that the operator can analyze it, and—based on the results of analysis—send control signals to the motor to control the wings.

Product Testing

The components of a prototype examined on a test range include the wireless camera's ability to send a signal to the ground base and to the motors' control circuitry, fabrication, the electronic components to ensure that they can survive gravitational forces

Component	Purpose
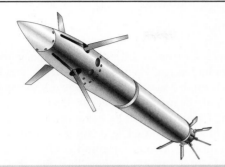	(a) Simulated model of projectile during stress analysis
	(b) Alaris 30 rapid prototype 3-D printer used to fabricate projectile components
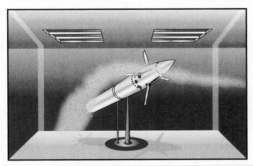	(c) Prototype during wind tunnel testing
	(d) Projectile (initial position) and final position after the wing is actuated (before and after); Based on Jay P. Wilhelm, Edward R Jackson, Patrick Browning, Wade Huebsch, Victor Mucino, and Mridul Gautam "Flight Simulation of a Hybrid Projectile to Estimate the Impact of Launch Angle on Range Extension," Proceedings of the ASME International Mechanical Engineering Congress & Exposition, November 2012, Houston, Texas.

TABLE 10.8 Illustrations of the Prototype under Modeling and Testing

during launch, and an antenna. The model's range, flight path, velocity, maneuverability, video feedback, and crash survivability are evaluated.

10.2.8 **Summary**

The case study of transformable UAV provides a typical example of how basic tools of product design can be systematically applied at each step of the product creation process. The design steps in this product go through concept development, choosing the best design amongst alternatives, evaluation of the merits of alternate design configurations, and initial performance estimation. This is followed by optimization of aerodynamic design, rapid prototyping, testing in a wind tunnel, and computational fluid dynamics (CFD). Other steps involve definition of avionics, ground control requirements for mission profile, demonstration of optical tracking capability, line of sight range verification of telemetry and terminal guidance RF links, matching the design with wind tunnel testing, procuring final components through supply chain, in-house survivability evaluation and hardening for gravitational force, detailed mechanical design and mechanical/structural analysis, safety analysis, and final testing. The results have demonstrated the usefulness of design in a real-life environment.

REFERENCES

1. Tarek Sobh, Khaled Elleithy, Jeongkyu Lee, Ali El-Rashidi, Jovin Joy, and Leon Manole, "Design and Implementation of the Wireless Camera and Control Components for a Transformable Unmanned Aerial Vehicle," *Journal of Intelligent & Robotic Systems* 66 (2012), 401–414.
2. Jay P. Wilhelm, Edward R Jackson, Patrick Browning, Wade Huebsch, Victor Mucino, and Mridul Gautam "Flight Simulation of a Hybrid Projectile to Estimate the Impact of Launch Angle on Range Extension," *Proceedings of the ASME International Mechanical Engineering Congress & Exposition, November 2012, Houston, Texas.*
3. Leon R. Manole, Ernest L. Logsdon, Jr., Mohan J. Palathingal, and Anthony J. Sebasto. Gun launched hybrid projectile. U.S. Patent US 8115149 B1, July 21, 2009.
4. D. Shetty and L. Manzione, "Micro UAVs—Design Trends Using Mechatronic Trends," *Proceedings of ASME International Mechanical Engineering Congress & Exposition, November 2012, Denver, Colorado.*
5. Shevell, Richard S. *Fundamentals of Flight, 2nd Ed.* Englewood Cliffs, NJ: Prentice Hall, 1989.

ACKNOWLEDGMENTS

The author would like to thank Mr. Leon Manole, Ernest Lonsdon, and Arthur Pizza from Armament Research, Development and Engineering Center (ARDEC) at Picatinny Arsenel; and Dean Lou Manzione, Luke Ionno, Tim Maver, Chris Bepko, Tom Lavoie, Claudio Campana, Dr. Eppes, and Dr. Milanovic from the University of Hartford, CT. In addition the assistance of the Connecticut Center of Advanced Technology (CCAT) is appreciated.

APPENDIX
Product Design Related Resource Material

Innovative, inventive solutions play an important part to improve product quality and productivity, lower manufacturing costs, and reduce defects, rework, and downtime. When inventive solutions are protected by patents, they enhance profitability and ensure product life longevity. Most engineering design problems fall into the patentable categories of utility patents or design patents.

Patents

Intellectual property deals with the protection of ideas with patents, copyrights, and trademarks. The United States Patent and Trademark Office (USPTO) is an agency of the U.S. Department of Commerce. A patent for an invention is the award of a property right to the inventor, issued by the U.S. Patent and Trademark Office. In general, the term of a new patent is 20 years from the date on which the application for the patent is filed in the United States. The intent of the right conferred by the patent is "the right to exclude others from making, using, offering for sale, or selling" the invention in the United States or "importing" the invention into the United States.

1. **Utility patents** may be granted to anyone who invents or discovers any new and useful process, machine, article of manufacture, or composition of matter, or any new and useful improvement thereof.
2. **Design patents** may be granted to anyone who invents a new, original, and ornamental design for an article of manufacture.
3. **Plant patents** may be granted to anyone who invents or discovers and asexually reproduces any distinct and new variety of plant.

There are three criteria for awarding a patent: (a) the invention must be new and novel, (b) the invention must be useful, and (c) the invention be nonobvious to a person skilled in the art covered by the patent. A key requirement is novelty. U.S. Patent and Trademark Office: http://www.uspto.gov/

Trademark

A trademark is a word, name, symbol, or device that is used in trade with goods to indicate the source of the goods and to distinguish them from the goods of others. Trademark rights may be used to prevent others from using a confusingly similar mark, but not to prevent others from making the same goods or from selling the same goods or services under a clearly different mark. Trademarks that are used in interstate or foreign commerce may be registered with the USPTO. The registration procedure for trademarks and general information concerning trademarks can be found in the separate book entitled "Basic Facts about Trademarks" (http://www.uspto.gov/trademarks/basics/Basic_Facts_Trademarks.jsp).

Copyright

Copyright is a form of protection provided to the authors of "original works of authorship" including literary, dramatic, musical, artistic, and certain other intellectual works, both published and unpublished. The 1976 Copyright Act generally gives the owner of copyright the exclusive right to reproduce the copyrighted work, to prepare derivative works, to distribute copies of the copyrighted work, to perform the copyrighted work publicly, or to display the copyrighted work publicly. The copyright protects the form of expression rather than the subject matter of the writing. As an example, a product can be copyrighted, but this would only prevent others from copying the description. It will not prevent people writing their own explanation of the product and using the product. Copyrights are registered by the *Copyright Office of the Library of Congress*.

REFERENCES

1. Andy Gibbs and Bob Dematteis, *"Essentials of Patents,"* (Hoboken, NJ: John Wiley & Sons, Inc., 2003).
2. Karl Ulrich and Steven Eppinger, *Product Design and Development*, 5th ed. (New York: McGraw-Hill, 2011).
3. Avital Fast, Devdas Shetty, et al., *"Wheelchair Attachments."* U.S. Patent 7,766,342 B2 filed on Apr 9, 2007 and issued on Aug 3, 2010.

Patent Sample

US007766342B2

(12) **United States Patent**
Fast et al.

(10) **Patent No.:** **US 7,766,342 B2**
(45) **Date of Patent:** **Aug. 3, 2010**

(54) **WHEELCHAIR ATTACHMENTS**

(75) Inventors: **Avital Fast**, Glen Cove Road, NY (US); **Devdas Shetty**, West Hartford, CT (US); **Moshe Raz**, Tel Aviv (IL); **Glora Rothman**, Hod Hasharon (IL)

(73) Assignees: **Montefiore Medical Center**, Bronx, NY (US); **The University of Hartford**, West Hartford, CT (US)

(*) Notice: Subject to any disclaimer, the term of this patent is extended or adjusted under 35 U.S.C. 154(b) by 166 days.

(21) Appl. No.: **11/697,934**

(22) Filed: **Apr. 9, 2007**

(65) **Prior Publication Data**

US 2008/0246251 A1   Oct. 9, 2008

Related U.S. Application Data

(60) Provisional application No. 60/790,596, filed on Apr. 7, 2006.

(51) **Int. Cl.**
A61G 5/06   (2006.01)
B62B 5/02   (2006.01)

(52) **U.S. Cl.** **280/5.28**; 280/5.22; 280/304.1; 280/DIG. 10; 74/2

(58) **Field of Classification Search** .... 280/3, 280/5.28, 5.3, 5.32, 6.155, 8.3, 43.13, 43.16, 280/86.1, 5.2, 304.1, DIG. 10; 74/2
See application file for complete search history.

(56) **References Cited**

U.S. PATENT DOCUMENTS

572,658 A * 12/1896   Marx 280/5.28

857,696 A	*	6/1907	Weinstein 280/5.28
2,123,707 A	*	7/1938	Bloch 16/44
3,133,742 A	*	5/1964	Richison et al. 280/5.28
3,279,812 A	*	10/1966	Rizzuto 280/5.2
3,438,641 A	*	4/1969	Bradley 280/5.28
3,494,440 A	*	2/1970	Hanson 180/8.2
4,326,622 A	*	4/1982	Ellzey 198/322
4,962,942 A		10/1990	Barnett et al.
5,338,048 A	*	8/1994	Medina 280/5.22
6,554,086 B1		4/2003	Goertzen et al.
2007/0095581 A1	*	5/2007	Chambliss et al. 180/8.2

* cited by examiner

Primary Examiner—Lesley Morris
Assistant Examiner—Wesley Potter
(74) *Attorney, Agent, or Firm*—Wiggin and Dana LLP; Gregory S. Rosenblatt

(57) **ABSTRACT**

Attachments for a manual wheelchair are provided for navigating a wheelchair over obstacles and uneven terrain, such as a typical curb on a street. The attachments provide for regulating the movement of the wheelchair as the wheelchair descends the curb and to prevent the wheelchair from flipping over during such movement. The attachments include a front caster wheel slider assembly on each side of the wheelchair, a follower wheel assembly and a track belt damping bar. The front caster wheel slider assembly includes a piston that will quickly push the front caster wheels down after they roll over the top edge of the curb. The follower wheel assembly acts as a sensor that will release the piston in the assembly after the front caster wheels roll over the top edge of the curb. As the back wheels of the wheelchair roll over and down the curb, the track belt damping bar slows the descent of the wheelchair.

7 Claims, 11 Drawing Sheets

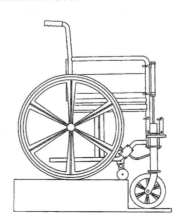

INDEX